Evolutionary Design and M

Springer

London
Berlin
Heidelberg
New York
Barcelona
Hong Kong
Milan
Paris
Singapore
Tokyo

I.C. Parmee (Ed.)

Evolutionary Design and Manufacture

Selected Papers from ACDM '00

With 182 Figures

Springer

I.C. Parmee, BSc, PhD
Engineering Design Centre, University of Plymouth, Drake Circus, Plymouth,
Devon, PL4 8AA, UK

ISBN 1-85233-300-6 Springer-Verlag London Berlin Heidelberg

British Library Cataloguing in Publication Data
Evolutionary design and manufacture
    1.Engineering design - Congresses 2.Manufacturing processes
    - Congresses 3.Evolutionary programming (Computer science)
    - Congresses
    620'.0042
    ISBN 1852333006

Library of Congress Cataloging-in-Publication Data
A catalog record for this book is available from the Library of Congress

Typesetting: Camera ready by contributors
Printed and bound by Athenæum Press Ltd., Gateshead, Tyne & Wear
34/3830-543210 Printed on acid-free paper SPIN 10761878

Preface

The fourth evolutionary/adaptive computing conference at the University of Plymouth again explores the utility of various evolutionary/adaptive search algorithms and complementary computational intelligence techniques within design and manufacturing. The content of the following chapters represents a selection of the diverse set of papers presented at the conference that relate to both engineering design and also to more general design areas. This expansion has been the result of a conscious effort to recognise generic problem areas and complementary research across a wide range of design and manufacture activity.

There has been a major increase in both research into and utilisation of evolutionary and adaptive systems within the last two years. This is reflected in the establishment of major annual joint US genetic and evolutionary computing conferences and the introduction of a large number of events relating to the application of these technologies in specific fields. The Plymouth conference remains a long-standing event both as ACDM and as the earlier ACEDC series. The conference maintains its policy of single stream presentation and associated poster and demonstrator sessions. The event retains the support of several UK Engineering Institutions and is now recognised by the International Society for Genetic and Evolutionary Computation as a mainstream event. It continues to attract an international audience of leading researchers and practitioners in the field.

Although evolutionary computing research and application provides the mainstay in the great majority of papers as is the tradition of the ACDM series, papers utilising neural computing technologies and related computational intelligence techniques within design/manufacture environments have also been included.

Contributions addressing aerospace, mechanical, thermal and structural engineering are included in addition to urban planning, food and chemical design and the design of networks. In terms of manufacturing processes cell formation, facility design, system control and robotics are addressed. In most cases results from application to or integration with real-world industrial problems are very much in evidence.

The true potential of evolutionary and adaptive computing within the design and manufacture field is now becoming very apparent with significant research effort moving away from relatively straightforward optimisation to investigate interactive exploratory systems, decision support, evolutionary and agent-assisted multi-objective strategies and problems relating to appropriate design representation. Algorithmic structure and development is also addressed with particular regard to the requirements of a design and manufacture environment.

The following chapters therefore provide information regarding current state-of-the-art with regard to the integration of evolutionary and adaptive computing with design and manufacture whilst also indicating future direction in terms of appropriate research and development.

I.C. Parmee
April, 2000

Organisation:

Conference Chairman:

I.C. Parmee — University of Plymouth

Scientific Committee:

T. Baeck	Informatik Centrum Dortmund, Germany
G. Bugmann	University of Plymouth, UK
B. Carse	University of West of England, UK
C. Coello Coello	LANIA, Mexico
K. Deb	Indian Institute of Technology, Kanpur
K. De Jong	George Mason University, USA
M. Denham	University of Plymouth, UK
T.C. Fogarty	Napier University, UK
D. Fogel	Natural Selection Inc, USA
M. Gen	Ashikaga Institute of Technology, Japan
J. Gero	University of Sydney, Australia
E. Goodman	Michigan State University, USA
D. Grierson	University of Waterloo, Canada
P. Hajela	Rensselaer Polytechnic Institute, USA
C.J. Harris	University of Southampton, UK
C. Hillermeier	Siemens Research, Germany
C. Hughes	Logica, Cambridge, UK
P. Husbands	University of Sussex, UK
A. Keane	University of Southampton, UK
F. Lohnert	Daimler Chrysler, Germany
M.L. Maher	University of Sydney, Australia
C. Moore	University of Cardiff, UK
J. Morris	University of Newcastle-upon-Tyne, UK
S. Patel	Unilever Research Laboratory, UK
C. Poloni	Universita of Trieste, Italy
W. Punch	Michigan State University, USA
M. Schoenauer	École Polytechnique, France
H.-P. Schwefel	University of Dortmund, Germany
E. Semenkin	Siberian Aerospace Academy, Russia
P. Sen	University of Newcastle-upon-Tyne, UK
A.E. Smith	University of Pittsburgh, USA
G.D. Smith	University of East Anglia, UK
J. Taylor	Kings College London, UK
G. Thierauf	University of Essen, Germany

Supporting Bodies:

The Institution of Electrical Engineers, UK;
The Institution of Civil Engineers, UK;
The Institution of Mechanical Engineers, UK;
The Institution of Engineering Designers, UK;
The British Computer Society;
Society for the Study of Artificial Intelligence and Simulation of Behaviour (AISB), UK;
International Society for Genetic and Evolutionary Computation (ISGEC);
European Network of Excellence in Evolutionary Computation (EVONET);
UK EPSRC Engineering Network in Adaptive Computing in Design and Manufacture (ACDM-net);
CASE Centre for Computer-Aided Engineering and Manufacturing, Michigan State University, USA;
Key Centre for Design Computing, University of Sydney.

ACDM Lecture:

Multi-objective Evolutionary Optimization: Past, Present and Future

Professor K. Deb
Indian Institute of Technology, Kanpur, India

Keynote Presentations:

Representation in Architectural Design Tools

Dr. U.-M. O'Reilly
MIT, Cambridge, USA

Adaptive Techniques for Evolutionary Topological Optimum Design.

Dr. M. Schoenauer
École Polytechnic, Paris, France.

Constrained Design: Using Constraints to Advantage in Adaptive Optimisation of Manufacturing Systems

Professor A.E. Smith
Auburn University, Alabama, USA

Contents

1. Engineering Design Applications

2. General Design Applications

3. Design Representation Issues

4. Manufacturing Applications

5. Multi-objective Satisfaction

6. Algorithm Comparison and Development

7. Neural Nets and Hybrid Systems

Chapter 1

Engineering Design Applications

Optimisation of Power Plant Design: Stochastic and Adaptive Solution Concepts
C. Hillermeier, S. Hüster, W. Märker, T.F. Sturm

Intelligent Searcher for the Configuration of Power Transmission Shafts in Gearboxes
S.D. Santillán-Gutiérrez, G. Olivares-Guajardo, V. Borja-Ramírez, M. López-Parra, L.A. González-González

Post-Processing of the Two-dimensional Evolutionary Structural Optimisation Topologies
H. Kim, O.M. Querin, G.P. Steven

Optimisation of a Stator Blade Used in a Transonic Compressor Cascade with Evolution Strategies
M. Olhofer, T. Arima, T. Sonoda, B. Sendhoff

Mixed-integer Evolution Strategy for Chemical Plant Optimization with Simulators
M. Emmerich, M. Grötzner, B. Groß, M Schütz

Advantages in Using a Stock Spring Selection Tool that Manages the Uncertainty of the Designer Requirements
M. Paredes, M. Sartor, C. Masclet

Optimization of Power Plant Design: Stochastic and Adaptive Solution Concepts

Claus Hillermeier[1], Steffen Hüster[2],
Wolfgang Märker[3] and Thomas F. Sturm[4]

[1] SIEMENS AG, Corporate Technology, D-81730 München, Germany;
Technical University of München, Dep. of Math., D-80290 München, Germany
email: Claus.Hillermeier@mchp.siemens.de

[2] Univ. of Augsburg, Dep. of Math., Universitätsstr. 14, D-86135 Augsburg, Germany
email: huester@gmx.de

[3] SIEMENS AG, Power Generation, Freyeslebenstr. 1, D-91058 Erlangen, Germany
email: Wolfgang.Maerker@notes.kwu.siemens.de

[4] SIEMENS AG, Corporate Technology, D-81730 München, Germany
email: Thomas.Sturm@mchp.siemens.de

Abstract. Optimizing the design of industrial plants requires making optimal structural decisions (concerning the selection and arrangement of various components) as well as optimizing all continuous design variables. This paper presents genetic and stochastic optimization strategies which are based on a representation of the plant by means of a modified decision tree. By taking into account hierarchical dependencies of decisions this representation guarantees that designs generated by mutation operators automatically comply with an important class of constraints. The method is explained and its potential is demonstrated with the example of an important industrial application problem: The design optimization of feed-water heater strings in fossil-fueled power plants. For the treatment of the structural and continuous design variables two strategies have been implemented and tested. The first approach considers structural decisions as the primary problem, which is solved by means of a Metropolis algorithm, and regards the optimization of the continuous variables as a subproblem, which is solved by Sequential Quadratic Programming for each generated plant structure. The second strategy is an evolutionary one-level algorithm which simultaneously optimizes both types of variables. In the design problem which is investigated here, the one-level Evolutionary computation algorithm performs slightly better than the hierarchical method. This result is explained by analyzing the objective function.

1 Introduction

The central issue of this paper is to solve a real-world task of industrial design, namely to find the optimal design of feed-water heater strings in fossil-fueled power plants. Optimization, here, is defined with respect to the aim of energy supply companies to generate electric power at minimal generating costs.

This application problem allows us to develop, investigate and demonstrate general concepts and algorithms for computing the optimal design of industrial plants (for a discussion of the potential of soft computing techniques for design problems see [10]). Industrial plants are typically composed of operating components which are linked by e.g. pipes or conveyor belts in order to facilitate flows of energy, of raw materials or of products under construction. The selection and arrangement of the various components together with the arrangement of their links determine the *structure* of the industrial plant. Apart from that, the design of the plant is specified by a set of parameters characterizing the features of the single components and links.

Thus, the task of optimizing the design of industrial plants typically requires optimal structural decisions as well as optimizing all continuous design variables. There are two principal approaches to optimizing structures by means of stochastic algorithms.

Approach (A), which is closely related to concepts of Genetic Programming (see [7]), starts from some minimal core structure and defines mutation operators which try to achieve improvements by gradually refining the plant structure. A slight modification of that approach is starting at an arbitrarily chosen feasible structure and defining two types of mutation operators: Those which refine and those which simplify the plant structure.

The alternative approach (B) is based on some structure of maximum complexity defined by a design expert. Here, the task of the optimization algorithm is to achieve improvements by gradually pruning parts of the structure.

In order to enable the construction of mutation (and, perhaps, recombination) operators, the plant structure has to be modelled by some mathematically well-defined representation. For both solution strategies (A) and (B) advantageous representations can be borrowed from graph theory. The graph representation matching approach (A) models plant components by nodes and flows of energy or matter by edges. In order to enable compatibility checks, features and parameters can be assigned to nodes and edges (see e.g. [6]).

Within approach (B) the maximum structure is determined. Thus, all structural decisions could in principle be modelled by binary (or, more generally, integer) numbers denoting the existence/non-existence (and, perhaps, position) of components and links. Such a binary encoding would, however, not take into account hierarchical dependencies of the decision variables. If some component is not existent in the current structure, all decision variables concerning its features and subcomponents are meaningless. An elegant way to automatically take these dependencies into account is a representation by means of some type of decision tree.

This paper reports a solution of the task of power plant design which is based on approach (B) and a representation of the plant by a modified decision tree. Alternative adaptive strategies for algorithmic power plant design are discussed in [3] and [2].

For the treatment of the structural and continuous design variables there are, again, two different ways. The first way considers the structural decisions as the primary problem and regards the optimization of the continuous variables as a sub-problem which is to be solved for each generated plant structure. The second strategy is a one-level algorithm which simultaneously optimizes both types of variables. In the present paper, both strategies are realized in the form of adaptive computing algorithms and compared by numerical tests.

The paper is organized as follows. Section 2 presents the application problem and formalizes it as a mathematical optimization task. The issue of Section 3 is to develop the modification of the decision tree concept which is required for its applicability to plant design. Section 4 tailors two different adaptive optimization algorithms for the given design task, Section 5 reports the results of numerical tests, and Section 6 draws conclusions.

2 The task of designing a feed-water heater string

2.1 Structure of fossil fueled power plants

Fossil power plants transform thermal energy produced by burning coal, gas or oil into electrical energy. The energy conversion is effected by water or steam which runs through a circular process absorbing heat energy and emitting mechanical energy.

Figure 1 shows a scheme of the basic elements of a power plant. In the steam generator, the so-called feed-water gets heated, vaporized and superheated. The resulting hot vapour under high pressure is denoted as live steam. It is conducted into turbines driving a generator that produces electrical power. By driving the turbines the pressure of the steam is reduced. Since the size of a turbine depends on the pressure of the live steam, different turbines are used for the high-, medium-, and low-pressure area. Having passed the low-pressure turbine, the steam is condensed. The feed-water pump gets it back to the pressure level of live steam, and the feed-water arrives again in the steam generator.

By *preheating the feed-water* before it enters the steam generator, the efficiency η of such an elementary water/steam cycle will be increased. For that purpose little portions of steam are extracted from different parts of the turbines and are used to raise the temperature of the feed-water with the help of heat-exchangers called feed-water heaters. These heaters are connected in series, forming the *feed-water heater string*. Figure 4(a) shows a modern steam power plant with feed-water preheating. In the top right part of the figure, the high-, medium-, and low-pressure turbine are sketched. Extraction pipes connect the turbines and the feed-water heater string shown at the bottom of the figure.

According to the state of matter of the heat-emitting medium, heat-exchanging components can be classified in the following way:

Desuperheater: The heat-emitting medium, i.e. the extracted steam, emits heat but remains in the vaporous state (vapour $\rightarrow$ vapour).

Condensing heater: The extracted steam condenses releasing the condensation heat. Thus, the heat-emitting medium enters as vapour and comes out as a liquid (vapour $\rightarrow$ water).

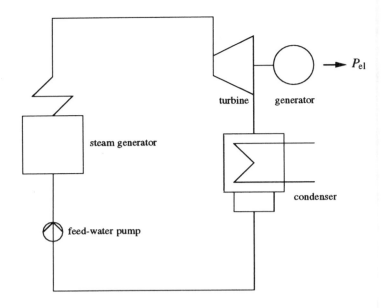

Figure 1: Water/steam circuit

Condensate cooler: Having left the condensing heater, the heat-emitting medium enters as (hot) water, emits further heat and leaves as water (water → water).

A feed-water heater consists at least of one condensing heater. It may be supplemented by a desuperheater and/or a condensate cooler. All heat-exchanging components that belong to one extraction may either be integrated in a single frame or may be installed separately. If installed separately, the components may be shifted along the feed-water heater string — desuperheaters may be moved from the position of the condensing heater towards the steam generator, and condensing heaters may be shifted towards the condenser. For example, in Figure 4(a) the two heat-exchangers on the right have a moved condensate cooler, and in Figure 4(b) the heat-exchanger on the left is a shifted desuperheater.

Instead of cooling the condensed water by a condensate cooler it may be directly pumped into the feed-water stream just before the condensing heater (see the second extraction in Figure 4(b)).

A heat-exchanger designed a bit differently is the feed-water storage tank (see e.g. the forth heat-exchanger from the left in Figure 4(a)). Here, the extracted vapour directly enters a tank filled with feed-water and condenses at the water surface. This tank also serves as a buffer for the feed water pump during operational malfunctions.

2.2 The mathematical optimization problem

2.2.1 Design variables

Designing the feed-water heater string means to specify the number and position of the steam extractions, the arrangement and characterization of the heat-exchanging components, and the connections between extractions and heaters.

The basic connectivity is determined by the laws of thermodynamics. On the one hand, the pressure of the steam monotonically decreases along the turbine flow, and the pressure of the extracted steam uniquely determines the position of the extraction and vice versa. The condensation temperature is a monotonic function of the pressure. Therefore, the temperature levels, at which the condensing heaters connected with the individual extractions operate, monotonically decrease from extraction to extraction. Since, on the other hand, the temperature of the feed-water has to be monotonically increased on its way from the condenser to the steam generator, the assignment of extractions to condensing heaters is determined.

The main degrees of freedom which are left for the optimization algorithm are the following:

(i) Number and position of the extractions. According to the concept of an expert-given maximum structure, nine extractions — numbered from 1 to 9 starting at the low-pressure turbine — are defined to be possible (3 for the low-pressure, 4 for the medium-pressure and 2 for the high-pressure turbine). For each possible extraction, the algorithm has to decide whether that extraction actually should be built. For some extractions i, in addition, the exact position or, equivalently, the extraction pressure p_i have to be determined.

(ii) For each existent extraction, the heat-exchanging components connected to it have to be specified. For that purpose one has to decide

- whether a desuperheater should be built,
- if yes, if it should be separated from the condensing heater,
- if yes, how far it should be shifted (by 0, 1, 2 or 3 positions),
- whether the extracted steam should, after condensation, be directly pumped into the feed-water stream,
- if no, whether a separate condensate cooler should be built.

The properties of a heat-exchanger depend on its terminal temperature difference which indicates the minimum difference in temperature between heat-emitting and heat-absorbing media and strongly influences the size of the heat-exchanger. Thus, for each separate heat-exchanging component its terminal temperature difference has to be determined.

2.2.2 Design constraints

A set or tuple of design variables has to meet a set of constraints in order to specify a working power plant.

The hierarchical dependencies of the discrete decisions can be seen directly from the listing of variables mentioned above. According to these dependencies, the existence/nonexistence decisions of the main components determine the existence/nonexistence decisions of the subcomponents (In that context, an extraction is considered as a component, and the heat exchangers connected to it are regarded as subcomponents). Position shifts and continuous parameters are only well-defined if the respective components exist.

Each feed-water heater string has to contain exactly one feed-water storage tank. Owing to technical reasons the corresponding extraction pressure has to be larger than 5 and smaller than 15 bars. Therefore, that steam has to be extracted from the medium-pressure turbine.

The feasible region of the continuous variables is, first of all, limited by box constraints. For the extraction pressures, these are given by the pressure intervals of the respective turbines. On top of that, the extraction pressures have to strictly increase along with their numbering. This condition can be expressed by a sequence of inequalities

$$p_{i+1} - p_i \geq \Delta p, \tag{2.1}$$

where i and $i+1$ are the integer labels of two neighbouring extractions and Δp denotes a minimum pressure gap.

Note that the number of box and inequality constraints depends on the structure of the heater string, i.e. on the current choice of the discrete decision variables.

2.2.3 The objective function

The optimization goal is defined by the interest of the energy supply companies to generate electrical current at minimal generating costs. Thus, the generating costs per kWh electric power serve as the objective function. These costs are composed of the costs caused by the investment and the costs owing to the operation of the plant which are mainly given by the fuel costs. This yields the following objective function

$$f = \underbrace{\frac{costs_{\text{invest}} \cdot a}{P_{\text{el}} \cdot h}}_{\text{Investment costs}} + \underbrace{\frac{costs_{\text{fuel}}}{\eta}}_{\text{Fuel costs}}, \quad \text{where} \tag{2.2}$$

$costs_{\text{invest}}$	$=$	total costs of the investment,
a	$=$	annuity factor comprising both the annual interest rate and some estimated price of maintenance,
P_{el}	$=$	produced electric power,
h	$=$	annual operating hours of the plant,
$costs_{\text{fuel}}$	$=$	fuel costs per kWh primary energy,
η	$=$	total efficiency of the plant.

From the viewpoint of design optimization, a, h and $costs_{fuel}$ are considered as constants, whereas the values of $costs_{invest}$, P_{el} and η depend nonlinearly on the design variables. In order to model these nonlinear dependencies properly, we employ the plant simulation program KRAWAL which has been developed by Siemens Power Generation and which calculates the thermodynamic equilibrium state as well as the geometric dimensions of the heating components which result from a given tuple of design variables.

3 Representation as a modified decision tree

As described in Paragraph 2.2.2, the discrete decision variables are subject to many constraints which are due to the hierarchical dependence of these decisions. By straightforwardly encoding the discrete decisions by a tuple of bits, the compatibility with the hierarchical constraints would have to be checked *a posteriori*, i.e. after some optimization algorithm has generated a new decision tuple, by means of Boolean expressions.

Computing time can be used more economically by representing the power plant design in such a way as to force compliance with the hierarchical constraints *a priori*. For that purpose the design representation has to be endowed with further information about the hierarchical dependencies of the design variables. We, therefore, represent a plant design by a modified decision tree.

The design variables are modelled by nodes, and each dependency relationship is modelled by an edge connecting the respective variable nodes. If some decision variable only exists if some higher rank variable has a special value, this value is assigned to the connecting edge. For example, the decision "separate desuperheater" is defined only if the variable "desuperheater existent" has the value 1, and the variable "pump condensate directly" has to be 0 to evoke a decision concerning "condensate cooler existent". There are no decision variables which depend on the value of any of the continuous variables. Therefore, continuous variables are always mapped to leaves of the decision tree. Since there are decisions of the same hierarchical level which can be made independently, the tree contains additional nodes to which no variable is assigned but which serve only for branching the tree. In the following we call nodes to which a decision variable is assigned *decision nodes* and nodes without an assigned variable *branching nodes*.

Figure 2 shows the resulting modified decision tree. Decision nodes are symbolized by empty circles, branching nodes by filled circles, and leaves which correspond to continuous variables by squares. To represent all possible values of the discrete variables, each feasible value of such a variable is modelled by an edge. This requires the definition of "empty nodes" (leaves) — symbolized by small filled squares — which serve only as end-points of such value-denoting edges.

With the help of Figure 2 it is easy to see the dependencies of the decision variables. For example, the variables coding the existence of the different extractions can be chosen independently. If an extraction does not exist, i.e. the value of this variable equals to 0, no further decisions for this extraction can be made. If an extraction exists, the decisions "desuperheater existent" and "pump condensate directly" as well

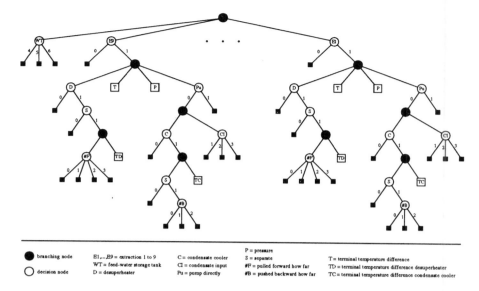

branching node   E1,...E9 = extraction 1 to 9   C = condensate cooler   P = pressure   S = separate   T = terminal temperature difference

WT = feed-water storage tank   CI = condensate input   #F = pulled forward how far   TD = terminal temperature difference desuperheater

decision node   D = desuperheater   Pu = pump directly   #B = pushed backward how far   TC = terminal temperature difference condensate cooler

Figure 2: Representation of hierarchical parameter dependencies as modified decision tree

as the continuous variables "extraction pressure" and "heater terminal temperature difference" are all independent from each other. The decision whether a desuperheater is separated from the condensing heater depends on the decision whether it is existent at all. Analogously, a condensate cooler may be used only if the condensate is not directly pumped back.

To find a tuple of values of the decision variables which represents a feasible power plant design, one has to choose the values of decision variables and to collect the dependent variables matching these values by a run through the decision tree which starts from its root and which obeys the following rules:

- In a branching node, *all* edges have to be followed one after the other.

- In a decision node, only that edge has to be followed which has the actually chosen value of the corresponding variable.

- Having reached a leaf, the next edge of the next (not yet completely treated) higher-ranking branching node has to be followed.

To demonstrate that prescription, Figure 3 shows a part of the decision tree where the edges collected by such a run through the graph are marked bold. In this sample run, the following values of the decision variables have been chosen: "pump condensate directly" = 0, "condensate cooler" = 1, "separate condensate cooler" = 0, "condensate input" = 1.

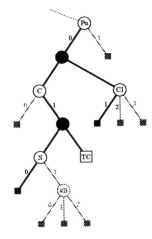

Figure 3: Feasible way(s) within the decision tree

4 Optimization strategies

4.1 Hierarchical optimization

Every feasible structure of the power plant is determined by a tuple $d \in D$, where D is a finite set of vectors of discrete decisions. Depending on d a vector $x \in R(d) \subseteq \mathbb{R}^{n(d)}$ of continuous variables has to be chosen to determine the properties of the power plant components. There, $n(d)$ denotes the number of leaves (coding continuous variables) collected by the run through the graph corresponding to d, and $R(d)$ denotes the corresponding feasible region as defined in Paragraph 2.2.2. This yields

$$S := \bigcup_{d \in D} \{d\} \times R(d)$$

as the search space of all feasible variable sets (d, x). Thus, the optimization problem

$$\min_{(d,x) \in S} f(d,x) \tag{4.3}$$

with the objective function f as defined in Paragraph 2.2.3 has an inherent hierarchical structure and can equivalently be put as

$$\min_{d \in D} \underbrace{\left(\underbrace{\min_{x \in R(d)} f(d,x)}_{=:f_d(x)} \right)}_{=:F(d)}.$$

The hierarchical optimization approach splits problem (4.3) into a discrete problem

$$\min_{d \in D} F(d) \tag{4.4}$$

and a continuous sub-problem

$$\min_{x \in R(d)} f_d(x). \tag{4.5}$$

Each objective function $f_d : R(d) \to \mathbb{R}$ is assumed smooth enough such that problem (4.5) can be solved by an appropriate method of nonlinear optimization. For the task of power plant design, the Sequential Quadratic Programming (SQP) method as e.g. explained in [4] has been chosen.

In order to treat the discrete problem (4.4), a Metropolis algorithm (see [8] or [11]) has been implemented. The operator "CreateNeighbour", whose task — similarly to the mutation operator of a genetic algorithm — is to create a slightly modified version of some decision vector d, works on the basis of the tree representation explained above. It creates a modification d' in the neighbourhood of the current decision vector d by running through the graph along the feasible way coded by d, changing each decision with some given mutation rate. Having generated a new tuple d', the leaves which are part of the corresponding feasible ways through the graph are collected to give the continuous variables of the subproblem (4.5). Afterwards, the feasible region $R(d')$ is defined by collecting the corresponding inequalities and box constraints, and the SQP-algorithm is started.

Below this hybrid method for the optimal design of feed-water heater strings will be referred to as M-SQP algorithm.

4.2 One-level optimization

An alternative strategy for solving problem (4.3) is to treat discrete and continuous variables in an analogous manner by a one-stage optimization algorithm. Such an algorithm has to cope with the special structure of the search space consisting of subspaces of variable dimensions.

The tree representation makes it possible to solve that task by means of an appropriately designed genetic algorithm (GA) (for an introduction to GAs see e.g. [5] or [9]; we are aware that our algorithm combines principles of GAs and Evolution strategies, but call it a GA for the sake of brevity). The GA which has been tailored for the present task uses a $(\mu + \lambda)$-selection strategy (see [1]), does without recombination, and employs a mutation operator which, similarly to the 'CreateNeighbour'-operator sketched above, works on the basis of the tree representation. Unlike the 'CreateNeighbour'-operation, mutation here includes all types of decision variables. It is controlled by mutation rates, each of which defines the probability for an individual decision variable to mutate. These mutation rates differ for discrete and continuous variables. For the latter, in addition, one has to specify the size σ of the mutation step.

For a general GA, where the individuals are encoded by binary strings, the mutation rate is usually the same for all variables. Owing to our tree representation, the mutation operator only has the chance to change "active" variables lying on the feasible way which is currently run through. Thus, the chance of a mutation depends on the hierarchical level of a decision variable. The deeper a decision node is in the tree hierarchy, the smaller is the chance to get acted on by the mutation operator. Since all decision nodes which have successor nodes are assigned to binary decision variables, the probability that a node on the hierarchy level δ lies on a feasible way is given by $\left(\frac{1}{2}\right)^{\delta-1}$ (see Figure 2). Here, the numbering of the hierarchy levels starts at the root and is increased by 1 each time a decision node is passed. In order to compensate for that decrease of the chance to be mutated, the mutation rate for a decision node of hierarchy level δ is calculated according to

$$\tilde{p}_m(\delta) = p_m \cdot 2^{\delta-1} \,,$$

where p_m denotes some given basic mutation rate.

When, during its run along the feasible way of the current individuum, the mutation operator has arrived at some decision node, it first determines whether a mutation is to be performed. For that purpose, a random number is drawn from a uniform distribution in the interval $[0, 1]$. If that number is smaller than $\tilde{p}_m$, the decision variable is mutated. The method of mutation is characteristic for each type of variable. For the integer variables, the new value is randomly drawn from the feasible interval according to a uniform distribution, for binary variables the bit is simply flipped, and to a continuous variable the mutation operator adds a random number drawn from a Gaussian distribution whose standard deviation σ is some given fraction of the feasible interval. Hereby, compliance with the box constraints is guaranteed by reflection of the new value at the box boundary, if necessary. The inequality constraints (2.1) are taken into account by adding appropriate penalty terms to the value of the objective function. During an optimization run both the basic mutation rate p_m and the stepsize (standard deviation) σ are decreased as a linear function of the generation number in order to induce a smooth transition from global optimization at the beginning to local optimization at the end of the run.

As the described GA performs the task of mixed-integer optimization, it will be referred to as MIGA in the following.

5   Results

In order to test and compare the performance of the developed algorithms, we consider two sample problems in the form of two different maximum structures which have been defined by an expert in power plant design. Example 1 has 8 potential discrete variables and 11 potential continuous variables, example 2 consists of 26 potential discrete variables and 18 potential continuous variables. The initial configuration (maximum structure) for the latter example is shown in Figure 4(a). For reasons of confidentiality, the objective function values, i.e. the costs for generating 1 kWh electric power, are expressed in some fictitious currency.

Example 1 has been optimized by 100 iterations of the M-SQP method and, alternatively, by 600 iterations of the MIGA method (with $\mu = 50$ and $\lambda = 15$) to get a comparable number of objective function evaluations. To optimize example 2, two runs have been made with each of both algorithms. The first run employed the same strategy parameters as the run for example 1, the second run employed 300 iterations

14

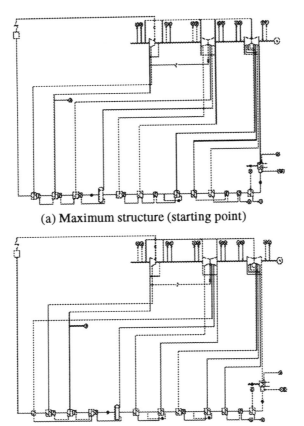

(a) Maximum structure (starting point)

(b) Optimal structure as calculated by the M-SQP-algorithm

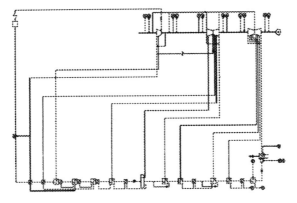

(c) Optimal structure as calculated by the MIGA-algorithm

Figure 4: Structure of a steam power plant with feed-water heater string

of the M-SQP method and 800 iterations of the MIGA method (with $\mu = 80$ and $\lambda = 40$).

The results of these optimization runs are summarized in the following table:

Method	f_{min}	# Function evaluations	Computing time
Example 1, objective function value (start): 2.41662330			
M-SQP	2.39118268	12181	12 h
MIGA	2.39079621	9050	9 h
Example 2, objective function value (start): 2.40341842			
M-SQP, 1st run	2.38700755	11419	11 h
MIGA, 1st run	2.38691590	10040	10 h
M-SQP, 2nd run	2.38652603	35026	33 h
MIGA, 2nd run	2.38559951	32080	30 h

The computation times refer to a SUN-ultra-1 workstation and are almost completely used up by the KRAWAL-simulations which serve for evaluating the objective function.

Not suprisingly, we see that both methods yield further improvements by an increase of computational effort. One can further see that, for a comparable number of objective function evaluations, the MIGA computes minima which are slightly better (with respect to the objective function) than those achieved by the M-SQP method. In order to obtain an explanation for this observation, let us have a closer look at the test results.

The structure of the minimum found by the M-SQP method for Example 2 is shown in Figure 4(b). This structure differs, indeed, in some parts from the starting structure. For example, the heat-exchanger at extraction 7 has got a desuperheater which is shifted by two positions.

Figure 5(a) shows the objective function values obtained during a run of the M-SQP method. The starting points for every SQP run are generated by the Metropolis algorithm and are marked by an asterisk. Each sequence of points following an asterisk reflects the gradual improvements gained by optimizing the continuous variables (Where the SQP stops after the first iteration, the corresponding continuous subproblem has been badly conditioned). As can be clearly seen, varying the plant structure has a more prominent effect on the values of the objective function than solving the continuous subproblem exactly. This explains why the MIGA method is more successful since, by construction, it invests a much greater portion of its computational efforts in structural variations than the M-SQP method.

Figure 4(c) shows the power plant structure of the minimum as it has been found by the MIGA method. Differently from the structure found by the M-SQP method, two desuperheaters are shifted, and the feed-water storage tank is connected to extraction 5. Both results, however, show that structures with shifted desuperheaters in the high-pressure area attain smaller values of the objective function. In order to validate this result from the point of view of real-world power plant design, the modelling of the investment costs for these components should be further refined.

16

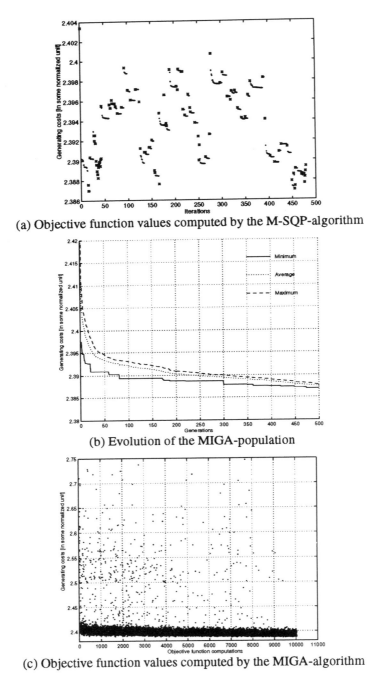

(a) Objective function values computed by the M-SQP-algorithm

(b) Evolution of the MIGA-population

(c) Objective function values computed by the MIGA-algorithm

Figure 5: Evolution of the objective function values during the optimization

The evolution of the population of the MIGA is shown in Figure 5(b). For every generation (at the moment immediately after the selection), the objective function value of the best and the worst individual are plotted together with the average over the population. During the optimization the distance between the three curves becomes smaller and smaller, i.e. the population gets more and more homogeneous. The objective function values of all generated individuals are shown in Figure 5(c). The stochastic mutation operator induces a large variance at the beginning of the optimization, where the search space is explored by the GA. During the optimization, the variance gets smaller indicating that the method concentrates on local minimization.

6   Conclusions

Global and adaptive optimization concepts have been presented which are able to solve a real-world task of industrial design: the optimal design of feed-water heater strings in fossil-fueled power plants. A key feature of the solution concept is the representation of the plant design by means of a modified decision tree. Thus, all constraints which are due to hierarchical dependencies between the decision variables are automatically taken into account by the encoding. For the mixed-discrete optimization problem two optimization strategies have been implemented and tested: A hierarchical strategy which treats discrete and continuous variables in a different way and which has been realized as a hybrid Metropolis-SQP method, and a one-level strategy realized by a mixed-integer GA (MIGA). Because of the character of the objective function for power plant design, the MIGA achieves slightly better results. More generally, one can conclude that for any given (mixed-integer) design problem the character of the objective function indicates which of the two strategies is more promising. If variations of the plant structure, i.e. variations of the discrete decision variables, have a much larger effect on the values of the objective function than variations of the continuous variables, then the one-level strategy is more promising. If, however, variations of discrete and continuous variables influence the objective function values in a similar way, the hierarchical approach is to be recommended.

From the viewpoint of real-world design, the minimum of the objective function, which is calculated by an optimization algorithm, is not the only criterion to assess the capability of that algorithm. Typically, there are design restrictions which cannot easily be modelled as constraints of the mathematical optimization problem. Thus, each solution calculated by the optimization algorithm has to be checked by a design expert for compliance with these restrictions. For the problem of power plant design such an unmodelled restriction is the necessity to put all the preheating components in a machine room of a given size.

As regards such "hidden" restrictions, the hierarchical M-SQP method has the advantage that different structural design solutions with optimized continuous parameters are computed. If the optimal solution cannot be realized because of unmodelled restrictions, then other solutions with different structures are offered. Compared with that, the individuals of the last generations of the mixed-integer GA differ only slightly from each other and have all the same values of the discrete variables owing to the selection pressure and to the reduced mutation variance at the end of the optimization run.

Acknowledgement

The authors gratefully acknowledge support by the German "Bundesministerium für Bildung und Forschung" in the framework of the project LEONET.

References

[1] Th. Bäck and H.-P. Schwefel. An overview of evolutionary algorithms for parameter optimization. *Evolutionary Computation*, 1/1993:1–23, 1993.

[2] K. Chen and I. C. Parmee. A comparison of evolutionary-based strategies for mixed discrete multilevel design problems. In *Proceedings of the Third International Conference on Adaptive Computing in Design and Manufacture*, pages 221–229. Springer Verlag, 1998.

[3] K. Chen, I. C. Parmee, and C. R. Gane. Dual mutation strategies for mixed-integer optimisation in power station design. In *Proceedings of the 1997 IEEE International Conference on Evolutionary Computations (ICEC'97)*, pages 385–390, 1997.

[4] R. Fletcher. *Practical Methods of Optimization*. John Wiley & Sons, New York, 1993.

[5] D.E. Goldberg. *Genetic Algorithms in Search, Optimization and Machine Learning*. Addison-Wesley, Reading/Mass., 1989.

[6] B. Groß, U. Hammel, P. Maldaner, A. Meyer, P. Roosen, and M. Schütz. Optimization of heat exchanger networks by means of evolution strategies. In *Proceedings of the 4th Internat. Conference on Parallel Problem Solving form Nature*, pages 1002–1011. Springer, 1996.

[7] J.R. Koza. *Genetic Programming: On the Programming of Computers by Means of Natural Selection*. MIT Press, Cambridge, MA, 1992.

[8] N. Metropolis, A. Rosenbluth, M. Rosenbluth, A. Teller, and E. Teller. Equation of state calculations by fast computing machines. In *Journal of Chemical Physics*, volume 21, 1953.

[9] Z. Michalewicz. *Genetic Algorithms + Data Structures = Evolution Programs*. Springer, Berlin, 2. edition, 1994.

[10] I. C. Parmee. Exploring the design potential of evolutionary/adaptive search and other computational intelligence technologies. In *Proceedings of the Third Internat. Conference on Adaptive Computing in Design and Manufacture*, pages 27–42. Springer Verlag, 1998.

[11] P.J.M. van Laarhoven and E.H.L. Aarts. *Simulated Annealing: Theory and Applications*. D. Reidel Publishing Company, Dordrecht/Boston/Tokyo, 1987.

Intelligent Searcher for the Configuration of Power Transmission Shafts in Gearboxes

S.D. Santillán-Gutiérrez, G. Olivares-Guajardo, V. Borja-Ramírez,
M. López-Parra & L.A. González-González.
Centro de Diseño y Manufactura, Facultad de Ingeniería, UNAM.
Laboratorios de Ingeniería Mecánica, Circuito Exterior C.U. México D.F. CP 04510.

1 Abstract

This paper reports on the foundation for the development of an evolutionary computing tool to help designers with the embodiment and detail design stages. The idea is to combine a database and a genetic algorithm to form an intelligent searcher. The case study under development is the configuration of power transmission shafts in gearboxes. At this stage the evolutionary computing tool would be focused on the selection process of involute splines for the shaft.

2 Terminology

A.- Addendum.
b.- Deddendum.
B.- Backlash.
C.- Distance between centres.
d.- Pitch diameter.
D_b.- Base diameter.
D_{Fe}.- Shape diameter of external involute.
D_{re}.- Internal diameter.
D_{ri}.- External diameter.
m.- Module.
π.- 3.1416.

M_w.- Transmission rate.
N.- Teeth number.
p.- Circular pitch.
P.- Diametral pitch.
R_b.- Base reference diameter.
t.- Tooth thickness.
m_w.- Transmission rate.
ϕ.- Pressure angle.
ϕ_D.- Involute angle.
ϕ_z.- Sevolute.
s.- Diameter for the major circumference for the tolerance class.

3 Definitions

- **Intelligent Searcher.**

An intelligent searcher is a computing system founded on machine learning tools. This system helps us to simplify in an important way the design problems that are related with catalogue search.

- **Catalogue Search.**

This is the type of search performed by selecting existing elements, from which we know the features and design procedures. The combinations of these elements generate the functional effect in a technical system.

- **Technical System.**

"The technical systems are the means by which energy in its various forms is manipulated in the solution of technical problems. The overall quality of a technical system is defined as the degree to which it meets the stated requirements and keeps within the imposed constraints, and is referred to as fitness for purpose (BSI 5750 (2))" (Aguirre, 1990).

4 Introduction

A genetic algorithm is a machine learning technique that is based on emulating the principles of natural selection proposed in the Darwinian theory of the natural selection for evolving species (Darwin, 1859). The theory is adapted to artificial scopes, where the information is manipulated as a basis for the process of natural selection.

Genetic algorithms (GA) were developed by John Holland in 1967 in the University of Michigan (Holland, 1967), after Holland adapted the selection processes of nature to artificial systems, for profiting the advantages of evolution. Genetic algorithms are considered to be a robust optimisation tool, because they can handle non-continuous solution landscapes of big size with many sub-optimal solutions.

In the next section we will be describing briefly the algorithm. Other descriptions can be found in the basic literature like Axelrod, 1987; Goldberg, 1989; Krishnakumar, 1993; Beasley, 1993; Koza, 1992.

Genetic algorithms have shown great potential to explore solution landscapes in non-linear problems and convergence to zones where there are possible solutions concentrated or *clustered* (De Jong, 1980). GA´s have shown several advantages when compared to hill climbing and other optimisation heuristic tools used for operations research. Consequently, they are considered adequate to be applied in design problems for configuration and conceptual design, especially when it is required to search for compatibility between the components of a system. Carlson (Carlson, 1996) published a paper on configuring hydraulic systems. Her work finds the elements and the optimisation behaviour of the system simultaneously. Thornton researched about solving the configuration of elements in a crankshaft-rod-piston assembly. (Thornton, 1996).

Although the results of these investigations are promising, further work is required, so that there would be different ways for using genetic algorithms during the configuration of mechanical systems. In the present paper the theoretical foundations for the application of the genetic algorithms are reported, applied to the configuration and selection of parts of power transmission shafts in gearboxes.

- *Previous Applications of the Genetic Algorithms.*

GAs have been usually used to solve complex problems of mathematical optimisation. Currently there are several systems that are being used for research and development. Moreover, there is some commercial software for specific applications of the GA to optimization problems. (Mill, 1996; Parmee, 1996; Carlson, 1994; Krishnakumar, 1993).

Some of the main applications of genetic algorithms in operations research are job shop scheduling (Jain, 1997), distribution systems (Lee, 1997), airline scheduling (Durand, 1996). In mechanical engineering they have been applied in assembly optimisation (Santillán, 1998) and in the hydraulic systems design (Carlson, 1996). In chemistry they have been applied in the information analysis of the waste flow in complex industrial multi-units. (Hugh).

- *Potential of genetic algorithms application to mechanical design.*

Considering the model of Pahl & Beitz (Pahl & Beitz; 1986),during the stage of conceptual design the ideas and operation principles acquire a specific form to configure a technical system.

The next stage is configuration design, which is used to choose the materials and shape of the technical system. If the designer is willing to be assisted by an intelligent searcher he/she will be considering to structure a data base containing material lists and operation principles for the proposed technical system.

Once the configuration design stage is done, detail design is undertaken. It requires consulting technical catalogues about all the elements that will form the operation principles of the proposed technical system.

The problem during these two stages is to take precise and opportune decisions. Sometimes the proposed solution is not the most convenient, and the process requires a iteration. It is at this point where the potential of the GA is evident, because they have the capacity to explore wide regions of the solution space in a short time, in order to provide clusters of solutions with good potential to fulfill the design specifications. Furthermore, they search more like an engineer, because they do not look for an optimal point or solution, but they look for sets which can be committed to multiple criteria satisfaction.

5 Objective

The aim is to present the theoretical foundations for the application of evolutionary computing techniques for designing mechanical systems through the use of intelligent searchers in interactive catalogues.

6 Hypothesis

Clusters of valid solutions for configuring technical systems can be found by using an intelligent searcher to help the designer. (The case study presented are shafts with involute splines). The solutions must satisfy design constraints and parameters defined in the PDS (Product design specifications). The intelligent searcher will assist by guiding the designer to find suitable clusters of involute splines.

However, it is the designer's decision to perform the final selection, from the results provided by the intelligent searcher.

7 Development

- *Intelligent searcher operation.*

The *intelligent searcher* system has a data base with relevant information for the problem of technical system configuration; in our case, for configuring a shaft with involute splines. Among others parameters that have been considered are: the manufacturer's number for the splines; the types of splines manufactured previously, suppliers' list, selection procedures, standards, materials and heat treatments. The company involved now handles over 170 combinations of involute splines.

The data base is the particular library of the GA and it is used to generate the individuals at the beginning of the GA. Each individual has in its genes a representation of a specific configuration for the shaft being designed. The process is supported by programs called filters, they only allow the generation of valid solutions, according to the particular set of restrictions for the problem.

During the selection process, individuals are chosen probabilistically. Those individuals showing good fitness are prone to be selected. In other words, the feasibility of solving the restrictions for the shaft configuration is a main factor for the selection process. The individuals selected will be used by the genetic algorithm to develop a new generation of shafts configurations.

General scheme:

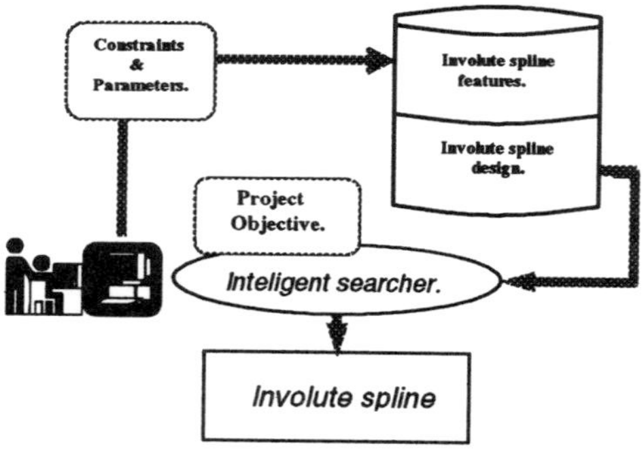

Figure 7.1 Representation of the operation of the intelligent searcher.

During the genetic material exchange stage, two different configurations represented by the character chains will form two completely new technical systems by exchanging part of their genes between them. This stage is crucial in the intelligent searcher because we cannot allow that the genetic algorithm generates invalid individuals as solution. For this reason, the splines are exchanged in corresponding pairs of male/female to make sure matching conditions for the shaft. This is achieved by incorporating connectivity restrictions combined with the reproduction operators (crossover and mutation), always taking in mind not to alter the content and meaning of the genetic information. Thus, we have a variant of the Simple Genetic Algorithm (SGA) reported by Holland and Goldberg (Holland, 1967; Goldberg, 1989).

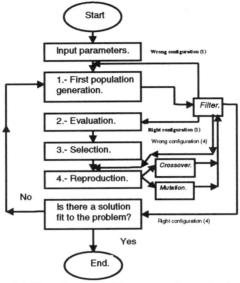

Figure 7.2 Flow diagram of the proposed genetic algorithm.

Whenever mutation is performed, the filters are applied to make sure a feasible solution is generated.

8 Case study

Variable transmissions using shafts of sliding gears, are very popular in the automotive industry. For power transmission, the involute spline shafts have milled slots around their circumference and along portions of their length, to achieve a sliding couple with the internal slots of the complementary piece, that could be a gear or something else.

Shafts with involute splines allow movements along the part, maintaining positive rotation and reference to the part coupled.

Figure 8.1 Involute spline on a shaft.

The slots of the involute splines can have straight side teeth or curved side teeth. They are ruled by the same guidelines and criteria of straight teeth gears. There are three different adjustments for coupling involute splines: sliding, firm and interference coupling. The adjustments are defined by combining the main dimensions of the involute splines of the shaft. Involute splines have their maximum strength at their base, that is the precise place for the maximum mechanical advantage, so that, they are preferred in applications for transmitting big loads with shock.

- *Theoretical justification.*

Solution space size.

The information of the involute splines is structured as shown in the Table 8.1 & 8.2

Part *	Teeth	Code *	Biggest Diameter	Smallest diameter	Pressure angle	Pitch
x45-1	15	83746	2.433	2.428	20	7 Esp.
y-3-5	26	35492	4.529	4.5265	20	6
z0-10	33	89708	2.181	2.171	37.5	16/32
c980	40	24365	3.45	3.44	30	12/24

Table 8.1 Information for processing.

Pitch Diameter	Base reference diameter*	Biggest circular thickness*	Biggest diameter between pins *	Smallest diameter between pins *	Pins *
2.1429	1.99987	0.2073	2.014	1.8734	0.1827
4.3333	3.9876	0.225	4.0756	3.8867	0.255
2.0625	1.74653	0.1281	1.9576	1.9089	0.10
3.3333	2.9645	0.1634	2.989	2.9633	0.13

Table 8.2 Information for processing.

Note: Sections with asterisk (*) were altered because of confidentiality.

The corresponding variables to involute splines are:

- Part.
- Maximum circular thickness.
- Diametral pitch.
- Code.
- Minimum diameter between pins.
- Pins.
- Biggest diameter.
- Circular pitch.

- Pressure angle.
- Teeth.
- Maximum diameter between pins
- Pitch diameter.
- Smallest diameter.
- Base reference diameter.
- Adjust.

Their functions are:

- ❖ $P = N/d$ — *Equation 1*
- ❖ $p \cdot P = \pi$. — *Equation 3*
- ❖ $C = \frac{1}{2}(d_{pinion} + d_{gear})$. — *Equation 5*
- ❖ $\phi = \cos^{-1}(R_b)/d$. — *Equation 7*
- ❖ $t = p/2$. — *Equation 9*
- ❖ $b = 1.25/p$. — *Equation 11*

- $p = \pi d/N$ — *Equation 2*
- $m = d/N$ — *Equation 4*
- $m_w = N_{gear}/N_{pinion}$. — *Equation 6*
- $B = 2(\Delta C)\text{sen}\phi$. — *Equation 8*
- $A = 1/p$. — *Equation 10*

- *Solution space.*

For developing an initial calculation, we consider that the following elements exist:

A_1 = 170 Parts. A_2 = 21 Teeth numbers.
A_3 = 25 Codes numbers. A_4 = 26 External diameters.
A_5 = 26 Internal diameters. A_6 = 4 Pressure angles.
A_7 = 16 Circular pitches. A_8 = 20 Pitch diameters.
A_9 = 25 Base diameters. A_{10} = 3 Backlashes.
A_{11} = 26 Major circular thickness. A_{12} = 26 Major diameters between pins.
A_{13} = 26 Minor diameters between pins. A_{14} = 19 Pins.
    So the number of possible solutions is:

❖ $A_1 \bullet A_2 \bullet ... \bullet A_{14} = 1.93 \times 10^{18}$.   *Equation 12*

Our chains for a shaft with four involute splines will have the following characteristics:

a a b b b c c c z z t a a b b b c c c z z t a a b b b c c c z z t a a b b b c c c z z t

{ Involute spline No 1 }{ Involute spline No 2 }{ Involute spline No 3 }{ Involute spline No 4 }

Each section of the code of the type a a b b b c c c z z t represents an involute spline shaft, the parts of the mentioned section are:

    - a a This is the first field of coding, representing the part number of a shaft spline.

    - b b b This is the second field of coding, representing the part number of a shaft spline.

    - c c c This is the third and last field of coding, representing the part number of a shaft spline.

    The fields a a b b b c c c link all the information of the design requirements, design parameters and restrictions involved. It also help us to represent the structure of the system with the support of Object Oriented Analysis (OOA) and Object Oriented Programming (OOP). Moreover, it defines the specific application of the involute spline. The variables a, b, c, could have a decimal value from zero (0) to nine (9). The processed information will be the one that is generated with decimals.

    - z z They represent the last two codes, representing of type involute spline involved.

    The field z z automatically links the code with the data base of the intelligent searcher, where there is the information developed by the company, together with their own characteristics. The z variables, could have a value from zero (0) to nine (9).

    - t Represents the involute spline type, that could be male or female. Hence the variable t could have a value of zero or one.

We can see the representation in Figure 8.2:

Modeling a shaft spline

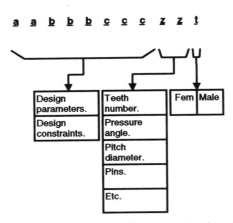

Figure 8.2 Representation of the processing functions.

- *Processing schemata.*

During this stage, the information content of the population is analysed, to understand the potential of performing the search and find clusters of high fitness (efficiency). The theory of *schemata* proposed by Holland (Holland, 1967), will allow us to see the probability of transmission of positive features from individuals of one generation to another.

In our case we have an alphabet of ten characters (digits from zero to nine) and a symbol of *do not* care, with a chain length of 44 characters, so, the number of possible schemata processed, according to Goldberg (Goldberg, 1989) is:

❖ $(n+1)^L = (n+1)^{44} = 11^{44} = 6.6264 \times 10^{45}$. *Equation 13*

where n is the number of characters (equal 10) and L is the chain length (equal 44).

- *Discrete solution field.*

The solution landscape for the present case study is discrete and non-continuous, because the variables processed during the design process of the involute spline are all discrete, because of reasons of geometry, design and manufacture.

9 Purposed implementation

- *Connectivity parameters and constraints.*

Among the parameters reducing the solution landscape size and simultaneously complicating the selection of valid individuals is *connectivity*. Although theoretically relations could be established between different involute splines that

28

physically are not compatible, because of machining, tolerances etc., they would form invalid individuals. For this reason, inside the filter programs there are routines that verify the corresponding data structure, to eliminate the possibility of physical incompatibilities during the selection. Other factors considered are constraints which originating from manufacturing, tools and materials, which actually are considered by the designers through empirical design guides. Inside the program, these guides will be incorporated either in the corresponding involute spline data structure key parameters or in the building procedure, that allow us to verify that these physical restrictions be considered in the crossover and selection process of the involute splines of the shafts.

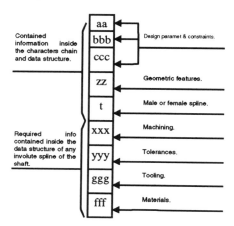

Figure 9.1 Proposed data structure.

- *Environment adaptation formulae (evaluation).*

The case study evaluation formulae are:

For flat root involute splines:

$$D_{Fe} = \frac{\sqrt{3N^2 + (N - 0.016P - 4.5)^2}}{2P}$$

(14)

$$\phi_z = \frac{D_{ri}}{D_b} - \left(\frac{s}{D} + inv\,\phi_D\right)$$

(15)

(Female involute spline)

For fillet root involute splines:

$$D_{Fe} = \frac{\sqrt{3N^2 + (N - 5.359)^2}}{2P} \qquad (16)$$

$$\phi_z = \frac{D_{re}}{D_b} - (\frac{t}{D} + inv\phi_D) + \frac{\Pi}{N} \qquad (17)$$

(Male involute spline)

For both cases:

$$\phi_z = sec\phi_z - (tan\phi_z - arc\phi_z). \qquad (18)$$

- *Experimental method.*

For the development of the application, it will look for the variable combination inside the genetic algorithm that delivers consistent solutions. In order to obtain it, the variations will be tested for the following operation parameters of the genetic algorithm: Generations number.

➢ Mutation probability.

➢ Crossover mechanism.

➢ Monitoring.

For the elements that are integral part of the case study, we will test with:

➢ Adaptation functions.

➢ Ponder over of the selection proceedings.

We will do different trial runs of the program, in order to monitor the relative performance of the results. Once the parameters of the rational use of the tool are defined, there will be a work stage with the designers from industry in order to confirm that they can evaluate and apply the tool effectively during the design process.

10 Expected Results

Once the programming stage is concluded, it is expected to obtain enough information about the convergence parameters for the problem. If the results are encouraging, it is anticipated that the designers of the participating company would adopt the tool for their day to day operation. Moreover, other parts of the gearbox would be treated in a similar way for the design stage.

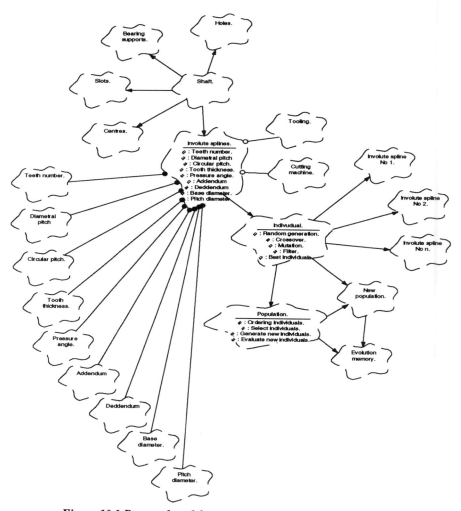

Figure 10.1 Proposed model using OOP and Booch representation.

11 Conclusions

The initial analysis of the problem for configuring shafts with involute splines using an intelligent searcher has been presented. The theoretical analysis have revealed a sizeable solution landscape, originating from discrete variables with non-linear behaviour, as a result of the different relations considered for the fitness function. The operational foundations of the intelligent searcher offer several possibilities for configuring and optimising simultaneously during the configuration stage, thus providing helpful design tools. However, these assumption are to be confirmed during the experimental stage presently being implemented.

12 Bibliography

[1] Aguirre, G.J. & Wallace, K.M. (1990). Evaluation of technical systems at the design stage. *International Conference on Engineering Design.* ICED.

[2] Axelrod, R. (1987). In L. Davis (Ed.), *Genetic algorithms and simulated annealing.* London: Pitman.

[3] Beasley D., et al. (1993). Overview of Genetic Algorithms, Research Topics. *University Computing.* London Whurr Publishers Ltd.

[4] Beasley D., et al. (1993). An Overview of Genetic Algorithms, Part 1 Fundamentals. *University Computing.* London Inter-University Committee on Computing.

[5] Bonomi, E., & Lutton, J.L. (1984). The N-city travelling salesman problem: Statistical mechanics and the Metropolis algorithm. *SIAM Review.*

[6] Brown Don R. (1993). Solving Fixed Configuration Problems with Genetic Search *Research in Engineering Design.* Springer-Verlag London-Limited.

[7] BSI 5750, Part 1 (1981), *Specification for Design, Manufacture and Installation,* British Standards Institution, London.

[8] Carlson. S.E., et al. (1994). Comparison of Three Non-derivative Optimization Methods with a Genetic Algorithm for Component Selection. *Journal of Engineering Design.* Journals Oxford Ltd.

[9] Carlson S.E. (1996). Genetic Algorithm Attributes for Component Selection. *Research in Engineering Design.* Springer-Verlag London-Limited.

[10] Chapman C.D., et al. (1994). Genetic Algorithms as an approach to Configuration and Topology Design. *Journal of Mechanical Design.* Vol 116.

[11] Coombs, S., & Davis, L. (1987). Genetic algorithms and communication link speed design: Constraints and operators. *Genetic algorithms and their applications: Proceedings of the Second International Conference on Genetic Algorithms.*

[12] Cramer, N.L. (1985). A representation for the adaptative generation of simple sequential programs. *Proceedings of an International Conference on Genetic Algorithms and Their Applications.*

[13] Darwin, Ch. (1889). *On the Origin of Species by natural selection.* 6th Edition. New York.

[14] De Jong, K.A. *A genetic-based global function optimization technique.* (Technical report No. 80-2). Pittsburg: Pittsburg University, Computing Sciences Department.

[15] Durand_N, Alliot_JM. (1996). Air traffic conflict resolution using genetic algorithms. *Nouvelle Revue Aeronautique Astronautique* No. 6.

[16] Goldberg, David E. (1989). *Genetic Algorithms in Search, Optimization, and Machine*

Learning. USA: Addison-Wesley.

[17] Hadamard, J. (1949). *The Psychology of Invention in the Mathematical Field.* Princeton, NJ: Princeton University Press.

[18] Holland, J.H. (1967). Nonlinear environments permitting efficient adaptation. In J. T. Tou (Ed.), *Computer and Information Sciences - II.* New York: Academic Press.

[19] Hugh, M. & Ben Jesson. *The analysis of waste flow data from multi-unit industrial complexes using genetic algorithms.*

[20] Jain AK. & Elmaraghy HA. (1997). Production scheduling/rescheduling in flexible manufacturing. *International Journal of production research.*

[21] Koza J.R. (1992). *Genetic Programming.* The MIT Press, Cambridge, Massachusetts. London, England.

[22] Krishnakumar, K. (1993). Genetic Algorithms A Robust Optimization Tool. *AIAA. 31st. Aerospace Sciences Meeting & Exhibit.* Reno. USA.

[23] Lee Y, et al. (1997). A genetic algorithm-based approach to flexible flow-line scheduling with variable lot sizes. *IEEE Transactions on Systems man and Cybernetics Part B- Cybernetics. Vol* 27, No.1.

[24] Mill F., et al. (1996). Shape and Topology optimization in engineering design with genetic algorithms. *Proceedings of ACED'96.* University of Plymouth. UK.

[25] Pahl G. & W. Beitz. (1986). *Engineering Design; a Systematic Approach.* 2nd edition. Berlin. Springer-Verlag.

[26] Parmee I.C. (1996). Towards an Optimal Engineering Design Process Using Appropriate Adaptive Search Strategies. *Journal of Engineering Design.* Vol. 7. No. 4. Journals Oxford Ltd.

[27] Pham D.T. / Yang Y. A genetic algorithm based preliminary design system. IMechE. *Proceedings of Mechanical Engineers.* Vol 207.

[28] Santillán S. (1998). Metaheuristic search using genetic algorithms for Boothroyd's Design for Assembly. *PhD. Thesis* Loughborough University 1998.

[29] Standard ANSI B92.1-1970. Involute Splines and Inspection.

[30] Thornton A.C. (1996). A Software Support Tool for Constraint Processes in Embodiment Design. *Research in Engineering Design.* Springer-Verlag London.

Post-Processing of the Two-Dimensional Evolutionary Structural Optimisation Topologies

H. Kim[†], O. M. Querin and G. P. Steven
Department of Aeronautical Engineering, Bldg. J07
University of Sydney NSW 2006 Australia
[†]email: alicia@aero.usyd.edu.au

Abstract. In this study, a post-processing algorithm for a two-dimensional topology is developed and investigated. It first converts the FE representation of an optimum design to the boundary representation. The spline control points are then determined and stored in a DXF file, which can be imported into most CAD/CAM packages. Although applied primarily to ESO-optimised structures, the algorithm is general enough for any FEA model. A few examples of the post-processing are displayed in this paper, some of which were manufactured. The post-processor minimises human-interaction, which also minimises the knowledge required by the user to carry out the optimisation and manufacturing of the design. The examples also demonstrate the effectiveness and user-friendliness of the algorithm.

1. Introduction

Many structural optimisation methods have been successful in providing the most structurally efficient design for a given engineering problem. With increasing computational power, these methods are becoming useful design tools, applied to realistic design problems. The most commonly employed tool for structural analyses during optimisation is Finite Element Analysis (FEA). FEA conducts its analysis by discretising the design into finite elements and hence the final optimum solution is often obtained by a group of regular-shaped elements. Whilst the topology can easily be identified visually on a computer screen, it is a cumbersome task to obtain the coordinates of smooth boundaries. However this boundary representation of a design is the data format that most CAD/CAM packages understand. Thus the FE representation of an optimum design must be converted to the boundary representation for the design to be communicated in a drawing or to be manufactured.

Evolutionary Structural Optimisation (ESO) is an optimisation method of an intuitive and heuristic nature. The original ESO concept was to slowly remove a small percentage of elements with low stress, approaching toward a fully stressed design. It is said that these lightly stressed elements are not efficiently utilised in carrying the applied load and hence can be removed with minimal effect on structural integrity. This process mimics nature's evolutionary process. Due to the simplicity of the concept, ESO has since been extended to accommodate other optimisation criteria such as stiffness and displacements, and applied to various problems including multiple loads, dynamic and buckling problems, [1].

ESO is comprised of two steps: FEA and optimisation. When a problem is specified, a FEA is conducted to the given design and determines the stress of each element. The design is then modified by removing elements according to the FEA results. FEA is applied again to the modified design and computes the new stress distribution, and the process is repeated until an optimum is found. Thus the final topology is represented by finite elements with jagged edges along the boundary, which requires excessive geometry interpretations to obtain smooth boundaries, often losing some optimal features. Figure 1.1 depicts a typical topology obtained by ESO - an optimum solution of a 1.6 aspect ratio cantilevered beam with a tip load. A topology is defined by a cluster of elements which in turn are defined by multiple nodes. Whilst the outlines or boundaries of a topology may be visually clear to the user, the ordered extraction of the boundary nodes and the conversion from FE to boundary representation is often a time-consuming and cumbersome task.

Figure 1.1 Optimum Cantilevered Beam of Aspect Ratio 1.6 by ESO

Most CAD/CAM packages employ boundary representation of a structure which consists of the coordinates of the boundary along the outlines of a topology, [2]. It is often these CAD/CAM packages from which the designers/engineers produce detailed drawings, virtual prototypes and/or tool paths for manufacturing. Thus the elemental representation of an optimal topology needs to be converted to boundary representation in order to manufacture or further modify the optimal structure.

This manuscript introduces an algorithm applicable to a FE representation of a topology to obtain smooth boundaries in a file format recognised by most CAD/CAM packages. Section 2 outlines a method which identifies the boundary node coordinates along the boundary. It then determines the coordinates of spline control points that can be used to smooth out the jagged edges of an optimal topology in section 3. These jagged edges are present simply due to the nature of the ESO method which removes whole elements, and the boundary needs to be smoothed-out in order to manufacture the structure. The spline control points are stored in a Drawing Interchange Format (DXF) file. DXF has become a standard file format to communicate true three-dimensional objects amongst CAD/CAM packages, [2]. Section 4 discusses the advantages and disadvantages of using DXF in more detail. This algorithm was applied to ESO optimised topologies, some of which were eventually manufactured. A few examples are given in section 5, followed by a conclusion.

2. Boundary Node Extraction

2.1. Problem Definition

When an optimisation problem is specified, a mesh is generated and the initial design is represented by the node coordinates of the elements. ESO removes elements by identifying which elements are needed and which are not. The node coordinates are stored in a random order and the elements which are defined by the nodes are also stored in a random order. Hence when an optimal structure is obtained, it is possible to visualise the topology as a cluster of elements on a computer screen, but there is little information on the connectivity between these elements for further modifications or for manufacturing.

A plate is defined by nodes, a typical continuum mechanics element for two-dimensional elasticity has four nodes. When an element is connected to another element, there is at least one node which is shared by two elements. When an element is on a boundary, there is at least one side of the element which is not connected to another element. With this information, an algorithm was developed in this research to determine the boundary node coordinates and rearrange them in an anti-clockwise manner. The following sections detail the algorithm.

2.2. Number of Boundaries

Any object has at least one boundary, ie. one that defines the outside shape. If a structure has a cavity, there is another boundary that defines the cavity. This boundary is not connected to the outside boundary and it can be identified as a separate boundary. Hence if there are N number of cavities, there are (N+1) boundaries that define a structure.

The proposed algorithm finds a boundary node, and searches the subsequent adjacent nodes in the anti-clockwise manner until all nodes on the boundary are considered. The algorithm then searches for another boundary node which was not included in the previous boundary and finds all the nodes which are on the new boundary. This process is repeated until all boundary nodes have been accounted for and this becomes the terminating condition for the boundary extraction routine. The number of boundaries can be counted at the end of this routine.

2.3. Finding a Boundary

The first step of the algorithm is to find the starting node of a boundary. For the first boundary, it finds a node on the outside boundary by determining the node at the bottom left hand corner. A boundary node is identified by a node on a boundary element which has at least one free side, ie. there is at least one edge which is not connected to another element. The algorithm then finds the boundary nodes adjacent to the previous nodes along the boundary until the starting node is met.

Once all nodes on the boundary have been determined, the algorithm finds another boundary. In order to achieve this, it searches through all boundary nodes and finds one that was not included in the previous boundaries. If another boundary is found, all the nodes on that boundary are determined. If all boundary nodes have been included in the previous boundaries, all the boundaries have been determined and the boundary node extraction routine is terminated.

2.4. Finding the Adjacent Nodes

In order to find the node which is adjacent to the current node on the boundary, the algorithm first searches for the boundary element that shares the current node. From this element, it finds the boundary edge of the plate and hence finds the next node coordinate. For example in figure 2.1, N1 and N2 have been accounted for and the current node is N2. The algorithm then finds boundary edges which contain N2, ie. E1 and E2. Thus the next node is either N1 or N3, however N1 has already been accounted for, thus N3 is obtained as the next node. This process is repeated until the next node found is the starting node.

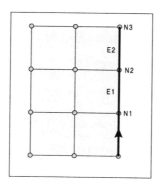

Figure 2.1 Boundary of a Structure

However there are some cases where the above method cannot be applied. Figure 2.2 displays such a case where all four adjacent nodes are on boundaries. N3 is connected to four other nodes which are all on the boundaries: N2; N4; N6 and N7. N2 has been accounted for previously hence it is excluded. N6 and N7 are on a different boundary and the algorithm must be able to recognise this in order to select N4 as the next node.

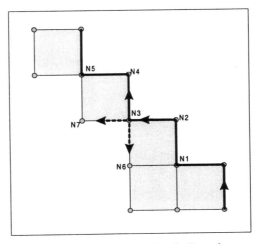

Figure 2.2 Another Example of a Boundary

In such a case, the coordinates of two plate centres, C1 and C2 are determined and the gradient of C1-C2 is computed, figure 2.3. Also the midpoints of the four boundary edges which include N3 are obtained to calculate the gradients between these points. For the above example, the midpoints are A, B, C and D as marked and thus three gradients calculated are of AB, AC and AD. Then the next boundary edge is the edge with a mid-point gradient value nearest to the plate centre gradient value. Comparing AB, AC and AD with C1-C2, it can be seen that AB would have the same gradient value as C1-C2. Thus the edge N3-N4 is selected as the next boundary edges and N4 becomes the next boundary node, and N6 and N7 are discarded.

This selection algorithm is repeated until the next boundary node is the starting node. At this point, all the nodes along the boundary have been selected and the algorithm searches for the new boundary using the method explained in section 2.3.

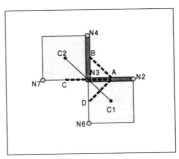

Figure 2.3 Boundary Node Selection of N4

2.5. Finding the Second Node

All boundary nodes are connected to two other nodes along the given boundary. The above method of finding the adjacent nodes in section 2.4 discards one of these two, which has already been accounted for. This applies for all nodes along the boundary except for the first boundary node, as neither of the adjacent nodes has been accounted for. It was proposed that all boundary nodes were to be arranged in the anti-clockwise direction, as it is an accepted standard direction for general engineering applications including FEA. Thus the second node must be the node adjacent to the first node in the anti-clockwise direction.

The algorithm first finds the two nodes adjacent to the first node. In figure 2.4, the starting node is N1, and N2 and N3 are the adjacent nodes. Then the cross-product of the vectors N1-N2 and N1-N3 is computed. If the cross-product is positive, then N2 becomes the next node, otherwise N3 becomes the next node.

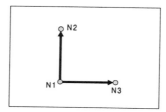

Figure 2.4 Determination of Second Node

2.6. Summary of Boundary Node Extraction Method

The methodology described in sections 2.2 to 2.5 are summarised in figure 2.5.

```
Let  N  be  total  number  of  boundaries,  then
initially N = 0;
Find the bottom left hand corner node, this node
becomes the First Node;
Do while (a new boundary exists),
    N = N + 1;
    Find the second node, this node becomes the
    Current Node (section 2.5);
    Do while (Next Node /= First Node),
        Find Next Node to Current Node (section
        2.4);
        Current Node = Next Node;
    End do;
    Find another new boundary (section 2.3);
End do;
```

Figure 2.5 Summary of Boundary Node Extraction Routine

3. Generation of Spline Control Points

3.1. Defining Segments

In an ESO optimised structure, the topology is described by horizontal and vertical lines, ie. segments due to the use of square elements, figure 1.1. Any straight lines at an angle or curves are therefore represented by jagged-edges.

Boundary representation defines a topology by the coordinates of the corners where the gradient of the curve changes. The boundary extraction routine described above offers all the node coordinates on the boundaries. However a straight line can be defined by two end points and it would be inefficient to store all nodes. Thus the boundary node coordinates are examined and only the two end coordinates of straight lines are stored and the segments of the topology are defined.

3.2. Spline Control Points

Holl [3] has attempted to develop an algorithm for fitting splines along the boundaries of the two-dimensional ESO topologies. Even though the algorithm largely depended on human interactions, spline-fitting using the mid-points of the horizontal segments as the control points produced reasonably good smooth topology. It was also found that this method preserves the total area of the topology after fitting the splines. It was proposed here to use the mid-points of the vertical segments as well as the horizontal segments.

The segments of the boundaries are identified by the end coordinates as described in the previous section. Using this information, the spline control points can be determined by computing the coordinates of the mid-points of each segment. These spline control points can be used to obtain a smooth geometry of the topology, by importing into any CAD/CAM software to generate splines or further manipulate the topology.

4. Drawing Interchange Format (DXF)

Computer-Aided Design (CAD) and Computer-Aided Manufacture (CAM) are some of the most popularly used tools in design. With the increasing use of computers and their networking capabilities, CAD/CAM has become a standard drawing tool in the engineering and design industry. Most CAD applications such as AutoCAD and Cadkey are able to display various file formats such as metafiles (WMF) and TIFF files as well as their own formats, eg. DRW in AutoCAD. In 1982, Autodesk introduced the Design Exchange Format (DXF) as a means of communicating design information between CAD and other applications [2]. DXF provides users an editable ASCII file and is capable of representing a complete 3D structure.

Due to the popularity of AutoCAD, the DXF format is widely supported by other CAD products as well as some word processors, desktop publishing and illustration tools on PC, MacIntosh and UNIX systems. Also many graphics applications such as Draw, Canvas and Autosketch are also able to import DXF files. Thus engineering drawings in DXF files can be transferred to any CAD and some non-CAD applications. Also the DXF format is able to represent full three-dimensional (3D) vector information and is therefore capable of communicating true 3D objects.

There are some disadvantages to using DXF files. Even though, a DXF file exists in both ASCII and binary formats, the DXF was originally developed in ASCII format, and this is still the most common form. The ASCII DXF format is slow to read and is not efficient for storage. Also DXF readers are often poorly implemented in some non-CAD applications. Nevertheless ASCII DXF files are still an accepted standard in exchanging engineering drawings.

It is for these reasons that the DXF ASCII was selected to store the coordinates of the spline control points. This allows further processing and/or manufacturing of the optimal topology, resulting in the ESO process becoming more easily applicable to design industry.

5. Examples

5.1. Short Cantilevered Beam

A classic short cantilevered beam was optimised, [1] and the optimum design is shown in figure 1.1. Applying the post-processor resulted in the final topology of figure 5.1. The tool path of this topology was generated and using a NC milling machine, the prototype was obtained, figure 5.2.

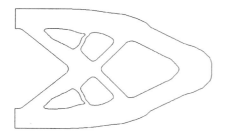

Figure 5.1 Optimum Design of the Cantilevered Beam

Figure 5.2 Final Product

5.2. Shape Optimisation of Animal Claw

The next topology was obtained from a shape optimisation of an animal claw using biological growth laws for multiple load cases, [4]. Figure 5.3 depicts the initial design with the boundary conditions and loading conditions. The initial design was constructed from a circular curvature. Two separate loads were applied at the tip: a pulling shear and a normal force.

Figure 5.4 was obtained as an optimum and the finite element representation of the optimum design was converted into the boundary representation of the spline control points. It was then imported into Mastercam as shown in figure 5.5, ready to be manufactured.

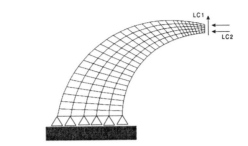

Figure 5.3 Initial Design

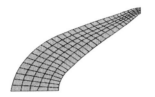

Figure 5.4 FE Representation   Figure 5.5 Optimum Design

5.3. Optimisation of Printed Circuit Board (PCB) Substrate

A printed circuit board (PCB) substrate was optimised for transient temperature field. There were 4 heat sources (F_1, F_2, F_3 and F_4) as shown in figure 5.6, and the non-design domain represents the area which is not to be modified, [5].

The optimum topology of figure 5.7 was obtained by applying thermal ESO. Applying the algorithm described in the previous sections extracted the boundary coordinates of the finite element representation. The DXF file of the spline control points were then imported into Mastercam and the splines were generated as displayed in figure 5.8, [6]. This can further be interpreted and modified if desired, or an engineering drawing can easily be produced from typical CAD/CAM packages. This boundary information can also be used to generate the Numerical Control (NC) machine tool paths for manufacturing the topology.

42

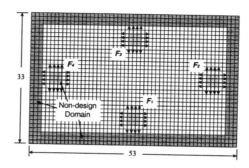

Figure 5.6 Initial design Domain of PCB Substrate

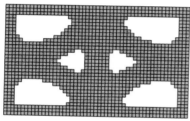

Figure 5.7 Finite Element Representation
of Optimum Topology

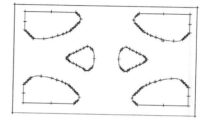

Figure 5.8 Optimum Topology

5.4. Optimisation of Bicycle Frame

A bicycle frame subjected to five multiple loading cases was optimised using Nibbling ESO, [1]. The initial design domain was as depicted in figure 5.9. The frame was fixed from all movements at the back wheel and only horizontal movement was allowed at the front wheel. There are three points of applications of loads: handle bar; saddle and bottom bracket. Five loading cases were considered: starting; climbing; speeding; braking and rolling, [7].

An optimum design for the problem was obtained as shown in figure 5.10. This finite element representation of the topology was then converted into a series of spline control points along the boundary, figure 5.11.

It was recognised that the boundary sections 1 and 2 marked in figure 5.11, are close to straight lines and may be represented by straight lines for ease of manufacturing. Thus three splines and two straight lines were fitted to the boundary and figure 5.12 was obtained. This final topology was 3-dimensionalised by applying an aerofoil cross-section and a plug was manufactured using a NC milling machine, figure 5.13. The composite bicycle frame thus was produced, figure 5.14.

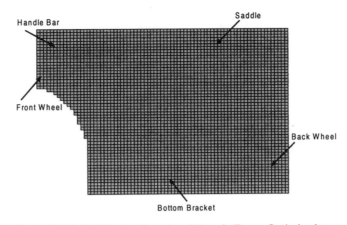

Figure 5.9 Initial Design Domain of Bicycle Frame Optimisation

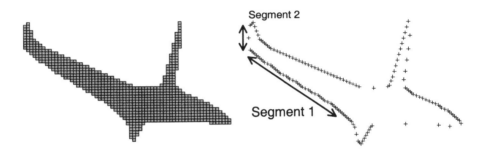

Figure 5.10 Optimum Bicycle Frame   Figure 5.11 Spline Control Points

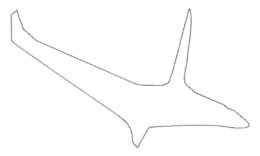

Figure 5.12 Final Optimum Topology

Figure 5.13 Bicycle Frame Plug   Figure 5.14 Manufactured Bicycle Frame

6. Conclusion

The post-processing algorithm discussed in this manuscript has three parts: Conversion of FE to boundary representation of a topology; determination of spline control points and importing into common CAD/CAM packages using DXF. The conversion of FE to boundary representation of a structure requires only the coordinates of the nodes and the element connectivity. This information is available for all FE models. Thus this conversion algorithm is applicable to any structures represented by two-dimensional finite elements. Storing the spline control point data in a DXF file provides a user the options of producing detailed drawings or manufacturing the model or further manipulating the optimum solution. As demonstrated through the various examples, this post-processing algorithm is robust and effective.

References
1.   Xie Y M, Steven G P, 1997. *Evolutionary Structural Optimization*. Springer-Verlag, Berlin.
2.   Autodesk, 1998. *AutoCAD Customization Guide*. u14.1.04.
3.   Holl N, 1997. Design and Manufacturing of Structures Using the ESO Method and 'Evolve'. BE thesis, University of Sydney, Australia.
4.   Querin O M, Lencus A, 1998. Optimisation of Structures Subject to Multiple Load Cases Using Biological Growth Laws. In: Steven G P, Querin O M, Guan H, Xie Y M (eds), 1998. *Proc. the Australasian Conference on Structural Optimisation*, Oxbridge, Sydney, Australia, pp 251-258.
5.   Li Q, Steven G P, Xie Y M, 1999. Thermoelastic Topology Optimization for Problems with Varying Temperature Fields. Submitted to *J Therm Stress*.
6.   CNC Software 1994. *Mastercam Design Version 5 User Guide*.
7.   Mechler J, 1993. Structural Optimisation of a Bicycle Frame. BE thesis, University of Sydney, Australia.

Optimisation of a Stator Blade Used in a Transonic Compressor Cascade with Evolution Strategies

Markus Olhofer[1]   Toshiyuki Arima[2]
Toyotaka Sonoda[2]   Bernhard Sendhoff[1]

[1]Honda R&D Europe (Germany)
Future Technology Research
D-63073 Offenbach/Main
Germany

[2]Honda R&D Co. Ltd.
Wako Research Centre
Wako-shi, Saitama, 351-0193
Japan

Abstract. The evaluation of fluid dynamic properties of various different structures in aerodynamic design optimisation is a computationally demanding process. For the application of evolutionary algorithms it would therefore be beneficial to restrict the population size to a minimum even if parallel genetic algorithms are employed. In this paper, we will show that specific evolution strategies can be successfully used for design optimisation even in the transonic regime with small population sizes and a full Navier Stokes solver for the evaluation. Furthermore, we analyse the self-adaptation properties of the evolution strategy.

1   Introduction

Evolutionary algorithms have recently been applied to the area of design optimisation, in particular to the air foil and wing design for aerodynamics [1–3] for a number of reasons. Firstly, gradient information, which would have to be estimated numerically, is not necessary during the optimisation and secondly, the stochastic component of evolutionary algorithms (EA) makes it possible to escape local minima. Furthermore, the population based approach inherent in EAs is very suitable for multi-objective optimisation [3,4]. On the other hand, the population size can pose a serious problem with regard to the computation time needed for one generation. In most applications parallel genetic algorithms are used, however, population sizes of the order 100 or more still seem infeasible for applications where a numerical solution of the Navier Stokes equation is needed. Indeed many applications reported in the literature use the Euler equation for computational fluid dynamics (CFD) calculations. However, as we will see in Section 2.1, the simplifications in the Euler method cannot be carried out for turbulent transonic flow, where viscosity and heat conduction have a large influence on the energy loss. Therefore, it would be beneficial if large populations could be avoided even if parallel algorithms are employed.

In this paper we will apply a specific version of the evolution strategy (ES), which extends the self-adaptation property of the standard ES by using a derandomised, cumulative step-size adaptation method. One of the main advantages of this approach is the possible reduction of the population size. In addition, for reasonably smooth fitness landscapes it was shown [5] that the ES outperforms the GA. According to our own experiments, the design optimisation task described in this paper can be regarded as reasonably smooth. In the following section, we will outline the encoding of the stator blade and describe the CFD simulation. In Section 3, we will briefly introduce the ES with particular attention paid to the lesser known extensions which allow the reduction of the population size. The results will be described in Section 4 together with an analysis of the self-adaptation of the strategy parameters during the optimisation. In Section 5 we will discuss our findings.

2   Encoding of the airfoil and CFD simulation

The encoding of the air foil geometry is one of the most critical steps in the optimisation since, together with the evolutionary operators it determines the modifications of the structure during the optimisation process and therefore strongly influences the evolution path. There are numerous constraints which should be fulfilled by the encoding, e.g.:

Completeness The encoding has to guarantee a maximal degree of freedom for the generated geometries, otherwise the underlying model introduces unnecessary constraints on the phenotype space. In the worst case it might not be possible to encode the optimal structures.

Causality Causality implies that small steps on the genotype space lead to small steps in the phenotype space [6]. This attribute is especially necessary for the adaptation of the strategy parameters in evolution strategies.

Compactness It is necessary to minimise the dimensionality of the encoding in order to reduce the dimension of the search space and in turn the calculation time.

Our initial encoding used an intuitive model, where each parameter had a specific aerodynamic meaning. It was based on engineering knowledge for the description of the stator blade and was designed to use heuristics for the optimisation. As such its applicability was confined to conventional designs. For a more complete and less biased description of the search space it was too restrictive.

When we devised an alternative model, it was our aim to comply with the above described conditions to a greater extent. The model used fulfils the demand for strong causality and is a trade-off between a high degree of freedom for the structure generation and a low dimensionality of the genotype space: The upper and lower side of the air foil are each described by five points fixing two 4th order polynomials. To reduce the dimensionality of the problem the x-coordinates of the points are fixed and only the y-coordinates are used as parameters for the optimisation. This method is feasible because of the special structure of the compressor cascade stator air foils. The leading and the trailing edges are described by a circle, which was given as a constraint for the design, see Figure 1. In this way, the encoding is reduced to 11 parameters, of which 10 represent the y-coordinates of the control points in a Cartesian coordinate system and one parameter determines the distance between two successive blades in a cascade.

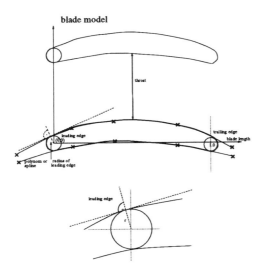

Fig. 1. The underlying air foil model based on two 4th order polynomials is fixed by 10 control points, which are the parameters of the model. The circles which depict the leading and trailing edges are generated in such a way that the polynomials are tangential to the circle. If this construction is not possible, then the geometry is marked as invalid.

2.1   CFD Simulation

In order to evaluate the quality of the generated geometries, a 2D Navier Stokes flow solver with a low-Reynolds number k-ϵ turbulence model is used [7]. The grid size is fixed during the optimisation to 191×55. With grid sizes which are equal or larger to this value, the calculation is nearly independent from the number of grid points for the range of geometry modifications and the necessary computation time is acceptable. However, it should be mentioned that the design of an universal grid, for which the numerical accuracy for all different geometries encountered during optimisation is roughly the same, is still an unsolved and interesting problem.

Additionally, we continuously monitored whether the flow solver converged. If no convergence could be achieved after a given time, the results were marked as invalid and the quality function was replaced by a sufficiently large penalty term. At the same time, we used the convergence monitor to stop the calculation in case of earlier convergence to reduce the computation time to a minimum. Due mainly to this technique the usage of the full Navier Stokes flow solver with a turbulence model becomes feasible. The need for the Navier Stokes solver as opposed to the simpler and computationally less demanding Euler method is demonstrated in Figure 2. In the correlation diagram, the x-coordinate describes the resulting quality value using the Euler method and the y-coordinate the value resulting from Navier Stokes method. A correlation cannot be seen, which clearly indicate that the simplifications in the Euler method cannot be made under these streaming conditions.

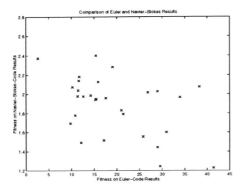

Fig. 2. Correlation diagram between the fitness values determined by the Euler and the Navier Stokes flow solver. The absolute difference between both solutions is not important for the optimisation, however, both results have to be strongly correlated in order not to mislead the search process during optimisation. The figure shows that this constraint is not satisfied, therefore, for the transonic regime the full Navier Stokes model has to be employed.

3 Evolution strategy with covariance matrix adaptation (CMA)

Standard evolution strategies have been described in several textbooks, e.g. [8], and therefore we will only briefly outline the lesser known extensions which have been proposed by Ostermeier [9] and Hansen [10,11]. Firstly, this is the derandomised strategy which reduces the stochastic influence on the self-adaptation of the strategy parameters. In the original mutative self-adaptation scheme, as proposed by Schwefel [8], both the strategy parameters, as well, as the objective parameters are subject to independent stochastic mutations. The idea behind the derandomised strategy is to use one stochastic source for both the adaptation of the objective and of the strategy parameters. In this case, the *actual* step length in the objective parameter space is used to adapt the strategy parameter, e.g. $\sigma(t)$ ($N(\mathbf{0}, \mathbf{1})$ denotes a random vector whose components are Gaussian distributed random variables with zero mean and variance equal to one):

$$\sigma(t) = \sigma(t-1) \exp\left(\frac{1}{d}\left(|z| - E[|N(\mathbf{0}, \mathbf{1})|]\right)\right), \quad z \sim N(\mathbf{0}, \mathbf{1}) \tag{1}$$

Equation (1) results in the following, simple, however successful, effect: If the mutation was larger than expected ($|z| > E[|N(0,1)|]$), then the strategy parameter is increased. This ensures that if this larger mutation was successful (i.e. the individual was selected), then such a larger mutation will again occur in the next generation, since $\sigma(t)$ was increased. The same argumentation holds if ($|z| < E[|N(0,1)|]$). Therefore, the self-adaptation of the strategy parameters depends more directly on the local topology of the search space (a topic which has received some criticism as being less "evolutionary" [8]).

The second method is the introduction of the cumulative step size adaptation. Whereas the standard evolution strategy extracts the necessary information for the adaptation of the strategy parameters from the population (ensemble approach), the cumulative step size adaptation relies on information collected during successive generations (time averaged approach). This leads to a reduction of the necessary population size. The main idea is to avoid strong correlations (positive or negative) in successive step sizes, because such cumulative steps can be more efficiently realized by single steps.

In the CMA algorithm, the full covariance matrix of the probability density function

$$f(z) = \frac{\sqrt{\det(C^{-1})}}{(2\pi)^{n/2}} \exp\left(-\frac{1}{2}(z^T C^{-1} z)\right). \tag{2}$$

is adapted for the mutation of the objective parameter vector. This has the advantage that the mutation direction is independent from the choice of the coordinate system and correlations between parameters can be represented. If the matrix B satisfies $C = BB^T$ and if $z_i \sim N(0,1)$ is a Gaussian distributed random variable with zero mean, then $B z \sim N(0,C)$. The adaptation of the objective vector can then be written as:

$$x(t) = x(t-1) + \delta(t-1) B(t-1) z, \quad z_i \sim N(0,1), \tag{3}$$

where $\delta(t-1)$ denotes the global step-size of the strategy. Thus, the overall mutation length can be adapted on a faster time scale (one parameter) than the direction which needs the adaptation of the covariance matrix. Since C^{-1} has to be positive definite with $\det(C^{-1}) > 0$, the different matrix entries cannot be determined independently and the detailed adaptation algorithm combined with the cumulative step-size approach is more involved, see [11,10,12] for a detailed description.

3.1　The quality function

The quality function of the objective parameter vector x consists of four terms. The first one depends on the pressure loss (ω), which is a direct measure of the energy loss and should be minimised. The second term measures the deviation of the averaged angle of the gas stream at the end of the blade

α_2 from the specified value $\tilde{\alpha}$, which is the target for the optimisation. A tolerance range ϵ is given for this difference. Finally, a third and a fourth term take geometrical constraints into account. The minimal diameter of the blade d_1 has to be larger than a given minimal thickness d_{min} and the maximal diameter d_2 of the blade has to be larger than a given value d_{max} to fix the minimal diameter at the thickest part, which is necessary for stability and manufacturing.

$$f(\boldsymbol{x}) = \eta_1 \cdot \omega + \eta_2 \cdot f_2(\alpha_2) + \eta_3 \cdot f_3(d_1) + \eta_4 \cdot f_4(d_2) \qquad (4)$$

$$f_2(\alpha_2) = \begin{cases} 0 & , \quad \tilde{\alpha} - \epsilon \leq \alpha_2 \leq \tilde{\alpha} + \epsilon \\ (\tilde{\alpha} - \alpha_2)^2 - \epsilon^2 & , \alpha_2 < \tilde{\alpha} - \epsilon \text{ or } \alpha_2 > \tilde{\alpha} + \epsilon \end{cases} \qquad (5)$$

$$f_3(d_1) = \begin{cases} 0 & , d_1 \geq d_{min} \\ (d_{min} - d_1)^2 & , d_1 < d_{min} \end{cases} \qquad (6)$$

$$f_4(d_2) = \begin{cases} 0 & , d_2 \geq d_{max} \\ (d_{max} - d_2)^2 & , d_2 < d_{max} \end{cases} \qquad (7)$$

In equation (4), η_i denote the factors which allow a weighting between different objectives; they have to be chosen manually. In order to include given constraints concerning outlet angle and geometry into the fitness function the nonlinear functions f_i are introduced. As long as the corresponding values are in the given tolerance range or larger than the minimal value the function is equal to zero. If the deviation angle is too large or the geometrical constraints are not fulfilled the corresponding terms act as a penalty term.

The weighting factors η_i are necessary to normalise the range of possible values for the different terms. Furthermore, they allow certain criteria to be prioritised. The pressure loss ω varies during the optimisation between 0.095 at the beginning and 0.055 at the end, see Figure 5. The factor for the outlet angle was set to $\eta_2 = 10$ and the factor for the geometric constraints was set to $\eta_3 = \eta_4 = 10^7$. If we take the range of the values of the angle and the thicknesses during the optimisation into account, the relation between the different terms can be estimated. The maximal difference between the outlet angle and the design value was 2.5°. The tolerance in equation (5) was set to $\epsilon = 0.3°$. The maximal differences between the stipulated and the measured minimal thickness at the thickest and the thinnest part were maximal 1% during the optimisation. Therefore, the relation between the terms in the order given in equation (4) can be estimated as (1 : 100 : 30 : 30) at the beginning of the optimisation. As a consequence of the small influence of the pressure loss and the high influence of the constraints, the shape is modified to an "allowed" shape in the first step. After the constraints for the outlet angle and the blade geometry have been fulfilled, minimisation of the pressure loss becomes the main target of optimisation. This has the advantage that during the whole optimisation valid shapes are created.

Alternatively, instead of combining the different terms into one quality function, which makes it necessary to fix the unknown parameters η_i, it would have been possible to strive for the determination of the Pareto set of possible solutions in the framework of a multi-objective approach to design optimisation. The main drawback is that in order to determine the Pareto set the population size has to be increased significantly.

4 Experimental Results

The presented results are calculated with a (1,12) evolution strategy with covariance matrix adaptation as described in Section 3. The population was initialised with an existing air foil geometry in order to start with an air foil whose quality can be calculated with the Navier Stokes flow solver. In Figure 3 the history of the overall quality during the optimisation process is shown. The number of quality evaluations (calculations) rather than generations is shown on the x-axis of all plots. Figure 5 and 6 show the development of the pressure loss (ω) and the deviation angle (α_2) during the optimisation. The outliers in Figure 5 correspond to solutions which did not converge and which were removed from the population. Since the deviation angle has a strong influence on the quality function (η_2 is comparably large), mainly (α_2) is optimised during the very first generations (~ 10). Thereafter, the deviation angles of the individuals fluctuate near the given tolerance range and the optimisation of the pressure loss becomes more important.

In order to observe the self-adaptation of the strategy parameters, the global step size δ and the condition of the covariance matrix during the optimisation are shown in Figure 7 and in Figure 8. The condition of a matrix is the relation between its largest and smallest eigenvalue. The eigenvalues are shown in Figure 4. If all eigenvalues are equal (thus the condition of the matrix is one), the iso-density lines of the probability density function for the mutation of the parameters are circles and no adaptation of the strategy parameters has taken place. The larger the difference between the eigenvalues the more pronounced is the "cigar-shape" of the iso-density lines and the more distinct is the preference to one direction. The magnitude of the mutation is controlled by the global step size.

For the structure optimisation task of the compressor cascade stator blade, we observe the following characteristics in the figures. After 50 to 100 calculations, which in the case of the used (1,12) ES corresponds to the 4th to 9th generation, the outflow angle lies within the tolerance range and the optimisation of the pressure loss starts. Overall, the self-adaptation of the covariance matrix during the whole evolution process is successful as is indicated by the trend of the condition of the matrix and the development of single eigenvalues in Figure 4. It is interesting to note that after generation 25 (300 calculations) the largest eigenvalue increases. This indicates that at this time adaptation of the covariance matrix has taken place and a direction for the search path has been found. At the same time, the global step size strongly increases by a factor of four. The behaviour of the covariance matrix monitored by the eigenvalues and the development of the global step size clearly indicates an adaptation of the strategy parameters to the local topology of the fitness function after about 25 generations. These observations imply that a significant adaptation of the covariance matrix occurs much earlier for this problem than the approximation of the adaptation time $O(n^2)$ given in [10,11]. After about 450 calculations the covariance matrix becomes unfavourable. Since the mutation rate is high according to the large global step size and since the largest eigenvalue of the covariance matrix is reduced, the diversity within the population is increased and no further progress is achieved at this point. As a result, the global step size becomes smaller and the variation in the eigenvalues indicates a new adaptation of the covariance matrix to the slightly different local topology of the fitness function[1]. Of course, we should be careful when drawing precise conclusions from the observations which only rely on one particular optimisation run (although similar results have been observed during other design optimisations of the stator blade). However, we can conclude that a self-adaptation of the strategy parameters indeed occurs and fortunately at an earlier stage than expected.

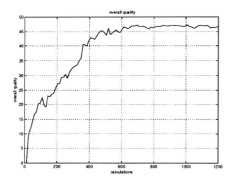

Fig. 3. The quality of the best individual in different generations.

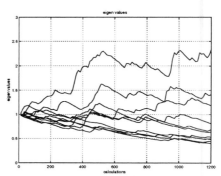

Fig. 4. The eigenvalues of the covariance matrix during optimisation.

5 Discussion and Conclusion

In this paper we successfully applied the derandomised evolution strategy with covariance matrix adaptation to the design of a transonic stator blade for a gas-turbine engine. With the CMA algorithm it was possible to reduce

---

[1] This behaviour is different from escaping local optima, where it is generally assumed that the step-size increases. In our case, we believe that the *new* adaptation occurs as a result of solely local properties, like for example if we overshot the minimum.

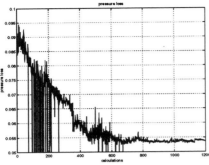

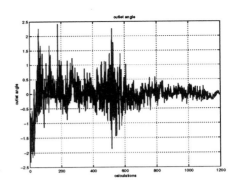

Fig. 5. The development of the pressure loss during optimisation

Fig. 6. The development of the deviation angle during optimisation

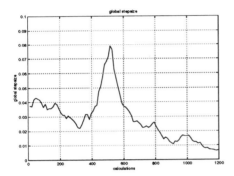

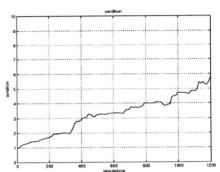

Fig. 7. Adaptation of the global step size parameter during optimisation.

Fig. 8. The condition of the covariance matrix during optimisation.

the population size considerably, so that together with a coarse parallelisation (twelve Sun workstations) and a dynamic termination of the simulation process, we were able to run the optimisation with a full Navier-Stokes solver and the k-ε turbulence model for 100 generations (this corresponds to 1200 calculations, compared to a typical population size of 100, only 12 generations would have been possible).

The resulting blade design is considerably different from the "standard" engineering design, which indicates that the evolutionary optimisation resulted in more than a fine tuning (two patents have been submitted in Japan for the optimised blades [13]). The stator blade is currently being built and tested in a wind tunnel at the Virginia Polytechnic Institute and State University; first results are very promising.

We also analysed the self-adaptation of the strategy parameters during evolution and ascertained that the optimisation strategy is very well adjusted to the local topology of the fitness landscape. In addition it also seems to be able to cope with changes in this topology during search.

We did not employ a multi-objective approach which might be one of the directions for future work. However, in particular with reference to 3D optimisation, it seems difficult to further increase the number of structure evaluations. Instead it seems likely that "real-world" applications of multi-objective optimisation will only be possible together with some kind of meta-model for evaluation.

Acknowledgements The authors would like to thank E. Körner and Y. Jin for support and stimulating discussions.

References

1. A. Vicini and D. Quagliarella. Airfoil and wing design through hybrid optimization strategies. In *16th Applied Aerodynamics Conference*. American Institute of Aeronautics and Astronautics, 1998. AIAA Paper 98-2729.
2. D.J. Doorly. Parallel genetic algorithms for optimization in cfd. In J. Périaux and G. Winter, editors, *Genetic Algorithms in Engineering and Computer Science*. Wiley, 1995.
3. S. Obayashi, Y. Yamaguchi, and T. Nakamura. Multiobjective genetic algorithm for multidisciplinary design of transonic wing planform. *Journal of Aircraft*, 34(5):690–693, 1997.
4. C.M. Fonseca and P.J. Fleming. Multiobjective optimization and multiple constraint handling with evolutionary algorithms – part II: Application example. *IEEE Transactions on Systems, Man and Cybernetics*, 28(1):38–47, 1998.
5. Th. Bäck and H.-P. Schwefel. An overview of evolutionary algorithms for parameter optimization. *Evolutionary Computation*, 1(1):1–23, 1993.
6. B. Sendhoff, M. Kreutz, and W. von Seelen. A condition for the genotype–phenotype mapping: Causality. In Thomas Bäck, editor, *Genetic Algorithms: Proceedings of the 7th Int. Conf. (ICGA)*, pages 73–80. Morgan Kaufmann, 1997.
7. T. Arima, T. Sonoda, M. Shirotori, A. Tamura, and K. Kikuchi. A numerical investigation of transonic axial compressor rotor flow using a low-Reynolds number k-ε turbulence model. *Journal of Turbomachinery*, 121:44–58, 1997. Transactions of the ASME.
8. H.-P. Schwefel. *Evolution and Optimum Seeking*. John Wiley & sons, New York, 1995.
9. A. Ostermeier. A derandomized approach to self adaptation of evolution strategies. *Evolutionary Computation*, 2(4):369–380, 1994.
10. N. Hansen and A. Ostermeier. Adapting arbitrary normal mutation distributions in evolution strategies: The covariance matrix adaption. In *Proc. 1996 IEEE Int. Conf. on Evolutionary Computation*, pages 312–317. IEEE Press, 1996.
11. N. Hansen and A. Ostermeier. Completely derandomized self-adaptation in evolution strategies. *Evolutionary Computation*, 2000. To appear.
12. M. Kreutz, B. Sendhoff, and Ch. Igel. *EALib: A C++ class library for evolutionary algorithms*. Institut für Neuroinformatik, Ruhr-Universität Bochum, 1.4 edition, March 1999.
13. M. Olhofer, Y. Yamaguchi, B. Sendhoff, and E. Körner. Stator blade of axial flow compressor. Japanese Patent No. 348577, December 1999. (in Japanese).

Mixed-Integer Evolution Strategy for Chemical Plant Optimization with Simulators

Michael Emmerich[*]   Monika Grötzner[‡]   Bernd Groß[‡]
Martin Schütz[*]

[*] Informatik Centrum Dortmund
Center for Applied Systems Analysis
Joseph-von-Fraunhoferstr. 20
44227 Dortmund, Germany
{emmerich,schuetz}@icd.de

[‡] Rheinisch-Westfälische Technische Hochschule Aachen
Lehrstuhl für Technische Thermodynamik
52056 Aachen, Germany
mg@ltt.rwth-aachen.de
bernd.gross@cognis.de

Abstract The optimization of chemical engineering plants is still a challenging task. Economical evaluations of a process flowsheet using rigorous simulation models are very time consuming. Furthermore, many different types of parameters can be involved into the optimization procedure, resulting in highly restricted mixed-integer nonlinear objective functions.

Evolution Strategies (ES) are a promising robust and flexible optimization technique for such problems. Motivated by a typical chemical process optimization problem, in this paper a non standard ES is presented, which deals with nominal discrete, metric integer and metric continuous parameters taken from limited domains. Genetic operators from literature are combined and adapted.

Experimental results on test functions and an application example – the parameter optimization of a HDA process – show the robust convergence behaviour of the algorithm even for small population sizes.

1   Introduction

The availability of high speed parallel computers has given way to new techniques of solving optimization problems. Evolution Strategies (ES) [8] utilize principles of biological evolution, like recombination, mutation and recombination, to search for a global optimum. As a flexible and robust search technique, they have been successfully applied to various real world problems.

This paper deals with a special variant of an ES for the optimization of metric continuous, metric integer and nominal discrete parameters taken from restricted domains. The design of this Mixed-Integer Evolution Strategy (MI-ES) has been motivated by optimization problems occurring in chemical plant design using rigorous process simulators. Therefore a typical chemical plant design problem (HDA process optimization) is introduced in the next section, followed by a detailed description of the MI-ES in section 3. Finally, in section 4 the typical convergence behavior of this algorithm is demontrated on three selected test functions, before a sketch of some test results on the HDA process optimization is given.

2 Application problem: HDA process optimization

As a typical example for a small chemical plant, the well documented Hydrodealkylation (HDA) process [1] is used in this paper. Although the MI-ES is of more general applicability, its design has been mainly motivated by applications of this kind.

The aim of the HDA process is the production of benzene from toluene. Unfortunately diphenyl is formed as an undesired side reaction. The reactions (1) and (2) takes place at a high temperature and pressure.

$$Toluene \ (C_7H_8) + Hydrogen \ (H_2) \rightleftharpoons Benzene \ (C_6H_6) + Methane \ (CH_4) \qquad (1)$$

$$2 \ Benzene \ (C_6H_6) \rightleftharpoons Diphenyl \ (C_{12}H_{10}) + Hydrogen \ (H_2) \qquad (2)$$

Figure 1 shows a possible flowsheet for the HDA process which serves in the following as reference configuration. There are many other possible configurations for the HDA process. Structural alternatives of the process can be represented by binary decision variables, selecting alternative subsystems (see e.g. [3]).

The toluene and hydrogen raw material streams are mixed with the recycle streams before they are fed into the reactor. The product stream leaving the reactor contains excess hydrogen and toluene, the main product benzene, some methane and the undesired byproduct diphenyl. With a partial condenser (*flash*) most of the hydrogen and methane is separated from the aromatics.

In order to prevent the accumulation of methane, which enters the process as an impurity of the hydrogen feed stream and which is also produced by reaction 1, a purge stream is required in the gas recycle loop.

It is not possible to separate all hydrogen and methane in the flash drum. Therefore most of the remaining amount is removed in a distillation column. The benzene product is then recovered in a second distillation column and

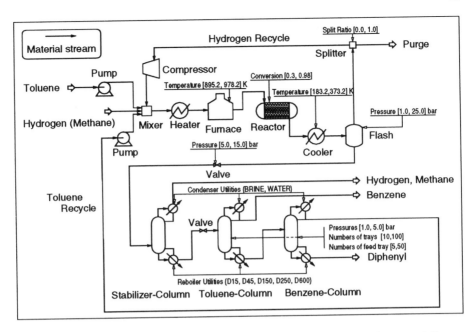

Figure 1: Flowsheet of the HDA process network (taken from [1]) with different chemical devices (unit operations) connected by material streams. The parameters involved in the optimization together with their domains are assigned to the devices.

finally the unwanted diphenyl is separated from the excess toluene which should be recycled for economical reasons.

Each chemical engineering device has a set of individual parameters which are responsible for the performance in the plant. Some of them are given by the problem definition, others can be obtained by calculations and some of them may serve as optimization parameters. Before an optimization of the flowsheet in question can be carried out all parameters that are to be optimized have to be defined. The candidates as parameters can be divided into three main classes:

- metric real parameters (e.g. pressure and temperature levels, conversion rate of the reactor),

- metric integer parameters which denote for example the number of trays in a distillation column or which defines at which tray a feed stream enters the distillation column.

- nominal discrete parameters which represent parameters for which no reasonable order can be defined (e.g. structural alternatives, type of utility streams used for heating and cooling).

When dealing with engineering problems like the optimization of chemical engineering plants, restrictions for the range of the parameters need to be de-

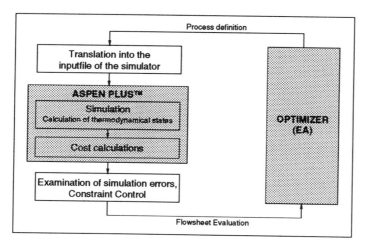

Figure 2: The interaction between the flowsheet simulator and the optimizer. After the time expensive simulation a cost function and a penalty term for the simulation errors is calculated. See [3] for more details concerning the interface to the simulator and the target function.

fined, which are for example given by technical or economical constraints. Furthermore, sometimes dependencies between parameters have to be respected in order to get feasible plants.

3   Mixed-Integer Evolution Strategy

As we have seen from the preceding problem description, the optimization of a chemical plant may be formalized in the following way:

$$f(r_1, \ldots, r_{n_r}, z_1, \ldots, z_{n_r}, d_1, \ldots, d_{n_d}) \to min$$

subject to

$$r_i \in R_i := [r_i^{min}, r_i^{max}] \subset \mathbb{R}, \ i = 1, \ldots, n_r$$

$$z_i \in Z_i := [z_i^{min}, z_i^{max}] \subset \mathbb{Z}, \ i = 1, \ldots, n_z$$

$$d_i \in D_i := \{d_{i,1}, \ldots, d_{i,|D_i|}\}, i = 1, \ldots, n_d$$

Here D_i denotes a set of nominal discrete values. There might be further equality and inequality constraints. The objective function is denoted here by f. In the context of Evolutionary Algorithms we will talk about f as the *fitness function*. In this paper it is assumed, that the fitness function may comprise a penalty term proportional to the number and severity of explicit and implicit constraints violations (c.f. Figure 2).

To solve this problem with an Evolution Strategy let us represent individuals as tuples taken from the search-space:

Algorithm: (μ, κ, λ) - Evolution strategy
$t := 0$
initialize Population $P(t) \in I^\mu$
evaluate the μ initial individuals applying fitness function f
while Termination criteria not fulfilled **do**
    recombine λ offspring individuals out of μ parents, by choosing randomly two
    individuals and recombine them, to obtain each of the offspring individual.
    set age of the λ offspring individuals to zero
    mutate the λ offspring individuals
    evaluate the λ offspring individuals
    select the μ best individuals for $P(t+1)$ from λ offspring individuals and μ parents
    with age lower than the maximal lifespan κ
    increase the age of the individuals in $P(t+1)$ by one
    $t := t+1$
end while

Figure 3: Main loop of a (μ, κ, λ)- ES.

$$I := R_1 \times \cdots \times R_{n_r} \times Z_1 \times \cdots \times Z_{n_z} \times D_1 \times \cdots \times D_{n_d} \times A_s$$

A_s denotes the domain of strategy parameters, defined as:

$$A_s = \mathbb{R}_+^{n_\sigma + n_\varsigma} \times [0,1]^{n_d}, n_\sigma \le n_r, n_\varsigma \le n_z$$

An individual of a population P(t) in generation t is denoted as:

$$\vec{a} = (r_1, \ldots, r_n, z_1, \ldots, z_{n_z}, d_1, \ldots, d_{n_d}, \sigma_1, \ldots, \sigma_{n_\sigma}, \varsigma_1, \ldots, \varsigma_{n_\varsigma} p_1, \ldots, p_{n_p})$$

The parameters $r_1, \ldots, r_n, z_1, \ldots, z_{n_z}, d_1, \ldots, d_{n_d}$ take part in the evaluation of the fitness function. They are called *object parameters*. The parameters $\sigma_1, \ldots, \sigma_{n_\sigma}$ are average step sizes for the real valued and $\varsigma_1, \ldots, \varsigma_{n_\varsigma}$ average step sizes for integer parameters and $p_1, \ldots, p_{n_p}$ are mutation probabilities for the discrete object parameters (c.f. Section 3.3).

The main loop of the Evolution Strategy is described in Figure 3. After a random initialization and evaluation of the first μ individuals, λ offspring individuals are generated step-wise with a recombination operator and a mutation operator. Then the fitness of the λ offsprings is evaluated. The selection operator chooses the μ best individuals out of the λ offspring individuals and the μ parental individuals, which do not exceed the maximal life-span (age), denoted as κ. As long as the termination criterion[1] is not fulfilled, these μ selected individuals form the parental generation for the next iteration loop.

---

[1] In most cases a maximal number of generations is taken as termination criterion.

The proposed algorithm is called a (μ, κ, λ)- ES and it has first been applied by Schwefel [8]. In many cases a ratio of $\mu/\lambda \simeq 1/7$ leads to a good performance of Evolution Strategies.

3.1 Initialization

The initialization operator generates the μ individuals of the start population. These individuals should be distributed uniformly over the search space, in order to explore a big ratio of the search space volume. Furthermore it should create feasible individuals, because an evaluation of infeasible individuals would be a waste of computing time.

Therefore we choose the initial object parameter values randomly distributed over the search space, which is defined by the parameter intervals in the case of continuous and integer parameters or the finite domain of values in the case of discrete parameters. In our application example additionally the restriction that the feed stage of a distillation column is always lower than its total height is taken into account. First the total height parameter is initialized in its feasible interval. This value is then an upper bound for the initialization interval of the feed stage, where this parameter is then initialized uniformly distributed.

3.2 Recombination

The recombination operator can be subdivided into two steps. The first step is to choose recombination partners. In this paper we choose two recombination partners $a_1 \in I$ and $a_2 \in I$ uniformly distributed from the parental generation for each of the offspring individuals. Then a random function $r : I \times I \mapsto I$ assigns an offspring individual to each pair of parents.

In the case of Evolution Strategies we distinguish between dominant and intermediate recombination [8]. In a *dominant recombination* the operator chooses with equal probability one of the corresponding parental parameters for each offspring vector position and in an *intermediate recombination* the arithmetic mean of both parental parameters is calculated.

The intermediate recombination is applied for all strategy parameters, and the dominant recombination for the object parameters. Therefore no difficulties arise, if we want to calculate some kind of mean value for the discrete or integer parameters.

In the application of the algorithm to the HDA process optimization, the feed stage height parameter should be kept lower then the total column height again. This problem was solved by applying a *coupled recombination*. This means that the parameters of the feed stage and the total height of a distillation column are both taken from the same recombination partner.

3.3 Mutation

For the parameter mutation, standard mutations for real, integer and discrete parameter types are combined, as they are described in [6, 7] and [8].

The choice of mutation operators was guided by the following guidelines.

- **Reachability:** Every point of the individual search space should be reachable by a chained application of the mutation operator.

- **Feasibility:** The mutation should produce feasible individuals.

- **Symmetry:** No additional bias should be introduced by the mutation operator.

- **Causality / Local Search:** Small variations of a parameter should occur more often than big variations. This may be achieved by the choice of a symmetric unimodal distribution for the variation of the parameters with a maximum at the original value (c.f. [8]).

- **Scalability:** There should be an efficient procedure, by which the strength of the impact of the mutation operator on the fitness values can be controlled.

- **Maximal Entropy:** If there is no additional knowledge about the objective function available the mutation distribution should have maximal entropy [6]. By this measure a more general applicability can be exspected.

The mutation of real parameters is achieved by the addition of a value, obtained by a normal distributed random function, to their old value. The corresponding standard deviations underlie the evolution process, thus are multiplied in each step with a logarithmic distributed random number. Schwefel [8] termed the resulting process as *self-adaptive*, because of the capability of the process to adapt the step-sizes to the local fitness landscape.

The normal distribution possesses the maximum entropy under all continuous distributions with finite variance on $\mathbb{R}$. The multidimensional normal distribution is symmetrical to its mean value and unimodal. The step-sizes represent standard deviations of the multi-dimensional normal distribution for each real valued variable. Thereby the requirement of scalability is achieved.

New integer parameters are chosen by the addition of a geometrically distributed random variable to their old value (c.f. [6]). The multidimensional geometric distribution is one of the distributions of maximum entropy for integer domains that have a finite variance. It is l_1-symmetrical to its mean value, unimodal and has an infinite support. Thereby symmetry and reachability of the mutation is achieved. The strength of the mutation for the integer parameters is controlled by a set of step-size parameters which represent the mean value of the absolute variation of the integer object variables.

Finally, a mutation of the discrete parameters is carried out with a mutation probability that is assigned to each parameter. The probability is a strategy parameter for each discrete variable (c.f. [7]). Each new value is chosen randomly (uniformly distributed) out of the finite domain of values. The application of a uniform distribution, is due to the principle of maximal entropy, since the assumption was made that there is no reasonable order defined between the discrete values.

To reason about requirements like symmetry and scalability we need to define a distance measure on the discrete sub-space. The assumption that there is no order, which can be defined on the finite domains of discrete values, leads to the application of the trivial distance measure $d : (D_1 \times \cdots \times D_{n_d}) \times (D_1 \times \cdots \times D_{n_d}) \mapsto \mathbb{N}_0$ defined as $d((u_1, \ldots, u_{n_d}), (v_1, \ldots, v_{n_d})) = \sum_{i=1}^{n_d} d_T(u_i, v_i)$ with $d_T(u_i, v_i) := \begin{cases} 0 & \text{if } u_i = v_i \\ 1 & \text{else} \end{cases}, i = 1, \ldots, n_d$.

The probability values should be chosen lower than 0.5. In this case the causality guideline will be met, if we choose a constant mutation probability p for each of the n positions of the discrete tuple, because the probability that the result of the mutation is a tuple $(u_1, \ldots, u_n)$ with distance $d_N((g_1, \ldots, g_n), (u_1, \ldots, u_n)) =: \alpha$ which can be calculated as:

$$P(m(g_1, \ldots, g_n) = (u_1, \ldots, u_n)) = p^{\alpha}(1 - p)^{n-\alpha}$$

If we insert values $p \leq 1/n$ an unimodal distribution for the variation will be achieved with a maximum at zero. Furthermore, it can be shown that by introducing this bound, small mutations of the discrete tupel concerning this distant measure are more likely to occur than big mutations.

A self-adaptation of the mutation probability for the discrete parameters may be achieved, by a logistic mutation of these parameters, generating new probabilities in the feasible domain. The logistic transformation function is recommended and discussed by Schütz [7].

If individual mutation probabilities $p_1, \ldots, p_{n_d}$ are applied, an analysis of the mutation requirements turns out to be more difficult. Empirical runs showed that the use of individual probabilities often leads to worse results. To understand this, we may consider a discrete variable that is changed by the mutation operator to its best value, because the probability for changing this variable has been mutated to a high value. Then the probability for disturbing this variable setting again in the next mutation is comparably high.

Since we have to keep integer and real process parameters in their feasible interval, the mutation operators need to be extended. Therefore a transformation function is applied in the mutation operators, that reflects parameters back into the feasible domain.

For the real parameters this is achieved by the *transformation function* $T^r_{[a,b]} : \mathbb{R} \mapsto [a, b], [a, b] \subset \mathbb{R}$ defined as:

$$T^r_{[a,b]}(x) := \begin{cases} x - 2kw & \text{if } a + 2kw \leq x < a + (2k+1)w \\ 2a - x - 2(k+1)w & \text{if } a + (2k+1)w \leq x < a + (2k+2)w \end{cases}$$

$$w := b - a, k \text{ integer}$$

Analogously a transformation function for integer parameters $T^z_{[a,b]} : \mathbb{Z} \mapsto [a, b], [a, b] \subset \mathbb{Z}$ is defined as follows:

$$T^z_{[a,b]}(x) := \begin{cases} x - 2kw & \text{if } a + 2kw \leq x < a + (2k+1)w \\ 2a - x - 2(k+1)w - 1 & \text{if } a + (2k+1)w \leq x < a + (2k+2)w \end{cases}$$

$$w := b - a + 1, k \text{ integer}$$

These transformation functions may be illustrated as a reflection at the interval boundaries. They are discussed in [2]. It cannot be avoided that a certain bias is introduced by the transformation function. But keeping the ratio between step-size and interval width low, the main characteristics of the unconstrained mutation are kept.

4 Empirical Results

Some selected empirical results demonstrate the convergence behaviour of the proposed MI-ES algorithm. In this paper test results for theoretical test functions are illustrated, instead of test runs on the application example. This gives the reader the opportunity to compare results of this EA with that of other optimization algorithms. Since rather simple test functions have been chosen, the experiments give hints on the minimal running time for practical problems of similar dimension.

The fitness history plots in Figure 4 show the behaviour of a $(3, 5, 10)$ MI-ES on a modified step function (f_1) and a weighted sphere function (f_2):

$$f_1(r_1, \ldots, r_{n_r}, z_1, \ldots, z_{n_z}, d_1, \ldots, d_{n_d}) := \sum_{i=1}^{n_r} \lfloor r_i \rfloor^2 + \sum_{i=1}^{n_z} (z_i \text{ div } 10)^2 + \sum_{i=1}^{n_d} (d_i \text{ mod } 2)^2,$$

$$f_2(r_1, \ldots, r_{n_r}, z_1, \ldots, z_{n_z}, d_1, \ldots, d_{n_d}) := \sum_{i=1}^{n_r} ir_i^2 + \sum_{i=1}^{n_z} iz_i^2 + \sum_{i=1}^{n_d} id_i^2,$$

$$f_3(r_1, \ldots, r_n, z_1, \ldots, z_n, d_1, \ldots, d_n) = \sum_{i=1}^{n} (\sum_{j=1}^{i} x_j + z_j + d_j)^2$$

$$r_i \in [0, 1000] \subset \mathbb{R}, z_i \in [0, 1000] \subset \mathbb{Z}, d_i \in \{0, \ldots, 4\}$$

For the test runs we chose 30 (9 and 60) dimensional functions with $n_r = n_z = n_d = n = 10$. The initial step-size for the real and integer values is set to 100 and the initial mutation probability to 0.1. The minimum of all functions can be obtained, setting all object parameters to zero.

The choice of the test functions has been motivated by the following: The weighted sphere model represents a function with an elliptical geometry. Experiments on this function can detect if a speed up can be achieved by the learning of individual strategy parameters for each parameter. Furthermore it is an example for a function with a simple quadratic and convex geometry. The step function has been chosen to demonstrate that the algorithm is capable to deal with large *plateaus* in the fitness landscape. The word plateau is used here for linked areas of neighboured solutions in the search space, that lead to the same fitness value. Such plateaus occur in practical applications for example when searching for feasible points, using penalty functions that are proportional to the number of violated constraints or simulation errors.

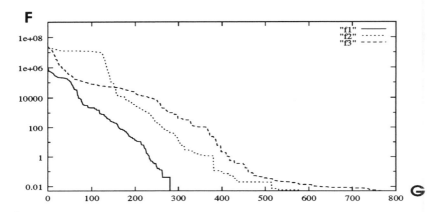

Figure 4: The plots depict typical fitness histories of f_1, f_2 and f_3 for $n_r = n_z = n_d = 10$. For each generation (G) the best fitness value (F) is shown.

The dimension of the search space and the size of the intervals and discrete sets is higher than that of the application example, but in a similar order of magnitude.

The function f_3 represents a general quadratic fitness function. In contrast to the first two fitness functions a strong interaction between all parameters can be obtained. The contour lines of this function have approximately the shape of ellipsoids. The condition number of the corresponding real quadratic problem - measuring the numerical difficulty - increases quadratically with the number of variables (c.f. [8], p. 326).

Test experiments show a robust convergence behaviour for $n_\sigma = 10, n_\varsigma = 10$ and $n_p = 1$ (c.f. Figure 4). The plots show the typical convergence behavior of the MI-ES on quadratic fitness functions, which is characterized by an exponential decrease of fitness values in the first generations. In this stage it is easy to find improvements for the integer and discrete values. After some time the search process nearly stagnates for f_2. The reason for this may be that in this stage the exact settings for the optimal integer and discrete parameters have to be found and kept. There is a high danger of disturbing the solution again, after mutating an integer or discrete variable. It can be conjectured that this disturbing effect is decreased by the application of recombination, extracting the similarities of good solutions, and the learning of a low mutation rate or step size.

Furthermore, table 1 shows a comparison of two MI-ES variants with small populations ($((1, 5, 3)$-ES and $(3, 5, 10)$-ES)) applied to the functions f_1, f_2 and f_3 for different numbers of object variables. The table illustrates the number of generations needed to obtain optima for different problem-sizes. Furthermore the comparison shows that the total running time can be effectively decreased by appliing parallel fitness evaluations. Besides, the use of larger population sizes seems to increase the convergence security.

Last but not least test runs on the application example have been per-

μ, λ, F	$1, 3, f_1$	$3, 10, f_1$	$1, 3, f_2$	$3, 10, f_2$	$1, 3, f_3$	$3, 10, f_3$
$n = 9$, Best	152	71	101	85	630	122
Median	304	84	316	101	729	644
$n = 30$, Best	1200	274	2049	543	2358	676
Median	1483	289	2380	561	3579	1091
$n = 60$, Best	5086	805	7826	1962	9256	8758
Median	12814	917	8452	2271	>100000	11405

Table 1: Comparison of different MI-ES algorithms: The number of generations to obtain the optimum for an $((1, 5, 3)$-ES and a $(3, 5, 10) - ES)$ on the test functions f_1, f_2 and f_3 for different number of variables ($n_r = n_z = n_d = 3(10, 20)$) has been measured. For each combination 5 test runs have been performed. The optimal function value had to be approached with an absolute accuracy of $\epsilon = 0.005$. Discrete recombination has been applied to the object variables and intermediate recombination to the strategy variables.

formed. The following set of parameters, illustrated in Figure 1, has been involved in the optimization: 8 continuous parameters, representing the pressures of the columns and the flash, the reactor conversion, split ratio, cooler temperature; 4 integer parameters, representing total number of column trays and feed tray numbers and finally 6 discrete parameters, representing the heating and cooling utilities.

In [2] a series of 16 test runs with the MI-ES for the application example is illustrated. After 100 generations of the $(3, 5, 10)$- MI-ES all test runs converged to the supposed optimum of the fitness function with a relative accuracy of 0.01. Fitness values have been obtained by a simulator based product price calculation combined with a sophisticated penalty function to handle constraints and simulation errors [3]. A discussion of the fitness landscape and detailed presentation of the obtained optimal plant is beyond the scope of this paper, which focused on the algorithmic methodology and not on the chemical engineering details. The interested reader is referred to [3] for a detailed discussion of the fitness function and to [2] for an illustration of the convergence behaviour on the application example.

It shall be noticed that a similar mixed-integer evolution strategy has been successfully applied on a set of nonlinear mixed-integer benchmark problems with up to 12 variables and constraints, taken from the domain of chemical engineering problems, showing a robust convergence behaviour. The results of this study are summarized in [3] and [4]. Furthermore, a comparison of variants of the proposed EA on further test functions can be found in [5].

5   Summary and outlook

In this paper a detailed decription of a comparably fast and robust Evolution Strategy for the parameter optimization problems with integer, nominal discrete and continuous parameters, which are restricted by intervals or finite domains, has been given. The design of the evolutionary operators has been oriented on guidelines like the maximum entropy principle for mutation, which shall guarantee an universal applicability. For the resulting MI-ES algorithm different operators from literature have been combined and adapted. The design has been motivated by the simulator based optimization of chemical plants. Results on two test functions demonstrate the robust convergence behaviour in high dimensional search spaces, which has also been observed for the optimization of an application problem - the HDA process optimization.

Further studies may explore the applicability for related problems. Besides, a more detailed analysis of the convergence behaviour on fitness functions with different shapes would be useful. The algorithm is designed for chemical engineering problems with only a few structural alternatives. If a high variety of structures is considered, things turn out to be by far more difficult. Here, the use of graph representations together with knowledge integration techniques combined with local parameter optimization (c.f. [2]), seems to be a promising approach.

References

[1] J.M. Douglas. *Conceptual Design of Chemical Processes*. MacGraw Hill, Boston, MA, 1988.

[2] M. Emmerich. Optimierung verfahrenstechnischer Prozeßstrukturen mit Evolutionären Algorithmen. Technical report, Department of Computer Science, University of Dortmund, 1999.

[3] B. Groß. *Gesamtoptimierung verfahrenstechnischer Systeme mit Evolutionären Algorithmen*, volume 608 of *3*. VDI-Verlag, Düsseldorf, 1999.

[4] B. Groß and P. Roosen. Total Process Optimization in Chemical Engineering with Evolutionary Algorithms. In *Comp. Chem, Engng. Suppl. ESCAPE 8, Vol 22*, pages 229–236, 1998.

[5] R. Olschewski. Evolutionäre Algorithmen für gemischt-gannzzahlige Optimierungsprobleme. Diploma thesis, University of Dortmund, Germany, 1999.

[6] G. Rudolph. An evolutionary algorithm for integer programming. In Y. Davidor, H.-P. Schwefel, and R. Männer, editors, *Parallel Problem Solving from Nature - PPSN III*, Lecture Notes in Computer Science, pages 139–148, Berlin, 1994. Springer.

[7] M. Schütz. Eine Evolutionsstrategie für gemischt-ganzzahlige Optimierprobleme mit variabler Dimension. Technical Report SYS-1/96, Department of Computer Science, University of Dortmund, 1996.

[8] H.-P. Schwefel. *Evolution and Optimum Seeking*. Sixth-Generation Computer Technology Series. Wiley, New York, 1995.

comment: Mutation of real valued parameters:

$N_c := N(0,1); \tau := \frac{1}{\sqrt{2*(n_r)}}; \tau' := \frac{1}{\sqrt{2\sqrt{n_r}}}$

if $n_\sigma = 1$ **then**

    $\sigma'_1 = \sigma_1 \exp(\tau N_c)$

else

    for all $i \in \{1, \ldots, n_\sigma\}$ **do**

        $\sigma'_i := \sigma_i * \exp(\tau N_c + \tau' N(0,1))$

    end for

end if

for all $i \in \{1, \ldots, n_r\}$ **do**

    $r'_i := T^r_{[r_i^{min}, r_i^{max}]}(r_i + \sigma'_{\min(n_\sigma, i)} * N(0,1))$

end for

comment: Mutation of integer parameters:

$N_c := N(0,1); \tau := \frac{1}{\sqrt{2*(n_z)}}; \tau' := \frac{1}{\sqrt{2\sqrt{n_z}}}$

if $n_\varsigma = 1$ **then**

    $\varsigma'_1 = \max(1, \varsigma_1 * \exp(\tau N_c))$

else

    for all $i \in \{1, \ldots, n_\varsigma\}$ **do**

        $\varsigma'_1 := \max(1, \varsigma_i * \exp(\tau N_c + \tau' N(0,1)))$

    end for

end if

for all $i \in \{1, \ldots, n_z\}$ **do**

    $u_1 := U(0,1); u_2 := U(0,1); s := \varsigma'_{\min(n_\varsigma, i)}$

    $p := 1 - \frac{s/n_z}{1 + \sqrt{1 + (\frac{s}{n_z})^2}}$

    $G_1 := \left\lfloor \frac{\ln(1-u_1)}{1-p} \right\rfloor; G_2 := \left\lfloor \frac{\ln(1-u_2)}{1-p} \right\rfloor$

    $z'_i := T^z_{[z_i^{min}, z_i^{max}]}(z_i + G_1 - G_2)$

end for

comment: Mutation of discrete parameters:

for all $i \in \{1, \ldots, n_p\}$ **do**

    $p'_i := T^r_{[p_{min}, p_{max}]}(\frac{1}{1 + \frac{1-p_i}{p_i} * \exp(-\tau' * N(0,1))})$

    if $U(0,1) < p'_{\min(n_p, i)}$ **then**

        choose a new element uniform distributed out of $D_i \setminus \{d_i\}$

    end if

end for

Figure 5: Mutation procedure performing a mapping of an individual $(r_1, \ldots, r_n, z_1, \ldots, z_{n_z}, d_1, \ldots, d_{n_d}, \sigma_1, \ldots, \sigma_{n_\sigma}, \varsigma_1, \ldots, \varsigma_{n_\varsigma} p_1, \ldots, p_{n_p}) \in I$ to an individual $(r'_1, \ldots, r'_n, z'_1, \ldots, z'_{n_z}, d'_1, \ldots, d'_{n_d}, \sigma'_1, \ldots, \sigma'_{n_\sigma}, \varsigma'_1, \ldots, \varsigma'_{n_\varsigma} p'_1, \ldots, p'_{n_p}) \in I$. In the algorithm the function $U(0,1)$ denotes an uniform distributed random variable between 0 and 1 and $N(0,1)$ the standardized normal distribution with mean 0 and variance 1.

Advantages in Using a Stock Spring Selection Tool that Manages the Uncertainty of the Designer Requirements

Manuel PAREDES, Marc SARTOR, Cédric MASCLET
LGMT, INSA, 135 avenue de Rangueil 31077 Toulouse Cedex 4
manuel.paredes@ insa-tlse.fr

Abstract. This paper analyses the advantages of using a stock spring selection tool that manages the uncertainty of designer requirements. Firstly, the manual search and its main drawbacks are described. Then a computer assisted stock spring selection tool is presented which performs all necessary calculations to extract the most suitable spring from within a database. The algorithm analyses data set with interval values using both multi-criteria analysis and fuzzy logic. Two examples, comparing manual and assisted search, are presented. They show not only that the results are significantly better using the assisted search but it helps designers to detail easily and precisely their specifications and thus increase design process flexibility.

1. Introduction

The creation of mechanical objects is often the end result of a long design process. Standard component selection is perhaps the simplest, but nonetheless an important, class of design decision problems as catalogues are becoming increasingly common and voluminous.

Let us analyze the method commonly used by design engineers to select stock springs in order to highlight the difficulties they encounter, the help they can find today and what could be added to improve it.

The usual method for selecting stock springs can be divided in three steps:

- Step 1 : evaluate, from the requirements, certain spring design parameters among those classified in the catalogue of the chosen spring manufacturer.
- Step 2 : find springs that are within parameter limits.
- Step 3 : calculate the operating parameters for each spring short-listed, so as to select one that satisfies the specifications.

Designers are confronted here with the following problem. On the one hand, when the specification is vague, it is difficult to choose the best spring from the large range available. When the specification is precise, on the other hand, choice of an appropriate spring becomes limited.

As paper-based methods are tedious and time-consuming, Yuyi [1] has implemented the first two steps of the search in an expert system but where step 3 has to be calculated. Technical literature provides mathematical methods to calculate the design parameters corresponding to the optimal design (Sandgren [2], Kannan and Kramer [3], Deb and Goyal [4]). These methods can be used in step 1 but problems have been simplified and the practical existence of the spring is never envisaged.

Text-only systems such as "SPEC" [5] have been developed which assist the designer during step 2. As with many computer-based methods, if the requirements are not specified precisely enough or lie outside the catalogue range, "all or nothing" search results can be obtained [6].

Finally, industrial software available for a designer during the spring definition work such as "Compression spring Software" from IST® [7], can be used in step 3.

The first drawback of the common component selection method is that it may take a long time when the requirements are advanced. The process time remains significant even when dealing with the simplest applications. The second disadvantage is the approximate nature of the procedure. Usually, a spring is selected without being sure of the pertinence of the choice. Generally, the designer cannot affirm that a spring best matching his specifications does not exist.

But the main drawback is probably the decrease in the design process flexibility as illustrated in Fig. 1. Once a stock spring has been selected, the designer is inclined to keep it unchanged for as long as possible to avoid the time-consuming work of making a new search. The choice is usually kept until the end of the design process. So, there is a need for software tools dedicated to reliable and less time-consuming catalogue search.

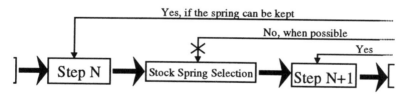

Fig. 1 Effects on the design process

This paper presents a tool armed at meeting designer expectations which also takes advantage of all specifications possibilities. The following capabilities, that are not usually presented, are proposed :
- taking uncertain parameters into consideration (data set with interval values),
- automatically performing all the necessary calculations (buckling, fatigue life...) to check that a spring satisfies the specifications,
- excluding "all or nothing" search results,
- introducing an objective function in order to propose the most suitable component.

This tool may be used even in the early design stages. To fit perfectly with the designer's incomplete knowledge, the method determines springs from a specification sheet where data can be uncertain. The associated algorithms select the best spring by calculating the operating parameters for a given objective. Using this kind of tool, the designer can express his specifications in a very formal and practical way. He can obtain search results instantaneously (number of springs available and the one selected). The present study deals only with helical compression springs with closed ends and with closed and ground ends.

2. A Stock Spring Selection Tool Working from Toleranced Specifications

First, spring characteristics are detailed in order to illustrate whether the tool can accept over-definite requirements. Then the resolution algorithm is described. Finally, the two different methods available to compare competing springs are presented.

2.1. Specifications for the Assisted Search

The parameters which define the spring geometry are: Do, D, Di, d, R, L0, Ls, n, z, p. Fig. 2 illustrates these parameters which characterize the intrinsic properties of the spring. A spring works traditionally between two configurations, one corresponding to the least compressed state W1, the other corresponding to the most compressed state W2. Thus the operating parameters which define the use of a spring are: P1, P2, L1, L2 and sh (see Fig. 3).

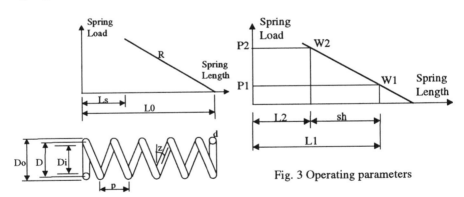

Fig. 3 Operating parameters

Fig. 2 Design parameters

Four independent design parameters have to be known in order to calculate the six others. When the design parameters are known, only two independent operating parameters (to be taken among P1, P2, L1, L2 and sh) are necessary to determine the two operating points W1 and W2. To express all the calculations inside the tool which is presented below, springs of the database are defined by Do, d, L0 and R and the chosen operating parameters are L1 and L2. Nevertheless, each particular parameter illustrated in Fig. 2 or Fig. 3 can be used by the designer to express his specifications.

The designer can decide on design and operating parameters by giving their bounds (lower and/or upper limits: $L0_s^L$, $L0_s^U$,.... $P1_s^L$, $P1_s^U$ ...) in the specification sheet (see Fig. 4). Each fixed parameter simply involves the specification of lower limit equal to upper limit. Moreover, the designer can define a number of other characteristics with interval values:

- Natural Frequency of surge waves
- Spring mass
- Overall space taken up when uncompressed (L=L0)
- Overall space taken up when compressed (L=L2).
- Internal energy during the operating travel

Designers can provide additional data to calculate other characteristics:

- The number of cycles (Ncycles) to calculate the fatigue life factor (to check that it is higher than unity)
- The end fixation factor (v) to calculate the buckling length and check that it is less than L2. [8]

The designer can also specify the material and the spring ends required.

Finally, to be able to select the most suitable spring, the objective function (maximize fatigue life, minimize mass, minimize L2...) has to be given.

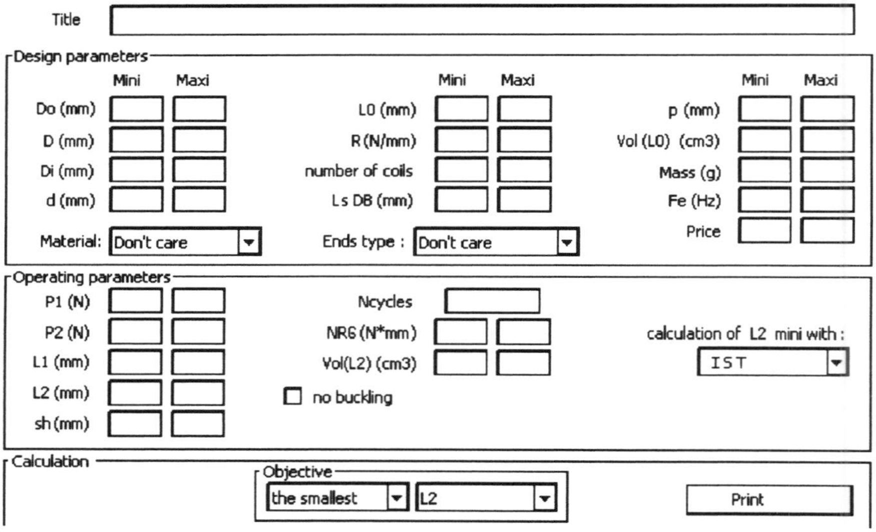

Fig. 4 Specification sheet

Any data not defined in the user specification sheet is set to a default value : 0 for a lower limit and 10^7 for an upper limit. Then the proposed resolution algorithm is performed.

2.2. Proposed Resolution Algorithm

General methods dedicated to component selection problems, as the one proposed by Bradley and Agogino [9], could be applied to the stock spring selection problem. Significant reduction in development costs and processing times can be obtained using a more direct method which is able to take advantage of the spring problem characteristics such as the COSAC system [10] developed at Bath University.

The method chosen here can be considered as the most reliable, since all springs are successively tested. The first spring of the catalogue is evaluated and set as the potential optimum. Then the second spring is evaluated and compared to it. If it is better, it becomes the new potential optimal spring. All the springs of the catalogue are thus evaluated and compared to the last potential optimal spring. When the end of the catalogue is reached, the spring that best matches the requirements is the potential optimal spring. This method provides an acceptable processing time (less than ten seconds) with the catalogue used in this paper which contains about five thousands references.

To evaluate a spring, its four associated design parameters are read from the database and the two operating parameters are automatically calculated in order to optimize the objective value (maximize P2, minimize P2, maximize L2 ...) [11]. All previously detailed design and operating parameters are then calculated. To fit with real-life industrial problems, other properties are added such as fatigue life, price, mass, buckling length or solid length. 23 criterions are thus calculated.

When all the spring criteria have been calculated, it remains to know how they fit the specifications. To manage the various needs of designers, two different analysis are proposed.

2.3. Comparing Springs Using Multi-criteria Analysis

In the first steps of the design cycle, when most part and shapes have not been chosen, specifications are often imprecise and constraint violations can be admitted. Multi-criteria analysis has been chosen to solve this problem. For each spring, the following equation is used to evaluate the constraint violations.

$$Violation = \frac{\sum_{c=1}^{Nc} K_c \times (Mark_c)}{\sum_{c=1}^{Nc} K_c}$$

The weighting coefficient K_c enables the relative influence of criteria to be adjusted. The mark for criterion c : $Mark_c$ (L_c, U_c, V_c) is calculated as follows :

L_c = Lower bound value of the specifications for criterion c (positive value)
U_c = Upper bound value of the specifications for criterion c
V_c = Criterion value of the spring

$$Mark_c = 0$$

```
IF V_c > U_c THEN
        IF U_c = 0 THEN
                Mark = V_c
        ELSE
                Mark = (V_c – U_c) / U_c
        END IF
END IF
IF V_c < L_c THEN
                Mark = (L_c – V_c) / L_c
END IF
```

To evaluate a spring, both objective function value (*Objective*) and constraints violation (*Violation*) values have to be taken into account.

The following equation has been selected.

$$Evaluation = Objective \times e^{a \times b \times Violation}$$

where

$a = 1$ if the objective function has to be minimized or $a = -1$ if the objective function has to be maximized.

b is the weighting violation coefficient, in our study, b = 100.

All the springs are then evaluated and the most interesting one according to the *Evaluation* value is selected.

2.4. Comparing Springs Using the Fuzzy Logic Analysis

The previous analysis often ends in proposing a spring that is close to certain limit values of the specifications. In the first steps of the design cycle, a spring near the centre of the solution domain can be the best choice, even if its objective function value is less interesting. To solve this kind of problem, fuzzy logic analysis is proposed here.

As the goal is to find springs within the limit values of the specifications, a basic comparison is first carried out on the number of constraints violated (*ncv*). A spring that has the lowest *ncv* value is definitely considered has better than the others. When both springs have the same *ncv* value, the comparison with fuzzy logic is performed. The first step is to evaluate how the tested spring matches the specifications. The comparison between this evaluation and that obtained for the potential optimal spring is made at the second step. In the third step, the comparison between the objective function values of these two springs is done. Then, a final comparison is made to select the new potential optimal spring.

2.4.1. Step 1 : How a Spring Matches the Specifications

First, an evaluation is made of each criterion. To evaluate how a criterion matches the specifications, the following method is used.

FuzzyMark$_c$ (L_c, U_c, V_c) = [VB, B, M, G, VG]$_c$ is calculated as follows.

IF U_c = 0 THEN

MarkU$_c$ = - V_c

ELSE

MarkU$_c$ = (U_c - V_c) / U_c * 100

END IF

MarkL$_c$ = (V_c - L_c) / L_c * 100

WorstMark$_c$ = Min(MarkU$_c$, MarkL$_c$)

WorstMark$_c$ is used to perform the evaluation of [VB, B, M, G, VG]$_c$

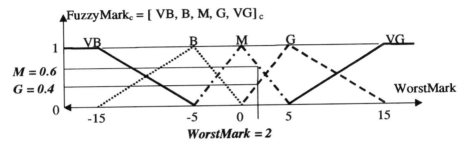

Fig. 5 FuzzyMark$_c$ evaluation

Then, the final values of VB (very bad), B (bad), M (medium), G (good), and VG (very good) are calculated using the formula :

$$[VB, B, M, G, VG]_{Spec} = \frac{\sum_{c=1}^{Nc} K_c \times FuzzyMark_c}{\sum_{c=1}^{Nc} K_c}$$

2.4.2. Step 2 : Comparison of Specifications

The comparison of the previous fuzzy values is performed using the Mamdani [12] definition for the AND connector as described in table 1.

Table 1 : Comparison of specifications

	Potential optimal spring				
Tested spring	VB (0)	B (0.50)	M (0.50)	G (0)	VG (0)
VB (0)	E (0)	S (0)	VS (0)	VS (0)	VS (0)
B (0)	I (0)	E (0)	S (0)	VS (0)	VS (0)
M (0.70)	VI (0)	I (0.50)	E (0.50)	S (0)	VS (0)
G (0.30)	VI (0)	VI (0.30)	I (0.30)	E (0)	S (0)
VG (0)	VI (0)	VI (0)	VI (0)	I (0)	E (0)

76

Then the Or connector (Mamdani) is used to obtain the value of
[VI (very inferior), I (inferior), E (equal), S (superior), VS (very superior)]$_{Spec}$.
Results are shown in table 2.

Table 2 : value of [VI, I, E, S, VS]$_{Spec}$

VI	I	E	S	VS
0.3	0.5	0.5	0	0

2.4.3. Step 3 : Comparison of Objectives

The comparison of the objective function values is made using the *ObjMark* value:
Objtop is the objective function value of the potential optimal spring.
Objective is the objective function value of the evaluated spring.
$ObjMark = 200 * (Objtop - Objective) / (Objtop + Objective)$
Then the comparison of the two objectives to calculate [VI, I, E, S, VS]$_{Obj}$
is made using the rule defined on fig 6.

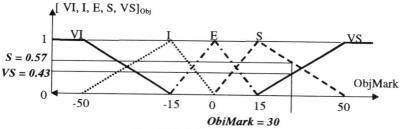

Fig. 6 [VI, I, E, S, VS]$_{Obj}$ value

2.4.4. Step 4 : Final Comparison

Springs are finally compared using both [VI, I, E, S, VS]$_{Obj}$ and [VI, I, E, S, VS]$_{Spec}$ with the same rules as in step 2 in order to calculate I, E, S values.

Table 3 : Final comparison

[VI, I, E, S, VS]$_{Obj}$	[VI, I, E, S, VS]$_{Spec}$				
	VI (0.3)	I (0.50)	E (0.50)	S (0)	VS (0)
VI (0)	I (0)	I (0)	I (0)	I (0)	E (0)
I (0)	I (0)	I (0)	I (0)	E (0)	S (0)
E (0)	I (0)	I (0)	E (0)	S (0)	S (0)
S (0.57)	I (0.30)	E (0.50)	S (0.50)	S (0)	S (0)
VS (0.43)	E (0.30)	S (0.43)	VI (0)	S (0)	S (0)

Table 4 : Value of I, E, S

I	E	S
0.30	0.50	0.43

For the case presented in the previous tables, as S value is superior to I value, the "old" potential optimal spring is superior to the tested spring. Thus, it is kept as the potential optimal one (otherwise the tested spring would have replaced it) and the next spring is tested.

3. Examples and Comparison Between Manual and Assisted Search

In order to compare the results between the manual and assisted search, the two examples have been performed using the same catalogue with both methods. Each assisted search has been carried out with the two different analysis. The average time of the first analysis is 3 seconds increasing to 6 seconds for the analysis using fuzzy logic.

3.1. A Spring for a Clamping Pin

3.1.1. Manual Search

A clamping pin has been custom-designed for an industrial manipulating robot. The time-consuming manual procedure resulted in the selection of the spring Do = 36.0 mm, d = 2.5 mm, L0 = 50 mm, R = 3.54 N/mm, L1 = 47 mm, **L2 = 36 mm**. (steel with closed and ground ends) from the paper catalogue. The result is shown in Fig. 7 and the clamping pin was added to the robot.

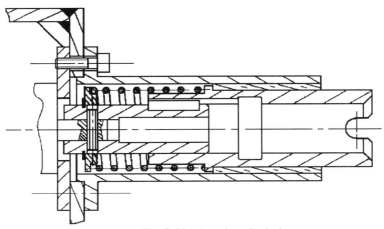

Fig. 7 Old clamping pin design

3.1.2. Assisted Search

During a reengineering procedure, it was decided to reselect a spring in order to reduce the main drawback of the clamping pin : its high axial length. Now, the proposed search tool can be used and the specifications can be detailed precisely.

Specifications : the maximum outside diameter of the spring is 38 mm, the minimum inside diameter is 27 mm, spring travel must be 11 mm, the maximum value of L1 is 50 mm, the maximum spring rate is 5.5 N/mm, the load P1 must be between 5 and 15N and the load P2 between 50 and 100N. The goal is to obtain the spring with the smallest value of L2.

According to the chosen objective, the algorithm calculates the operating parameters of each spring in order to have the minimum operating length L2 while satisfying the specifications.

Results : there are 7 springs that fit the given specifications.
The first analysis proposes a spring that is close to specification requirements (in terms of R) :
Do = 32.0 mm, d = 2.2 mm, L0 = 25 mm, R = 5.78 N/mm, L1 = 22.4 mm, **L2 = 11.4 mm**. (steel with closed and ground ends).
Using fuzzy logic the following spring is selected :
Do = 32.0 mm, d = 2.2 mm, L0 = 32 mm, R = 4.34 N/mm, L1 = 28.54 mm, **L2 = 17.54 mm**. (steel with closed and ground ends).

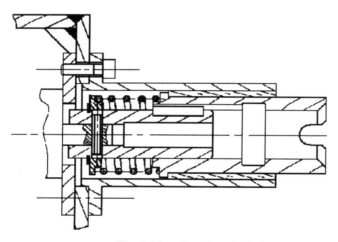

Fig. 8 New clamping pin design

As the properties of the two proposed springs are automatically calculated, the designer can easily choose the one that best matches his specifications. In fact, the spring found with the first method fits the geometrical parameters and has a law *Violation* value (= 0.051), whereas its L2 value is much lower than the one selected by the fuzzy logic method. This spring is chosen for the new design described in Fig. 8. The assisted search proposes springs significantly better than those found with the paper search, leading to useful changes in the clamping pin design.

3.2. A Spring for an Axial Displacement Sensor

3.2.1. Manual Search

In this example, the manual search in the catalogue led to the following spring :
Do = 12.5 mm, d = 1.25 mm, L0 = 100 mm, R = 0.8 N/mm, L1 = 93.75 mm,
L2 = 33.75 mm (**P2 = 53N** , steel with closed and ground ends).

3.2.2. Assisted Search

Once again, the assisted search allows to the following requirements to be
expressed :

Specifications : the maximum outside diameter of the spring is 13 mm,
the minimum inside diameter is 5 mm, spring travel must be 60 mm, the
minimum P1 value is 3 N and the length L2 must be between 30 and 45 mm. The
goal is to obtain the spring with the smallest value of P2.

According to the chosen objective, the algorithm calculates the operating
parameters of each spring in order to have the minimum operating load P2 while
satisfying the specifications.

Results : there are 14 springs that match the given specifications.
Using the first analysis, a spring that fits the specifications (*Violation = 0*) is
selected : Do = 11.0 mm, d = 0.9 mm, L0 = 100 mm, R = 0.3 N/mm, L1 = 90
mm, L2 = 30 mm (**P2 = 21N** , steel with closed and ground ends).
Fuzzy logic analysis proposes another spring that fit the specifications :
Do = 11.0 mm, d = 1 mm, L0 = 100 mm, R = 0.374 N/mm, L1 = 92 mm,
L2 = 32 mm (**P2 = 25.4N** , stainless steel with closed and ground ends).

In this case, the spring found by fuzzy logic analysis is within the
geometrical constraints. In order to obtain a reliable design, this spring is
included in the mechanism. Once again, the assisted search results in the choice
of a much better spring than the one obtained by the paper based method.

4. Conclusion

Paper-based methods for selecting stock springs are tedious, time consuming and
decrease design process flexibility. A stock spring selection tool managing
uncertain parameters and including all the necessary calculations in order to
suggest the most suitable spring is proposed. It has been developed and tested for
one year in collaboration with a spring manufacturer. It has shown that this kind
of tool changes the designer's approach during the catalogue search. With the
assistance of the proposed tool, the designer can specify his needs and quickly
choose the spring that best matches his requirements. Finally, this type of tool
increases design process flexibility as the component choice is made easier and
more efficient.

Acknowledgements

The financial and technical support of the spring manufacturer « Ressorts VANEL » is gratefully acknowledged.

References

1. YUYI L, KOK-KEONG T and LIANGXI W, 1995. Application of Expert System for Spring Design and Procurement. *Springs* march : 66-80. From the Spring Manufacturer Institute, http :\\www.smihq.org

2. SANDGREN E, 1990. Nonlinear Integer and Discrete Programming in Mechanical Design Optimization. *ASME Journal of Mechanical Design* 112 : 223-229.

3. KANNAN B.K. and KRAMER S.N., 1994. An Augmented Lagrange Multiplier Based Method for Mixed Integer Discrete Continuous Optimization and its Applications to Mechanical Design. *ASME Journal of Mechanical Design* 116 : 405-411.

4. DEB K and GOYAL M, 1998. A Flexible Optimization Procedure for Mechanical Component Design Based on Genetic Adaptive Search. *ASME Journal of Mechanical Design* 120 : 162-164.

5. G.B. Innomech, 1994. SPEC Select, Spring Selection.

6. HARMER Q.J., WEAVER, P.M., WALLACE, K.M., 1998. Design-led component selection. *Computer-Aided Design*, 30 : 391-405.

7. IST®, Institute of Spring Technology, Henry Street Sheffield S3 TEQ United Kingdom, http :\\www.istec.demon.co.uk

8. WAHL A. M., 1963. *Mechanical Springs*. McGraw-Hill, New York.

9. BRADLEY S.R. and AGOGINO A.M., 1994. An Intelligent Real Time Methodology for Component Selection: An Approach to Managing Uncertainty. *ASME Journal of Mechanical Design* 116 : 980-988.

10. VOGWELL J and CULLEY S.J., 1991. A strategy for selecting engineering components. *Proc. Imeche, Journal of Manufacturing systems* 205 : 11-17.

11. PARDES M, SARTOR M and MASCLET C, 2000. Stock Spring Selection Tool. *Springs* Winter. From the Spring Manufacturer Institute, http :\\www.smihq.org

12. MAMDANI E.H., 1974. Application of fuzzy algorithms for control of simple dynamic plant. *Proc. Institution of electrical engineers, Control and science* 121 : 1585-1588.

Chapter 2

General Design Applications

Designing Food with Bayesian Belief Networks
D. Corney

Flexible Ligand Docking Using a Robust Evolutionary Algorithm
J.-M. Yang, C.-Y. Kao

Network Design Techniques Using Adapted Genetic Algorithms
M. Gen, R. Cheng, S.S. Oren

Designing Food with Bayesian Belief Networks

David Corney

Computer Science Dept., University College London
Gower Street, London, WC1E 6BT
D.Corney@cs.ucl.ac.uk

Sira Technology Centre
South Hill, Chislehurst, Kent, BR7 5EH
David.Corney@ptp.sira.co.uk

Abstract. The food industry is highly competitive, and in order to survive, manufacturers must constantly innovate and match the ever changing tastes of consumers. A recent survey [1] found that 90% of the 13,000 new food products launched each year in the US fail within one year. Food companies are therefore changing the way new products are developed and launched, and this includes the use of intelligent computer systems. This paper provides an overview of one particular technique, namely Bayesian Belief networks, and its application to a typical food design problem. The characteristics of an "ideal" product are derived from a small data set.

1. Introduction

Bayesian Belief Networks are graphical models that encode probabilistic relationships between variables of interest. They have become increasingly popular within the AI community since their inception in the late 1980's [2] [3], due to their ability to represent and reason with uncertain knowledge. They have been used successfully in expert systems, decision support systems and diagnostic systems, among others. Figure 1 shows a typical network, described in more detail in part 3.

Historically, one of the first applications of Bayesian networks was to medical diagnosis. For example, a Bayesian network system has been developed from a database containing descriptions of many symptoms and associated diseases [4]. By entering a brief description of a patient's symptoms, the system can deduce likely causes, i.e. diseases. The system was designed as a decision-support system for use by medical experts, and as a teaching aid.

Bill Gates recently described Microsoft's competitive advantage as being its expertise in Bayesian networks [5]. Microsoft have been actively recruiting experts in the field since the early 1990's, and have become a significant research force. They have also released software components using Bayesian networks, such as the Office Assistant and the grammar checker, both in Office 97. Further applications of

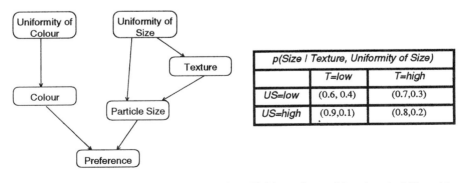

Figure 1: A hypothetical Bayesian network   Table 1: A conditional probability table

Bayesian networks include robot guidance [6], software reliability assessment [7], data compression [8] and fraud detection [9].

A great deal has been achieved with Bayesian networks, and (the author believes) they can and will be applied to product design. Products are artefacts purchased and used because of their properties and functions [10]. They are designed to meet the end-users' requirements, whether this means a car must be fast, a mouse-trap must be "better", or a plate of food must taste nice. Because of their ability to learn, adapt and explain, intelligent systems such as Bayesian networks can aid product designers in their work.

The next section describes the nature of the data used in food design work. This is followed by an overview of Bayesian networks, and descriptions of how the models can be built and used. Finally, some experimental results are presented.

2. Food Data

When designing new food products, companies typically obtain data from three sources: sensory panels, preference panels, and instrumental data. The nature of instrumental data is product-specific, and is not covered here in detail, but may include digital images, acoustic imaging or chemical fingerprinting.

The *sensory panel* is a group of typically 10-20 people, selected and trained for several months. The panel derives their own descriptors of product attributes, which can then be systematically used to describe different varieties of the product. The panel typically produces between 8 and 20 descriptors after discussion and analysis. Members of the panel are then presented with a variety of different products, selected to represent a wide range of flavours, colours, etc. They then measure each sample by ranking it for each descriptor. The ideal sensory panel should produce absolutely consistent and uniform results, allowing the panel to be treated as an instrument. In practice, human perception is neither absolute nor constant.

The *preference panel* is a larger group of untrained people, typically 50-500 potential consumers, who are bought in "off the street" specifically for the trials. They are individually presented with a few samples and are then asked to rank each one on a simple preference scale. No training is given and no discussion between panellists is allowed, so the results will be entirely subjective, and vary from panellist to panellist. The relatively large panel size should smooth out any unwanted discrepancies.

Once the preference panel has classified the samples, the sensory panel data is re-examined, to determine which sensory attributes best distinguish the different preferences. For example, suppose the preference panel gave two samples significantly different grades. Then if the sensory panel gave both of them the same grade for some measure, e.g. shape, then this attribute is a poor predictor of quality. If a correlation can be found between the one or more of the sensory panel attributes and the preference panel scores, then this can be used to guide future product design and marketing.

The entire data-gathering process is very expensive and very time-consuming, and depends on human perception, which lead to the most striking and important features of the data sets: they are small and sparse, and contain uncertainty. The data used here is described in section 5 and included as an appendix.

3. Features of Bayesian Networks

Bayesian Belief Networks are graphical representations of the joint probability distributions over a set of discrete variables, and incorporate conditional independence assumptions. They consist of a directed acyclic graph (DAG) such as the simple model shown in Figure 1, and a set of conditional probability tables, such as Table 1. In the graph, each node represents a variable and the arcs between nodes specify the independence assumptions between the variables. More precisely, each variable is "conditionally independent of any combination of its non-descendants, given its parents" [8]. Thus Figure 1 shows, for instance, that given "colour", then "uniformity of colour" has no influence over any variable.

One conditional probability table is determined for each node, defining the probability of the variable being in each possible state, given each of the possible states of its parent node(s). If a node has no parents, the unconditional probabilities are used instead.

Table 1 shows the conditional probability distribution for the "particle size" node, conditional upon its parents, namely "uniformity of size" (US) and "texture" (T). Each cell in the table has two numbers, the probability that the particle size is low (i.e. "small") and the probability that it is high (i.e. "large").

Bayesian networks have a number of features that make them suitable for product design, as shown in Table 2 and discussed in the remainder of this section.

3.1 Explaining away observations

"Explaining away" can be defined as "a change in the belief in a possible explanation if an alternative explanation is actually observed" [11]. The standard example of explaining away is the lawn sprinkler: suppose we observe that the lawn is wet one morning. There are two possible causes: either it rained or the sprinkler was left on. Our belief in both of these explanations increases. We then observe that our neighbour's lawn is also wet, and so deduce that it rained last night. Because we now believe that the wet lawn was caused by the rain, we no longer have any reason to believe that the sprinkler was left on, so we should retract that belief [12].

In the case of food modelling, if we know that sweet foods are generally preferred, and we have a particular sample that is both sweet and popular, then our simple model gives us no reason to believe its colour will affect its popularity. More traditional rule-based expert systems fail to cope with this type of situation, because the systems are *modular*, meaning that the rules are fired with no reference to the

Explaining away	Make effective use of all available information
Bi-directional Inference	Can diagnose what causes high preference
Complexity	Can scale up to represent complex models
Uncertainty	Can deal with uncertainty in the data
Confidence values	Provide confidence measures on results
Readability	Produce graphical, transparent models
Prior Knowledge	Can incorporate expert knowledge

Table 2: Features of Bayesian Networks

context of other rules or the source of the data. The conditional probabilities in the Bayesian network models encapsulate the desired effect.

3.2 Bi-directional Inference

Many intelligent systems (e.g. feedforward neural networks, fuzzy logic) are strictly one-way in the sense that when a model is given a set of inputs it can predict the output, but not vice versa. The question one really wants to ask "What features would a product have, if it had a high preference score?" This inverse problem can be solved by bi-directional modelling, where inputs can be used to predict outputs, and outputs can be used to "predict" or diagnose inputs.

Bayesian networks can do this within a single structure because variables are *not* specified as being solely for input or for output. By applying Bayes' theorem, the direction of the relationship can be reversed. For example, given the rule "*If (product is sweet) then (product is preferred)*" and given the fact "*product is sweet*" we can obviously deduce that it is preferred. However, with a Bayesian system, we might observe (or hypothesise) that "*product is preferred*" and deduce that this preference must be caused by its sweetness, i.e. that "*product is sweet*". In other words, while many systems can perform induction, Bayesian networks can also perform *abduction*.

3.3 Complexity

The independence assumptions expressed by the graph mean that fewer parameters need to be estimated because the probability distribution for each variable depends only on the node's parents. This independence assumption allows us to factorise the network, considering each node and its parents in isolation from the rest of the model. This means that far fewer parameters are needed to fully specify the relationships between the variables, than would be required by a fully connected network, or any other global, "unfactorable" model.

Similarly, when learning the structure of the graphs, the search can be local, with the optimal set of parents for each node being selected independently of the rest of the model. Thus even very complex models can be discovered without suffering from a combinatorial explosion. These efficiencies are particularly important when only small data sets are available, as is often the case with food design. The K2 algorithm described later relies on this feature.

3.4 Uncertainty

There are many sources of uncertainty, such as distortion, incompleteness and irrelevancy [11]. Consider asking a group of preference panellists, "How much do each of you like product X?" However much time is spent defining or describing the word "like", there is no guarantee that any two subjects will actually use the same scale to measure the product on, irrespective of personal differences in taste. Furthermore, experimental results show that individual subjects will give the same product different scores at different times, depending on the context, their mood, etc. The same problem occurs with sensory panel data. In common with all Bayesian systems, Bayesian networks model "degrees of belief", equivalent to probabilities, rather than a crisp true/false dichotomy. This means that uncertainty can be handled effectively, and explicitly represented.

3.5 Confidence Values

The output of any Bayesian model is a probability distribution, rather than simple scalar or vector. For example, whereas a neural network might predict a scalar preference score of say, 0.753, a Bayesian belief network might give an output in the form: $p(low) = 0.28$; $p(high)=0.72$. This sort of information can be used as a measure of confidence in the result, which is essential if the model is going to be used for decision support.

3.6 Readability

When a (human) designer produces technical drawings and reports, the aim is to aid manufacture, sales, marketing and so on. When computers are being used to generate the designs automatically, it is important that they are still *readable*. No one is going to invest a great deal of time, money and expertise developing a product if they cannot see why it will be good. Due to their graphical nature, Bayesian networks provide a transparent model, although very complex systems may require networks too large to be comprehensible.

3.7 Prior Knowledge

It is impossible to avoid the use of prior knowledge when building models. By defining the bounds of the solution space, the representation used, the scoring measure used and so on, the analyst will inevitably introduce biases. Bayesian approaches make these prior assumptions explicit and formal. The size of the data sets also influences the use of prior knowledge. Because food design data sets are typically small, little information is contained in them, so the use of alternative, non-electronic sources of information (i.e. experts) is significant. This could be in the form of selecting nodes, sub-graphs or even entire graphs, if these are known to be important.

4. Bayesian Belief Networks Theory

Having described many of the features of Bayesian networks, it is now time to describe some of the processes involved in building and using them. There are three problems that must be solved: defining the graphical structure (B_s), defining the parameters in the form of the conditional probabilities (B_p), and finally using the models to make predictions. Further details of learning both the structure and parameters can be found in [13], and making predictions (inference) is covered in [12].

4.1 Defining the Structure

The graph consists of two parts: a collection of nodes and a collection of arcs joining them. In graph theory, these are known as vertices and edges respectively. In some cases, suitable expert knowledge may be available to allow the entire structure to be defined by hand, with the expert stating which variables are relevant, and how they interact. More often however, such knowledge will be unavailable, or at best, imperfect.

The total space of all legal (i.e. directed, acyclic) graphs over a set of nodes is greater than exponential in the number of nodes. Therefore, in all but the simplest cases, an exhaustive search is impossible, requiring the use of heuristics. A number of search algorithms have been used, a selection of which are listed in Table 3.

Method	Comment	References
K2	Finds parents for each node via a greedy search.	[15]
Genetic algorithms	Constraints are similar to the Travelling Salesman Problem. Used to provide node ordering for K2.	[16] [17]
Branch-and-Bound	Often used in AI to limit the combinatorial explosion, e.g. during feature selection.	[18]
Structural EM	Learns structure and parameters with a modified Expectation Maximisation (EM) algorithm.	[19]

Table 3: Structure search techniques

With any search technique, we need some way of determining the quality, or fitness, of a Bayesian network. Given that we are trying to model some data, the direct way of considering this is to ask "How well does the data fit this model?" The Bayesian approach to this problem is to assume that the data was actually *generated* by the model, and then reverse the question to ask "How likely is it that this model produced the data?" This reversal is possible using Bayes' Theorem [14].

The experiments described later use the "K2" algorithm proposed by Cooper and Herskovits [15], and outlined here. Cooper and Herskovits show that the ideal model, i.e. that which maximises the posterior probability of the network structure given the data, $p(B_s|D)$, also maximises the joint probability, $p(B_s, D)$. This is easier to calculate, and they derive a polynomial-time function of this joint probability, using the frequency of variable instantiations in the data set. This gives a straightforward way of quantifying the goodness of fit between the model and the data, and therefore defines a fitness function for the models. We now have to search through the model space to find a good network structure.

To make the search tractable, the search space is limited by making a number of assumptions. K2 assumes that: all the variables are discrete; all the cases are independent given the model; there is no missing data; there is no prior knowledge regarding likely structures. In the current work, these present no problems: the variables can easily be discretised; there are no dependencies between the cases; the cases are complete; and there is no knowledge about the structures. K2 requires a fixed ordering of the nodes, such that each node will only be considered as a possible parent of nodes that appear later in the ordering. The algorithm also requires a maximum fan-in value, i.e. an upper bound on the number of parents any single node may have. Finally, it requires a complete database of cases.

By definition, nodes depend only on their parents; K2 makes use of this by searching for the optimum set of parents of each node independently, before finally constructing the network. The algorithm proceeds by considering each node in turn, and defining an initially empty set of parents for that node. Every possible parent is then considered, and the parent that maximises the K2 score is added to the node's parent set. Further parents are considered within the constraints of node ordering and maximum fan-in, until no further additions improve the fitness score. Then the parents of the next node are considered. The end result is a list of parents for each node. This list is sufficient to completely define the structure of a Bayesian network.

4.2 Defining the Parameters

Once the graph has been defined, the only remaining parameters are the conditional probabilities for each node. Remembering that each node depends only on its immediate parents, we need only estimate $p(v|\pi_v)$ for each node, where v is the variable (node), and π_v is the set of parents of the node. All the variables must be discrete in order for the propagation and inference algorithms to work (see below), so continuous data must be converted to discrete values prior to use.

The simplest way of estimating the probabilities is to use the frequency with which each configuration of variables is found in the data. As the number of data points observed increases, this frequency will tend towards the true probability distributions; however, the small data sets typical of food design studies tend to be very sparse when considered this way, as many configurations will not have been observed. An alternative approach therefore is to initially assume a particular distribution (e.g. uniform) and then update this to encapsulate the information contained in the data. This can be done using the Expectation Maximisation (EM) algorithm [20], optionally combined with an equivalent sample size [21].

4.3 Inference

Given a complete model, defining both the structure and the conditional probabilities, we can begin to make predictions. If the values of some variables are known ("observed"), then the probabilities of the remaining variables can be calculated. This is done by fixing the states of the observed variables, and then propagating the beliefs around the network until all the beliefs (in the form of conditional probabilities) are consistent. Finally, the desired probability distributions can be read directly from the network. The standard propagation algorithm is due to Lauritzen and Spiegelhalter [2].

5. Data, Experiments and Results

Two experiments are described here. The first uses the K2 algorithm to build Bayesian networks, and compares their accuracy at predicting preference scores against two alternative models. The second experiment uses one such Bayesian network to estimate the characteristics of the "perfect" product under consideration.

The data used throughout the remainder of this paper was provided by Unilever Research, and consists of a preference score ("P1") and eight sensory panel scores ("S1" to "S8"). A total of just 20 records were available, each record being a complete set of data for a single sample of the food. The exact nature of the food is commercially confidential, but the samples were carefully selected to represent the full range of varieties of the product.

The raw data is included in Appendix 1. Each value was converted to a binary score, by assigning the lowest ten scores of each attribute to the class "low" and the ten highest scores to the class "high". A more precise model could be obtained by discretising each attribute into more than two states, but the data would become extremely sparse.

5.1 Performance Measures

Two measures of performance were used: predictive accuracy and joint probability. To calculate the predictive accuracy of each model, the preference (P1) was treated as a target class, which the sensory scores (S1-S8) were used to predict. The accuracy score is simply the proportion of records that were assigned to the correct preference class (high or low). However, if the same data is used to both build and test the model, the resultant score tends to underestimate the model's true accuracy [14]. This is particularly true when small data sets are used (as in the current work), because models will tend to overfit the data, fitting both the underlying distribution and the inherent noise. To avoid this, leave-one-out cross-validation is often used. Thus given 20 records, we construct 20 models, each being built using a different subset of 19 records, and each being tested on the remaining record. This produces 20 accuracy scores, the mean of which is a good estimate of the model's true accuracy.

The maximum likelihood model is that which maximises $p(B_s|D)$, which [15] show is proportional to $p(B_s,D)$, the joint probability of the network and the data. Therefore in this work, this joint probability was calculated for every Bayesian network and the naïve Bayes classifier (described below). The entire data set was used to build each model and thus to calculate this probability score without the need for any cross-validation. This is because when performing Bayesian inference, complex models have lower prior probabilities than simple models, giving Bayesian techniques a built-in safeguard against overfitting. Note that no equivalent score exists for standard neural networks.

5.2 Model Performance

As an initial study, three techniques were compared to see which could most accurately predict preference from the sensory scores. Table 4 summarises the results.

The three techniques used were a neural network, a naïve Bayes classifier, and a Bayesian belief network. The neural network was a standard MLP, with 8 inputs (the sensory data), 5 hidden nodes and one output (the preference score). Leave-one-out cross-validation was used to measure the neural networks' accuracy at predicting the preference score, so each evaluation cycle actually consisted of building and testing 20 neural networks. Of 100 cycles, the mean accuracy was 0.796. [14] describes neural networks in more detail, as well as several cross-validation techniques.

The naïve Bayes classifier is a special case of Bayesian network that treats one variable as a target class. It assumes that all the other variables depend only on this class, being conditionally independent of each other. Here, we treat preference (P1) as the target class and assume that the eight sensory scores are independent. This produces a network where every sensory node has exactly one arc, which leads from the preference node, as shown in Figure 2. Using leave-one-out cross-validation gave an estimated accuracy of 0.80 for the naïve Bayes classifier. The log

	Accuracy		Log Joint Probability	
Neural Network	0.796	(0.05)	N/A	
Naïve Bayes Classifier	0.800	(0.00)	-123.08	(0.00)
Bayesian Belief Network	0.800	(0.00)	-107.51	(1.03)

Table 4: Comparison of models.
"Accuracy" is the estimated accuracy with which the model predicts P1 given S1-S8.
"Log Joint Probability" is the log of the joint probability of the model and the data,
ln $p(B_s,D)$. The standard deviation of each score is shown in brackets.

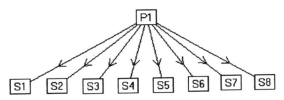

Figure 2: Naïve Bayes Classifier

joint probability was -123.08. Note that the corresponding variances shown in the table are zero, because the network structure is fixed, so the accuracy score has no variance.

The Bayesian belief networks used in this study were generated using the K2 algorithm, and Figure 3 shows one such network. The K2 algorithm was executed 100 times, with randomly generated node orderings. In each case, the joint probability $p(B_s, D)$ was calculated, and had a mean of -107.51. A separate experiment repeatedly used K2 with leave-one-out cross-validation. In every case, K2 selected the same two variables (S5 and S6) as parents, and so gave the same accuracy score of 0.80. This shows that (at least for this data set) the ordering of the nodes presented to K2 is not critical.

These results show that Bayesian belief networks, neural networks and naïve Bayes classifiers are equally effective at the specific task of predicting product preference from sensory panel scores.

Note that the Bayesian belief networks are constructed to model the *entire* data set, rather than just one relationship within it. In contrast, the other two techniques used here explicitly build models that are designed to predict preference. The final column of Table 4 shows that as well as making equally good preference predictions, the Bayesian network models the data more closely than the naïve Bayes classifier, as indicated by the higher log probability value. This suggests that the assumptions made by the naïve Bayes classifier are invalid, and therefore that the sensory panel variables are *not* independent.

5.3 Belief Propagation

If Bayesian networks are no more accurate than simpler alternatives, why use them? As outlined in section 3, Bayesian belief networks have many attractive features, including abduction: the ability to diagnose likely causes of an observed effect. In the current work, this is estimating the most likely characteristics of a hypothetically perfect product.

The Bayesian network shown in Figure 3 is used here to demonstrate how

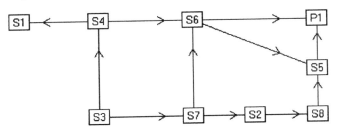

Figure 3: Bayesian network for sensory data

predictions can be made from limited observations. The parameters (probabilities) were defined using the data frequencies only, and the Lauritzen and Spiegelhalter algorithm [2] was used to propagate several observations and to make predictions.

Graph (a) in Figure 4 show the effect of observing, for some hypothetical sample, that the value of S1 is low. The chart shows the nine variables used in the model (Figure 3) with the first bar of each pair showing the *prior* probability of the variable having a "high" value, and the second bar showing the corresponding *posterior* probability, after the observation and belief propagation. For example, the prior belief that any given sample would have a high S1 measure is roughly 0.5, while the posterior is 0.0 - we are stating the level is low, so the probability of it being high is zero. The probability of a high S4 has increased from 0.2 to 0.5, suggesting that S1 and S4 are inversely correlated to some extent, so that a low S1 tends to "cause" a high S4. Finally, the probability of a high preference score (P1) decreases from 0.4 to 0.2, suggesting that the preference panel dislike products with a low S1 score.

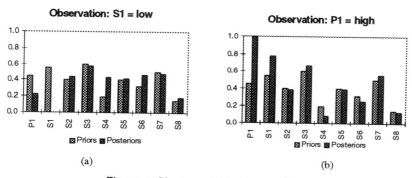

Figure 4: Single variable observations

Chart (b) shows the effect of observing a high preference score (P1). Here we are asking, "What features must a product have in order to be preferred?" The observation leads to an increase in the belief that the sample will have a high S1 score and a low S4 score. The other variables are largely unchanged, suggesting they have little direct influence over preference. The posterior probabilities here describe the "perfect" product according to the model.

6. Conclusions

Bayesian Belief Networks are a valuable addition to the product designers' toolkit. They are powerful tools for developing graphical models from a combination of data and expertise. They can be built from modest data sets, with or without background knowledge, and yet are scaleable because they are afford local optimisation. The results here show that they are as accurate as neural networks, but with the advantage of being reversible. This allows probabilistic predictions of optimal designs to be made, and these models are now being used to aid consumer preference modelling.

7. References

[1] AAFC 1991. "A Profile of the Canadian Speciality Food Industry" Market Report produced by the Canadian Department of Agriculture and Agri-Food.

[2] Lauritzen, S L, Spiegelhalter, D J 1988. "Local computations with probabilities on graphical structures and their application to expert systems" *Journal of the Royal Statistical Society*, Vol. 50 No. 2 pp.157-224.

[3] Pearl, J 1988. *Probabilistic Reasoning in Intelligent Systems: networks of plausible inference* Morgan Kaufmann.

[4] Barnett, G O, Famiglietti, K T, Kim, R J, Hoffer, E P, Feldman, M J 1998. "DXplain on the Internet" in *American Medical Informatics Association 1998 Annual Symposium*

[5] Helm, L 1996. " Improbable Inspiration" *Los Angeles Times*, October 28, 1996.

[6] Berler, A, Shimony, S E 1997. "Bayes Networks for Sonar Sensor Fusion" in Geiger, D, Shenoy, P (eds) 1997. *Proceedings of the Thirteenth Conference on Uncertainty in Artificial Intelligence* Morgan Kaufmann.

[7] Neil, M, Littlewood, B, Fenton, N 1996. "Applying Bayesian Belief Networks to Systems Dependability Assessment" in *Proceedings of Safety Critical Systems Club Symposium, Leeds, 6-8 February 1996* Springer-Verlag.

[8] Frey, B J 1998. *Graphical Models for Machine Learning and Digital Communication* MIT Press.

[9] Ezawa, K J, Schuermann, T 1995. "Fraud/Uncollectible Debt Detection Using a Bayesian Network Based Learning System: A Rare Binary Outcome with Mixed Data Structures" in Besnard, P, Hanks, S (eds) *Proceedings of the Eleventh Conference on Uncertainty in Artificial Intelligence* Morgan Kaufmann.

[10] Roozenburg, N F M, Eekels, J 1995. *Product Design: Fundamentals and Methods* Wiley.

[11] Krause, P, Clark, D 1993. *Representing Uncertain Knowledge: An Artificial Intelligence Approach*, Intellect Books.

[12] Jensen, F V 1996. *An Introduction to Bayesian Networks* UCL Press.

[13] Heckerman, D 1995. "A Tutorial on Learning With Bayesian Networks", Microsoft Research report MSR-TR-95-06..

[14] Bishop, C M 1995 *Neural Networks for Pattern Recognition* Oxford University Press

[15] Cooper, G F, Herskovits, E 1992. "A Bayesian Method for the Induction of Probabilistic Networks from Data" *Machine Learning* Vol. 9 pp. 309-347.

[16] Larranaga, P, Kuijpers, C M H, Murga, R H, Yurramendi, Y 1996. "Learning Bayesian network structures by searching for the best ordering with genetic algorithms" *IEEE Trans on Systems, Man and Cybernetics-A* Vol. 26 No. 4 pp.487-493.

[17] Etxeberria, R, Larranaga, P, and Pikaza, J M 1997. "Analysis of the behaviour of the genetic algorithms when searching Bayesian networks from data", *Pattern Recognition Letters* Vol. 18 No 11-13 pp 1269-1273

[18] Narendra, P M, Fukunaga, K 1977. "A Branch and Bound Algorithm for Feature Subset Selection" *IEEE Transactions on Computers*, Vol. 26, No. 9, pp. 917-922.

[19] Friedman, N 1998. "The Bayesian Structural EM Algorithm" in Cooper, G F, Moral, S (eds) *Proceedings of the Fourteenth Conference on Uncertainty in Artificial Intelligence* Morgan Kaufmann.

[20] Dempster, A P, Laird, N M, Rubin, D B 1977. "Maximum Likelihood from Incomplete Data via the EM Algorithm with discussion", *Journal of the Royal Statistical Society,* Series B, Vol. 39 pp.1-38.

[21] Mitchell, T M 1997 *Machine Learning* McGraw-Hill

8. Appendix: Data Set

The table below contains the raw data used in this work. Each row represents a record for a single sample. Column "P1" is the preference score, the remaining columns being eight sensory scores.

This data is also available in ASCII format from this URL:
http://www.cs.ucl.ac.uk/staff/D.Corney/FoodDesign.html

	P1	S1	S2	S3	S4	S5	S6	S7	S8
1	3.329	6.050	2.560	6.373	2.649	3.587	1.670	6.230	1.012
2	3.700	5.659	2.577	4.579	3.377	5.278	3.119	2.457	1.206
3	4.004	5.442	1.495	8.175	2.384	4.315	2.133	8.669	0.930
4	2.400	6.185	1.607	7.763	1.948	1.646	3.435	8.374	0.940
5	3.109	4.391	1.916	6.748	3.628	4.220	2.206	6.995	0.982
6	8.253	7.848	2.687	8.258	1.482	9.606	0.992	4.881	1.165
7	5.160	5.834	1.536	8.588	2.348	7.116	1.228	8.902	1.025
8	4.240	6.506	1.854	6.325	2.267	4.370	1.866	5.510	1.010
9	1.784	1.854	3.400	3.739	7.736	1.378	8.607	6.603	1.046
10	6.262	8.248	0.848	8.857	2.042	7.375	1.087	8.896	0.935
11	2.087	1.920	4.680	2.627	8.152	2.797	5.017	4.837	1.068
12	5.287	8.073	1.016	8.598	2.189	4.294	1.329	8.635	0.951
13	6.180	7.023	2.279	6.321	2.148	6.683	1.213	5.041	1.016
14	2.538	5.836	2.033	6.886	2.227	2.735	2.982	7.033	1.070
15	7.987	8.312	2.535	8.848	1.273	8.034	1.132	5.984	1.029
16	3.587	7.289	1.411	7.863	1.510	2.969	1.508	7.854	0.987
17	5.131	3.586	6.424	1.943	2.106	8.186	1.018	5.313	2.731
18	2.211	5.210	1.765	7.862	2.087	1.512	2.466	9.283	1.022
19	7.298	7.147	3.889	6.666	2.109	8.161	1.022	3.699	1.028
20	5.318	7.501	1.667	7.670	1.338	4.870	1.075	6.958	1.053

Acknowledgements

Unilever Research Ltd. have generously sponsored this work, and have provided data and advice throughout. The research was undertaken within the Postgraduate Training Partnership established between Sira Ltd and University College London. Postgraduate Training Partnerships are a joint initiative of the Department of Trade and Industry and the Engineering and Physical Sciences Research Council.

Flexible Ligand Docking Using a Robust Evolutionary Algorithm

Jinn-Moon Yang[1] and Cheng-Yan Kao[2]

Department of Computer Science and Information Engineering,
National Taiwan University, Taipei 106, Taiwan
e-mail:{moon,cykao}@csie.ntu.edu.tw

Abstract. A flexible ligand docking protocol based on evolutionary algorithms is investigated. The proposed approach incorporates family competition and adaptive rules to integrate decreasing-based mutations and self-adaptive mutations to act global and local strategies respectively. The method is applied to a dihydrofolate reductase enzyme with the anti-cancer drug methotrexate and two analogues of antibacterial drug trimethoprim. Conformations and orientations close to the crystallographically determined structures are obtained, as well as alternative structures with low energy. Numerical results indicate that the new approach is very robust. The root mean square derivation of the best docked lowest-energy structure with respect to the corresponding crystal structure is 0.67 Å.

1 Introduction

The docking problem is the computational prediction of the structures of ligand-protein complexes from the conformations of the flexible ligand and protein molecules. Minimization of the energy of intermolecular interactions of the docking process is an important method for drug design, called structure-based drug design [1] for identification of the lead compounds. This method has been of increasing interest because of the availability of high-resolution structures of enzymes of critical metabolic pathways. With these structures, computer-based methods can be used to identify or design ligands that posses good structural and chemical complementarity to active sites of the enzyme. The combination of molecular structure determination and computation is emerging as an important tool for drug design and discovery.

The methods for automated docking can be divided into two broad categories: matching methods and conformational search methods. The former attempts to find a good docking based on the geometry of a rigid docking molecular and receptor. In this treatment, the enzyme and ligand are taken as rigid objects and the search is reduced to finding energetically or geometrically favorable configurations of the ligand within the active site of the enzyme [2]. Unfortunately, this approach is likely to fail when the bound conformation of the ligand is unknown. Even with the rigid-body restriction, the number of possible ligand orientations is enormous and the computational problem belongs to the class known as NP-complete problem.

Conformational search methods are often to dock conformationally flexible ligand by employing a simulation or optimization method to search through the space of ligand-receptor configurations. These approaches, such as simulated annealing [3], [4] and genetic algorithms [5], [6], have the potential to identify a greater number and variety of known ligands. More recently, evolutionary algorithms [5], [7] have become a popular choice in molecule docking applications [5] and performed better than simulated annealing for some application domains [8].

Currently, there are about three main independently developed but strongly related implementations of evolutionary algorithms: genetic algorithms [9], evolution strategies [10], and evolutionary programming [11]. For genetic algorithms, both practice and theory, entails disadvantages of applying binary-represented implementation to global optimization. The coding function of genetic algorithms may introduce an additional multimodality, making the combined objective function more complex than the original function. To achieve better performance, real-coded genetic algorithms [8] have been introduced. In contrast, evolution strategies and evolutionary programming [12] mainly use real-valued representation and focus on self-adaptive Gaussian mutations. This type of mutation has succeeded in continuous optimization and has been widely regarded as a good operator for local searches. Unfortunately, experiments [13] show that self-adaptive Gaussian mutation leaves individuals trapped near local optima for rugged functions.

In this paper a new method called family competition evolutionary algorithm (FCEA) is proposed for docking conformationally flexible ligands into a rigid enzyme. The long-term objective of this work is to develop a method to conformationally screen and evaluate conformationally flexible ligands as potential lead compounds. We are seeking a method that is both rapid and reasonably accurate. The proposed method is applied to search the active site of an enzyme and then to search the conformational and configuration search space of known small ligands within the active site.

FCEA is a multi-operator approach which combines three mutation operators: decreasing-based Gaussian mutation, self-adaptive Gaussian mutation, and self-adaptive Cauchy mutation. It incorporates family competition and adaptive rules for controlling step sizes to construct the relationship among these three operators. In order to balance the exploration and exploitation, each of operators is designed to compensate for the disadvantages of the other. FCEA has been successfully applied to global optimization [13] [14].

The rest of this paper is organized as follows. Section 2 describes the flexible ligand docking problem, the scoring function, and the representation. Section 3 introduces the evolutionary nature of FCEA. In Section 4 we test FCEA on three flexible ligand docking problems. Conclusions are drawn in Section 5.

2 Problem description

The molecular docking problem is a problem of molecular recognition between two moleculars. In general, the recognition criteria can be given by the force field. Recognition processes are determined by the structure (including electrostatics) of molecular surfaces. Interaction of two molecular surfaces is a complex event involving molecular flexibility, induced fit, solvent, entropy, and hydrophobic, van der Waals interaction, and electrostatic interactions. This problem is the core problem in computational drug design. Broadly stated, molecular docking algorithms seek to predict the bound conformations of two interacting moleculars which are a small-molecule ligand and its receptor. The search spaces involved in docking are enormous, particular if a molecular is taken into account.

We adopt an AMBER-type potential function [15], [16] to score the different ligand orientations with the underlying assumption that the correct ligand binding conformation corresponding to the minimum of this function. For flexible ligand docking, the scoring function must consider the intraligand energy and the interaction energy between the ligand and the receptor. Our scoring function is

$$E_{tot} = E_{inter} + E_{intra} \tag{1}$$

where E_{inter} and E_{intra} are the intermolecular and intramolecular energy, respectively. In our scoring function, we use the Lennard-Jones 6-12 potential function to represent the energy of interaction between the ligand and the receptor. The Lennard-Jones equation is as follows [17]:

$$E_{inter} = \sum_{l=1}^{lig} \sum_{r=1}^{rec} \left[\frac{A_{lr}}{d_{lr}^{12}} - \frac{B_{lr}}{d_{lr}^6} + 332.0 \frac{q_l q_r}{\epsilon d_{lr}} \right] \tag{2}$$

where A_{lr} and B_{lr} are the nonbonded parameters, ϵ dielectric constants, d_{lr} is the distance between atoms l and r, q_l and q_r are the point charges of the atoms in the ligand and receptor respectively, and 332 is a factor that converts the electrostatic energy into kilocalories per mole. In this function, the first and second summation simulate the repulsive and attractive term in van der Waals interaction energy. The third summation simulates the electrostatic interaction between each pair of atoms.

The intramolecular energy of the ligand is

$$E_{intra} = \sum_{l<l'}^{lig} \left[\frac{A_{ll'}}{d_{ll'}^{12}} - \frac{B_{ll'}}{d_{ll'}^6} + 332.0 \frac{q_l q_l'}{\epsilon d_{ll'}} \right] \tag{3}$$

where the van der Waals and electrostatic sums over the one-four and higher interactions. In AMBER, certain atom pairs are models using explicit hydrogen bonds to fine-turn the hydrogen bond distance and energy. In our implementation we do not consider a specific hydrogen bond energy.

In the following, we give an outline of docking computational procedure and the representation of our algorithm. Given two molecules which consist of a number of atoms defined by their three dimensional coordinates, one defines the drug molecule, the other defines the receptor molecule. The basic ligand structure is presumed to be known (i.e. the atomic connectivities and positions), but the actual binding conformation of the ligand, described by a set of rotatable bonds, is taken as unknown. Both the binding conformation and the orientation of the binding conformation of the ligand with the protein are to be determined. Intuitively, the 3-dimensional location of the drug relative the protein, its three rotational angles relative to 3 axes, and its rotatable bonds are all adjustable.

In FCEA, each chromosome represents a ligand in a particular orientation and conformation space related to protein. That is, these adjustable variables are encoded as a chromosome:

$$(x_1, x_2, x_3, x_4, x_5, x_6, x_7, \cdots, x_n) \tag{4}$$

where x_1 , x_2 and x_3 represent the position of the drug molecule relative to the centroid of the receptor; x_4 , x_5 and x_6 are the rotational angles of the drug; and from x_7 to x_n are the twisting angles of the rotatable bonds inside the drug molecule. Generally, Docking problems can be broken down into three well-defined steps: defining the molecular of interest; modeling the protein and its interactions with solvent and the drug molecule; and performing the conformational and orientation search to find low-energy states of the system that correlate with the actual bindings model. In the coming section, our FCEA is described and implemented to execute these steps for the flexible ligand docking problems.

3 The Family Competition Evolutionary Algorithm

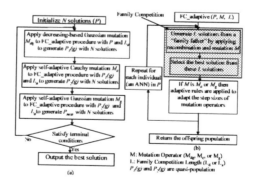

Fig. 1: Overview of our algorithm: (a) FCEA (b) FC_adaptive procedure

In this section, we present the detail of the family competition evolutionary algorithm (FCEA) for the flexible ligand docking problems. The basic structure of the FCEA is as follows (Fig. 1): N solutions are generated as the initial population. Each solution is presented by an encoded vector shown as Eq. 4. The initial ligand conformations are generated by randomizing the encoded vector, given that the center of mass of each ligand conformation must lie anywhere within the rectangular. Rigid rotation and rotatable dihedral angles are uniformly randomized between 0 and 360 degrees.

FCEA consists of three main stages in every iteration: decreasing-based Gaussian mutation, self-adaptive Cauchy mutation, and self-adaptive Gaussian mutation. Each stage is realized by generating a new quasi-population (with N solutions) as the parent of the next stage. As shown in Fig. 1, these stages differ only in the mutations used and in some parameters. Hence we use a general procedure "FC_adaptive" to represent the work done by these stages.

The FC_adaptive procedure employs four parameters, namely, the parent population (P, with N solutions), mutation operator (M), selection method (S), and family competition length (L), to generate a new quasi-population. The main work of FC_adaptive is to produce offspring and then conduct the family competition. Each individual in the population sequentially becomes the "family father." With a probability p_c, this family father and another solution randomly chosen from the rest of the parent population are used as parents to do a recombination operation. Then the new offspring or the family father (if the recombination is not conducted) is operated on by a mutation. For each family father, such a procedure is repeated L times. Finally L children are produced but only the one with the lowest objective value survives. Since we create L children from one "family father" and perform a selection, this is a family competition strategy. We think this is a way to avoid the prematureness but also keeps the spirit of local searches.

After the family competition, there are N parents and N children left. Based on different stages (or say the parameter S of the FC_adaptive procedure), we employ various ways of obtaining a new quasi-population with N individuals. For both Gaussian and Cauchy self-adaptive mutations, in each pair of father and child the individual with a better objective value survives. This procedure is called "family selection." On the other hand, "population selection" chooses the best N individuals from all N parents and N children. With a probability P_{ps}, FCEA applies population selection to speed up the convergence when the decreasing-based Gaussian mutation is used. With probability ($1-P_{ps}$), family selection is still considered. In order to reduce the ill effects of greediness on this selection, the initial P_{ps} is set to 0.05, but it is changed to 0.5 when the mean step size of self-adaptive Gaussian mutation is larger than that of decreasing-based Gaussian mutation. Note that the population selection is similar to ($\mu + \mu$)-ES in the traditional evolution strategies. Hence, through the process of selection, the FC_adaptive procedure forces each solution of the starting population to have one final offspring. Note that we create LN offspring in the procedure FC_adaptive but the size of the new quasi-population remains the same as N.

For all three mutation operators, we assign different parameters to control their performance. Such parameters must be adjusted through the evolutionary process. We modify them first when mutations are applied. Then after the family competition is complete, parameters are adapted again to better reflect the performance of the whole FC_adaptive procedure.

Regarding chromosome representation, we present each solution of a population as a set of four n-dimensional vectors $(x^i, \sigma^i, v^i, \psi^i)$, where n is the number of adjustable variables of a ligand and $i = 1, \ldots, N$. The vector x represents the adjustable variable vector, shown in Eq. 4 to be optimized; and σ, v, and ψ are the step-size vectors of decreasing-based mutations, self-adaptive Gaussian mutation, and self-adaptive Cauchy mutation, respectively. In other words, each solution x is associated with some parameters for step-size control. In this paper, the initial x_1, x_2, and x_3 are randomly chosen from the feasible box, and the others, from x_4 to x_n, are uniformly randomized between 0 and 360 degrees. The initial step sizes σ, v, and ψ are 0.8, 0.2, 0.2, respectively. For easy description of operators, we use $a = (x^a, \sigma^a, v^a, \psi^a)$ to represent the "family father" and $b = (x^b, \sigma^b, v^b, \psi^b)$ as another parent (only for the recombination operator). The offspring of each operation is represented as $c = (x^c, \sigma^c, v^c, \psi^c)$. We also use the symbol x_j^d to denote the jth component of an individual d, $\forall j \in \{1, \ldots, n\}$. In the rest of this section we explain each important component of the FC_adaptive procedure: recombination operators, mutation operations, and rules for adapting step sizes (σ, v, and ψ).

3.1 Recombination Operators

FCEA implements modified discrete recombination and intermediate recombination [10]. Here we would like to mention again that recombination operators are activated with only a probability p_c. The adjustable variable x and a step size are recombined in a recombination operator.

Modified Discrete Recombination: The original discrete recombination [10] generates a child that inherits genes from two parents with equal probability. Here the two parents of the recombination operator are the "family father" and another solution randomly selected. Our experience indicates that FCEA can be more robust if the child inherits genes from the "family father" with a higher probability. Therefore, we modified the operator to be as follows:

$$x_j^c = \begin{cases} x_j^a \text{ with probability } 0.8 \\ x_j^b \text{ with probability } 0.2. \end{cases} \tag{5}$$

Intermediate Recombination: Intermediate Recombination is defined as:

$$x_j^c = x_j^a + (x_j^b - x_j^a)/2, \text{ and} \tag{6}$$
$$w_j^c = w_j^a + \beta(w_j^b - w_j^a)/2, \tag{7}$$

where w is σ, v or ψ based on the mutation operator applied in the family competition. For example, if self-adaptive Gaussian mutation is used in this FC_adaptive procedure, x in (6) and (7) is v. We follow the work of the evolution strategies community [10] to employ only intermediate recombination on step-size vectors, that is, σ, v, and ψ. To be more precise, x is also recombined when the intermediate recombination is chosen.

3.2 Mutation Operators

Mutations are main operators of FCEA. After the recombination, a mutation operator is applied to each "family father" or the new offspring generated by recombination. In FCEA, the mutation is performed independently on each vector element of the "family father" by adding a random value with expectation zero:

$$x_i' = x_i + wD(\cdot), \tag{8}$$

where x_i is the ith component of x, x_i' is the ith component of x' mutated from x, $D(\cdot)$ is a random variable, and w is the step size. In this paper, $D(\cdot)$ is evaluated as $N(0,1)$ or $C(1)$ if the mutations are, respectively, Gaussian mutation or Cauchy mutation.

Self-Adaptive Gaussian Mutation: We adapted Schwefel's [10] proposal to use self-adaptive Gaussian mutation in global optimization. The mutation is accomplished by first mutating the step size v_j and then the adjustable variable x_j:

$$v_j^c = v_j^a \exp[\tau' N(0,1) + \tau N_j(0,1)], \tag{9}$$
$$x_j^c = x_j^a + v_j^c N_j(0,1), \tag{10}$$

where $N(0,1)$ is the standard normal distribution. $N_j(0,1)$ is a new value with distribution $N(0,1)$ that must be regenerated for each index j. For FCEA, we follow [10] in setting τ and τ' as $(\sqrt{2n})^{-1}$ and $(\sqrt{2\sqrt{n}})^{-1}$, respectively.

Self-Adaptive Cauchy Mutation: We define self-adaptive Cauchy mutation as follows:

$$\psi_j^c = \psi_j^a \exp[\tau' N(0,1) + \tau N_j(0,1)], \tag{11}$$
$$x_j^c = x_j^a + \psi_j^c C_j(t). \tag{12}$$

In our experiments, t is 1. Note that self-adaptive Cauchy mutation is similar to self-adaptive Gaussian mutation except that (10) is replaced by (12). That is, they implement the same step-size control but use different means of updating x. Cauchy mutation is able to make a larger perturbation than Gaussian mutation. This implies that Cauchy mutation has a higher probability of escaping from local optima than Gaussian mutation does.

Decreasing-Based Gaussian Mutations: Our decreasing-based Gaussian mutation adjusts the step-size vector σ with a fixed decreasing rate $\gamma = 0.95$ as follows:

$$\sigma^c = \gamma \sigma^a, \tag{13}$$
$$x_j^c = x_j^a + \sigma^c N_j(0,1). \tag{14}$$

Previous results [13] demonstrated that self-adaptive mutations converge faster than decreasing-based mutations but, for rugged functions, self-adaptive mutations more easily trap into local optima than decreasing-based mutations.

3.3 Adaptive Rules

The performance of Gaussian and Cauchy mutations is largely influenced by the step sizes. FCEA adjusts the step sizes while mutations are applied (e.g. (9), (11), and (13)). However, such updates insufficiently consider the performance of the whole family. Therefore, after family competition, some additional rules are implemented:

1. **A-decrease-rule:** Immediately after self-adaptive mutations, if objective values of all offspring are greater than or equal to that of the "family parent," we decrease the step-size vectors v (Gaussian) or ψ (Cauchy) of the parent:

$$w_j^a = 0.95 w_j^a, \tag{15}$$

where w^a is the step size vector of the parent. In other words, when there is no improvement after self-adaptive mutations, we may propose that a more conservative, that is, smaller, step size tends to make better improvement in the next iteration. This is inspired from the 1/5-success rule of $(1+\lambda)$-ES [10].

2. **D-increase-rule:** It is difficult, however, to decide the rate γ of decreasing-based mutations. Unlike self-adaptive mutations which adjust step sizes automatically, its step size goes to zero as the number of iterations increases. Therefore, it is essential to employ a rule which can enlarge the step size in some situations. The step size of the decreasing-based mutation should not be too small, when compared to step sizes of self-adaptive mutations. Here, we propose to increase σ if one of the two self-adaptive mutations generates better offspring. To be more precise, after a self-adaptive mutation, if the best child with step size v is better than its "family father," the step size of the decreasing-based mutation is updated as follows:

$$\sigma^c = \max(\sigma^c, \beta v_{mean}^c), \tag{16}$$

where v_{mean}^c is the mean value of the vector v; and β is 0.2 in our experiments. Note that this rule is applied in stages of self-adaptive mutations but not of decreasing-based mutations.

4 Molecular binding experiments

Our primary interest is the use of the FCEA to optimize the adjustable variables of the ligand. The accuracy of our FCEA incorporating flexible ligand docking is examined and summarized for three test systems in Table 1. We bind a real receptor molecule, dihydrofolate reductase enzyme(DHFR) with three drug molecules, the methotrexate(MTX), and two analogues(inhibitor 91 and inhibitor 309) of trimethoprim. Methotrexate is an anti-cancer drug which is used clinically to cure patients, and trimethoprim is an anti-bacterial drug.

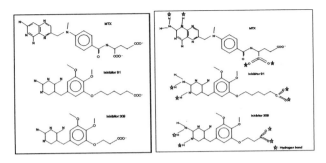

(a) The structures and the rotatable bonds

(b) The positions of the hydrogen bonds

Fig. 2: The structures and the rotatable bonds of MTX, inhibitor 91, and inhibitor 309, and the positions of the hydrogen bonds of these drugs with protein DHFR.

There has been some research that analyzes the binding structures of DHFR with the methotrexate molecule [12], [18]. The internal degrees of freedom of the docking MTX, inhibitor 91, and inhibitor 309 into the binding region of DHFR are 18, 19, and 18, respectively, because the rotatable bonds of these three ligands are 12, 13, and 12, respectively. These structures of rotatable bonds are the bold lines shown in Fig 3.2(a).

Table 1: The results of FCEA for docking MTX, 91, and 309 into DHFR (energy in kcal/mol)

Protein/ Ligand	# of rotatable bonds in Ligand	Degrees of freedom	Lowest Energy	Average Energy	Best # of Hydrogen[a]	Average # of Hydrogen[a]
DHFR-MTX	12	18	-111.5	-99.33	4	2.75
DHFR-91	13	19	-71.62	-66.17	4	3.48
DHFR-309	12	18	-66.07	-61.61	4	3.8

[a]The largest distance between atoms formed a hydrogen bond is $\leq 2.4\text{Å}$

Table 2 indicates the settings of FCEA parameters, such as family competition length and crossover rate. The population size is 50. FCEA stops if it exceeds a maximal number of function evaluations, or the individuals in the whole population are unable to be improved within five generations. The maximal number of function evaluations is 500,000. FCEA tests on these three docking problems with 20 independent runs.

The experimental results are given in Table 1. In addition to the binding energy, the existence of hydrogen-bonds is another criterion that determines the goodness of fitting between molecules. According to the results presented in [19] which has shown the binding structure of the MTX with DHFR, there

Table 2: Parameter settings of FCEA

parameter name	the value of parameter
population size	50
recombination rate p_c =0.2	
family competition $L = 3$	
step sizes	$v_i = \psi_i = 0.2$, $\sigma_i = 0.8$
decreasing rate	γ=0.95

is a pocket that exists in DHFR. Inspect all the testing cases, all generated drug structures are bound to this pocket, because hydrogen bonds are formed in this position. Fig. 3.2(b) shows the positions of these hydrogen bonds between molecules. Fig. 4.1(b) shows that FCEA is able to bind the ligand MTX into DHFR. Inspecting all the docked structures generated by FCEA, the four hydrogen bonds exist between molecules for the best solutions for three testing ligands according to Table 1.

Fig. 4.1(a) compares the difference between MTX crystal structure and the structure of our experimental results. The results of 20 docked runs of MTX into DHFR; 12 solutions are within 1.5 Å rmsd of the crystal structure of MTX. The best solution is 0.67 Å. From the analysis of hydrogen bonds between ligands and DHFR, and the comparison of the crystal structure, we claim that FCEA is able to find good fit molecular structures between ligands and DHFR. Our FCEA is a very useful tool for docking flexible ligands. Table 3 shows the

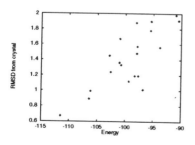

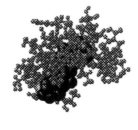

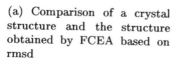

(a) Comparison of a crystal structure and the structure obtained by FCEA based on rmsd

(b) The docked structure

Fig. 1: The results of FCEA for docking the MTX into DHFR. The large balls in (b) are the atoms of the MTX and the others are the atoms of the receptor(DHFR).

comparison of FCEA with GAs, i.e. DIVALI [20] and DOCK [18], and evolutionary programming [12] on docking flexible MTX into DHFR. To achieve fair comparison, we use the rmsd between the crystal structure and the structures obtained by these four approaches. Except evolutionary programming [12], the others use an AMBER-type scoring function. Evolutionary programming, incorporating a special scoring function with a large population size 1000, has the smallest rmsd value among comparative approaches. FCEA is more stable than GAs in this case. We observe that FCEA is very competitive with states of the art evolutionary algorithms.

Table 3: Comparison FCEA with previous approaches on docking flexible MTX into DHFR based on rmsd by comparing the crystal structure with these structures obtained by four approaches.

FCEA		DIVALI [20]		DOCK [18]		EP [12]	
best	average	best	average	best	average	best	average
0.67 Å	1.37 Å	1.1 Å	2.24 Å	0.6 Å	$\approx 2.4 A^a$	$\approx 0.35 A^b$	$\approx 1.0 A^b$

$\dagger^a$ the value is approximated according to the Fig.3 in [18].
$\dagger^a$ the values are approximated according to the Fig.1(a) in [12].

5   Conclusions

This study has demonstrated that FCEA is a robust approach to explore the orientational and conformation space of a flexible ligand with a receptor. With FCEA, the location and the orientation of the ligand are not restricted to a portion of the active site. For FCEA, decreasing-based mutations with a large initial step size is a global search strategy; self-adaptive mutations with family competition procedure and replacement selection are local search strategies. These mutation operators can closely cooperate with one another.

Experiments on three test cases verify that the proposed approach generates low-energy solutions docked into a receptor. These solutions are compared to the crystal structure with an rmsd accuracy of 0.67 Å/atom. We believe that the flexibility and robustness of FCEA make it a highly effective tool for the flexible ligand docking problems.

In future, we will take our FCEA in three directions: (1) develop flexibility in the receptor; (2) improve the scoring function; and (3) study a more diverse set of ligand-protein complex to determine the limits of our FCEA for structure prediction.

References

1. I. D. Kuntz. Structure-based strategies for drug design and discovery. *Science*, 257:1078–1082, 1992.

2. I. D. Kuntz and et al. A geometric approach to macromolecular-ligand interactions. *Journal of Computational Biology*, 161:269–288, 1982.

3. D. S. Goodsell and A. J. Olson. Automated docking of substrates to proteins by simulated annealing. *Proteins Structure Function and Genetics*, 8:195–202, 1990.

4. C. J. Sherman, R. C. Ogden, and S. T. Freer. *De Novo* design of enzyme inhibitors by monte carlo ligand generation. *Journal of Medicinal Chemistry*, 38(3):466–472, 1995.

5. D. E. Clark and D. R. Westhead. Evolutionary algorithms in computer-aided molecular design. *Journal of Computer-Aided Molecular Design*, 10:337–358, 1996.

6. R. S. Judson et. al. Docking flexible molecules: A case study of three proteins. *Journal of Computational Chemistry*, 16(11):1405–1419, 1995.

7. W. E. Hart. Comparing evolutionary programs and evolutionary pattern search algorithms: A drug docking application. In *Proc. Genetic and Evolutionary Computation Conf.*, 1999.

8. G. M. Morris and et al. Automated docking using a lamarckian genetic algorithm and and empirical binding free energy function. *Journal of Computational Chemistry*, 19:1639–1662, 1998.

9. D. E. Goldberg. *Genetic Algorithms in Search, Optimization and Machine Learning*. Addison-Wesley Publishing Company, Inc., Reading, MA, USA, 1989.

10. T. Bäck, F. Hoffmeister, and H-P. Schwefel. A survey of evolution strategies. In *Proc. Fourth Int. Conf. on Genetic Algorithms*, pages 2–9, 1991.

11. D. B. Fogel. *Evolutionary Computation: Toward a New Philosophy of Machine Intelligent*. NJ:IEEE Press, Piscataway, 1995.

12. D. K. Gehlhaar and et al. Molecular recognition of the inhibitor ag-1343 by hiv-1 protease: conformationally flexible docking by evolutionary programming. *Chemistry and Biology*, 2(5):317–324, 1995.

13. J.-M. Yang, C.-Y. Kao, and J.-T. Horng. A continuous genetic algorithm for global optimization. In *Proc. of the Seventh Int. Conf. on Genetic Algorithms*, pages 230–237, 1997.

14. J.-M. Yang and C.-Y. Kao. An evolutionary algorithm for synthesizing optical thin-film designs. In *Parallel Problem Solving form Nature-PPSN V, (Lecture Notes in Computer Science, vol. 1498)*, pages 947–958, 1998.

15. S. J. Weiner and et al. A new force field for mocular mechanical simulation of uncleic acids and proteins. *Journal of the American Chemical Society*, 106:765–784, 1984.

16. S. J. Weiner, P. A. Kollman, D. T. Nguyen, and D. A. Case. An all atom force field fo simulations of proteins and uncleic acids. *Journal of Computational Chemistry*, 7:266–278, 1986.

17. N. Pattabiraman et. al. Computer graphics and drug design: Real time docking, energy calculation and minimization. *Journal of Computational Chemistry*, 6:432–436, 1985.

18. C. M. Oshiro, I. D. Kuntz, and J. S. Dixon. Flexible ligand docking using a genetic algorithm. *Journal of Computer-Aided Molecular Design*, 9:113–130, 1995.

19. D. A. Matthews and et al. Dihydrofolate reductase from lactobacillus casei : X-ray structure of the enzyme methotrexate nadph complex. *Journal of Biological Chemistry*, 253(19):6946–6954, 1978.

20. Ajay K. P. Clark. Flexible ligand docking without parameter adjustment across four ligand-receptor complexes. *Journal of Computational Chemistry*, 16(10):1210–1226, 1995.

Network Design Techniques Using Adapted Genetic Algorithms

Mitsuo Gen[*], Runwei Cheng[†] and Shmuel S. Oren[‡]

[*]Visiting Prof.;1999-2000, Dept of Industrial Engg. and Operations Res.
University of California, Berkeley, CA 94720 USA
E-mail:gen@ieor.berkeley.edu and
[*]Department of Industrial and Information Systems Engineering
Ashikaga Institute of Technology, Ashikaga 326-8558, Japan
E-mail: gen@ashitech.ac.jp

[†]Department of Industrial and Information Systems Engineering
Ashikaga Institute of Technology, Ashikaga 326-8558, Japan
E-mail: runweicheng@hotmail.com

[‡]Department of Industrial Engineering and Operations Research,
University of California, Berkeley, CA 94720 USA
E-mail: oren@ieor.berkeley.edu

Abstract. In recent years we have evidenced an extensive effort in the development of computer communication networks, which have deeply integrated in human being's everyday life. One of the important aspects of the network design process is the topological design problem involved in establishing a communication network. However, with the increase of the problem scale, the conventional techniques are facing the challenge to effectively and efficiently solve those complicated network design problems. In this article, we give out our recent research works on the network design problems by using genetic algorithms (GAs), such as, multistage process planning problem, fixed charge transportation problem, minimum spanning tree problem, centralized network design, and local area network design. All these problems are illustrated from the point of genetic representation encoding skill and the genetic operators with hybrid strategies. Large quantities of numerical experiments show the effectiveness and efficiency of such kind of GA-based approach.

1. Introduction

Genetic algorithms are one of the most powerful and broadly applicable stochastic search and optimization techniques based on principles from evolution theory (Holland, 1975). In the past few years, the genetic algorithms community has turned much of its attention toward the optimization of network design problems (Gen and Cheng, 1997, 2000). This paper is intended as a text covering applications of GAs to some difficult-to-solve network design problems inherent in industrial engineering and computer communication network (Gen and Kim, 1998).

2. Adaptation of Genetic Algorithms

Genetic algorithms were first created as a kind of generic and weak method featuring binary encoding and binary genetic operators. This approach requires a modification of an original problem into an appropriate form suitable for the genetic algorithms, as shown in Figure 1. The approach includes a mapping between potential solutions and binary representation, taking care of decoders or repair procedures, etc. For complex problems, such an approach usually fails to provide successful applications.

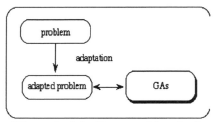

Figure 1. Adapting a problem to the genetic algorithms

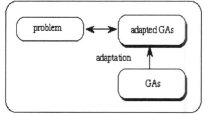

Figure 2. Adapting the genetic algorithms to a problem

To overcome such problems, various non-standard implementations of the genetic algorithms have been created for particular problems. As shown in Figure 2, this approach leaves the problem unchanged and adapts the genetic algorithms by modifying a chromosome representation of a potential solution and applying appropriate genetic operators. But in general, it is not a good choice to use the whole original solution of a given problem as the chromosome because many real problems are too complex to have a suitable implementation of genetic algorithms with the whole solution representation. Generally, the encoding methods can be either direct or indirect. In the direct encoding method, the whole solution for a given problem is used as a chromosome. For a complex problem, however, such a method will make almost all of the conventional genetic operators unusable because a vast number of offspring will be infeasible or illegal. On the contrary, in the indirect encoding method, just the necessary part of a solution is used as a chromosome. A decoder then produces the solution. A decoder is a problem-specific and determining procedure to generate a solution according to the permutation and/or the combination of the items produced by genetic algorithms. With this method, the genetic algorithms will focus their search solely on the interesting part of solution space.

A third approach is to adapt both the genetic algorithms and the given problem, as shown in Figure 3. A common feature of combinatorial optimization problems is to find a permutation and/or a combination of some items associated with side constraints. If the permutation and/or combination can be determined, a solution then can be derived with a problem-specific procedure. With this third approach, genetic algorithms are used to evolve an appropriate permutation and/or combination of some items under consideration, and a heuristic method is subsequently used to construct a solution according to the permutation and combination.

This approach has been successfully applied in the area of industrial engineering and has recently become the main approach for the practical use of genetic algorithms in the network design and optimization (Cheng, 1997).

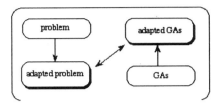

Figure 3. The third approach: adapt both the genetic algorithms and the problem

3. Multistage process planning problems

Multi-stage Process Planning (MPP) problem is abundant among manufacturing systems. It provides a detailed description of manufacturing capabilities and requirements for transforming a raw stock of materials into a completed product through multi-stage process. The MPP problem can be classified into two types (Chang and Wysk, 1985): variant MPP and generative MPP. Recently the generative MPP problem has received more attention since it can accommodate both general and automated process planning models, which is particularly important in flexible manufacturing systems (Kusiak and Finke, 1988).

The MPP system usually consists of a series of machining operations, such as turning, drilling, grinding, finishing and so on, to transform a part into its final shape or product. The whole process can be divided into several stages. At each stage, there is a set of similar manufacturing operations. The MPP problem is to find the optimal process planning among all possible alternatives given certain criteria such as minimum cost, minimum time, maximum quality or under multiple of these criteria. Figure 4 shows a simple MPP problem by the means of network flow.

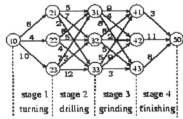

Figure 4. Flow network for a simple MPP problem

For an n-stage MPP problem, let s_k be some state at stage k, $D_k(s_k)$ be the set of possible states to be chosen at stage k, $k = 1, 2, \ldots, n$, and let x_k be the decision variable to determine which state to choose at stage k, obviously $x_k \in D_k(s_k)$, $k = 1, 2, \ldots, n$. Then the MPP problem can be formulated as follows:

$$\min_{\substack{x_k \in D_k(s_k) \\ k=1,2,\dots,n}} V(x_1, x_2, \dots, x_n) = \sum_{k=1}^{n} v_k(s_k, x_k) \qquad (1)$$

where $v_k(s_k, x_k)$ represents the criterion to determine x_k under state s_k at stage k, usually defined as a real number such as cost, time or distance. The problem can be rewritten as a dynamic recurrence expression and then approached by Shortest Path Method or Dynamic Programming. Obviously, with increased problem scale, many stages and states must be considered, which will greatly affect the efficiency of any solution method.

Awadh, Sepehri, and Hawaleshka proposed a binary string based-GA to deal with the MPP problem (Awadh, Sepehri, and Hawaleshka, 1995) and Zhou and Gen extended a state permutation encoding-based GA approach with multiobjecitves (Zhou and Gen, 1997a). The problem was encoded as a permu tation of states. In such an encoding, the position of a gene indicates the stage and the value of the gene indicates a state for an operation chosen at that stage. For example, [3 1 2] is an encoding for a feasible solution to the instance given in Figure 4, which represents the processing plan as $(s_{10}, s_{23}, s_{31}, s_{42}, s_{50})$. This encoding method is able to generate all feasible individuals through genetic operations such as crossover or mutation. Initial population is generated randomly with the number of all possible states in corresponding stage. Only neighborhood search-based mutation was used. Replacing the mutated gene with all its possible states forms a neighbor to a chromosome. Figure 5 shows an example for this mutation operation.

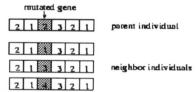

Figure 5. Illustration of mutation with neighborhood search.

When apply genetic algorithm in multiobjective context, a crucial issue is how to evaluate the credit of chromosomes and then how to represent the credit as fitness values. Zhou and Gen used a weighted-sum approach, which assigns weights to each objective function, combines the weighted objectives into a single objective function, and takes its reciprocal as the fitness value:

$$eval(v) = 1 / \sum_{k=1}^{p} w_k V_k(x_1, x_2, \dots, x_n) \qquad (2)$$

where v denote the chromosome, p the number of objectives, w_k the weight coefficient assigned to objective V_k, which is determined by the *adaptive weights approach* (Gen and Cheng, 2000). In the adaptive weight approach, weights will be adaptively adjusted according to the current generation in order to obtain a search pressure towards to the positive ideal point. Because some useful information from the current population is used to readjust the weights at each generation, the approach gives a selection pressure towards to positive ideal point.

4. Fixed charge transportation problem

The fixed charge transportation problem (fc-TP) is an extension of the transportation problem (TP) and many practical transportation and distribution problems can be formulated as this problem. For instance, in a transportation problem, a fixed cost may be incurred for each shipment between a given plant and a given consumer and a facility of a plant or warehouse may result in a fixed amount on investment. The fc-TP problem is much more difficult to solve due to the presence of fixed costs, which cause discontinuities in the objective function. Given m plants and n consumers, the problem can be formulated as follows:

$$\textbf{fc-TP:} \quad \min f(x) = \sum_{i=1}^{m} \sum_{j=1}^{n} (f_{ij}(x) + d_{ij} g(x_{ij})) \tag{3}$$

$$\text{s. t.} \quad \sum_{j=1}^{n} x_{ij} \le a_i, \quad i = 1, 2, \ldots, m \tag{4}$$

$$\sum_{i=1}^{m} x_{ij} \ge b_j, \quad j = 1, 2, \ldots, n \tag{5}$$

$$x_{ij} \ge 0, \quad \forall i, j \tag{6}$$

where $x = [x_{ij}]$ is the unknown quantity to be transported from plant i to consumer j, $f_{ij}(x)$ is the objective function of shipping, and

$$g(x_{ij}) = \begin{cases} 1, & \text{if } x_{ij} > 0 \\ 0, & \text{otherwise} \end{cases} \tag{7}$$

where d_{ij} is the fixed cost. Many solution procedures have been proposed for the fixed charge transportation problem range from exact solution algorithms to heuristic methods. Recently, Gottlieb and Paulmann (1998) proposed a genetic algorithm based on permutation representation for this problem. Sun, Aronson, Mckeown, and Drinka (1998) proposed a tabu search method. Since the solution of the problem has a network structure characterized as spanning tree, Gen and Li proposed a spanning tree-based genetic algorithm (Gen and Li, 1998, 1999). Figure 6 shows a simple example of transportation alternatives, expressed as a spanning tree. A transportation alternative can be encoded by a Prüfer number as shown in Figure 7. The detail of the decoding procedure from a Prüfer number to a transportation tree was given in the book (Gen and Cheng, 2000).

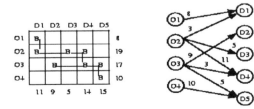

Figure 6. A transportation alternative with a spanning tree structure

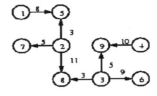

P(T) = [5 9 2 3 2 6 3]

Figure 7. A transportation alternative and its Prüfer Number encoding

Because a transportation tree is a special type of spanning tree, the Prüfer number encoding may correspond to an infeasible solution. Gen and Li designed a criterion for checking the feasibility of chromosomes. One-point crossover and inversion mutation were used to explore new solutions.

To demonstrate the effectiveness and efficiency of the spanning tree based genetic algorithm for the problem, Gen and Li carried out numerical experiments and compared with the matrix-based genetic algorithm (Vignaux and Michalewicz, 1991). For the small-scale problem, there was no obvious difference on results. For the larger scale problems, the spanning tree-based genetic algorithm can get the optimal solutions with much less CPU time than can the matrix-based genetic algorithm (Li, Gen, and Ida, 1998; Li, 1999).

5. Minimum spanning tree problem

A spanning tree structure is the best topology for telecommunication network designs, which usually consists of finding the best way to link n nodes at different locations. They may be the host, concentrators, multiplexors, and terminals. In a real-life network optimization situation, a spanning tree is often required to satisfy some additional constraints, such as the edge or capacity on a node. A tree structure network with the constraint on the edges is denoted as the degree-constrained minimum spanning tree problem (dc-MST) (Narula and Ho, 1980).

Considering an undirected graph $G = (V, E)$, let $V = \{1, 2, ..., n\}$ be the set of nodes and $E = \{(i, j) \mid i, j \in V\}$ be the set of edges. For a subset of nodes $S (\subseteq V)$, define $E(S) = \{(i, j) \mid i, j \in S\}$ be the edges whose end points are in S. Define the following binary decision variables for all edges $(i, j) \in E$.

$$x_{ij} = \begin{cases} 1, & \text{if edge } (i, j) \text{ is selected in a spanning tree} \\ 0, & \text{otherwise} \end{cases} \quad (8)$$

Let w_{ij} be the fixed cost related to edge (i, j), the problem can be formulated as follows:

$$\textbf{dc-MST:} \quad \min \quad z(x) = \sum_{i=1}^{n-1} \sum_{j=2}^{n} w_{ij} x_{ij} \quad (9)$$

$$\text{s. t.} \quad \sum_{i=1}^{n-1} \sum_{j=2}^{n} x_{ij} = n - 1 \quad (10)$$

$$\sum_{i \in S} \sum_{j \in S, j>1} x_{ij} \leq |S| - 1, S \subseteq V \setminus \{1\}, |S| \geq 2 \quad (11)$$

$$\sum_{j=1}^{n} x_{ij} \le b_i, i = 1,2,..., n \qquad (12)$$

$$x_{ij} = 0 \text{ or } 1, \quad i = 1, 2, ..., n - 1, \quad j = 2, 3, ..., n \qquad (13)$$

where b_i is the constrained degree value for node i. Inequality (12) is the constrained degree on each node. Equality (10) is true of all spanning trees. The dc-MST problem is NP-hard and there are no effective algorithms to deal with it. Zhou and Gen proposed a genetic algorithms approach to solve this problem (Zhou and Gen, 1997b and 1999; Zhou, 1999). There are two factors that should be taken into consideration if we want to keep the tree topology in genetic representation: one is the connectivity among nodes; the other is the degree value of each node. Therefore, a two-dimension structure was used as the genetic representation. One dimension encodes a spanning tree; another dimension encodes degree value. For an undirected tree, we can take any node as the root node of it and all other nodes are regarded as being connected to it hierarchically. Figure 8 illustrates an example of this degree-based permutation.

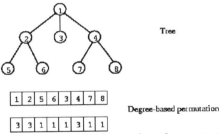

Figure 8. A rooted tree and its degree-based permutation encoding

In order to keep the degree constraint and connectivity between nodes, the genes in the degree dimension need to satisfy the following conditions: For an n-node tree, the total degree value for all nodes is $2(n-1)$. Suppose that d_{used} is the total degree value of the nodes whose degree value in degree dimension have been assigned and d_{rest} is the total lower bound of the degree values for all those nodes whose degree value in degree dimension have not been assigned. Then the degree value for the current node in degree dimension should hold: no less than 1. The degree value for the current node together with the number of the rest nodes should hold: no less than d_{rest} and no greater than $2(n-1) - d_{used}$.

Because this encoding is essentially a permutation one, uniform crossover and insertion mutations were adopted. Especially the insertion mutation plays a very important role for the dc-MST problem as it always keep the individuals as tree structure and evolves them to the fitter tree structures. This operator selects a string of genes (branch of a tree) at random and inserts it in a random gene (node). The operation is illustrated in Figure 9.

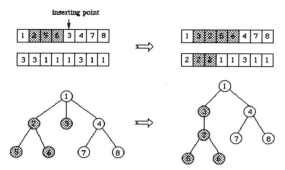

Figure 9. Illustration of insertion mutation

6. Centralized network design problem

A centralized network is a network where all communication is to and from a single site. In such networks, terminals are connected directly to the central site. Multipoint lines are used, where groups of terminals share a tree to the center and each multipoint line is linked to the center site by one edge only. This problem is also denoted as the capacitated Minimum Spanning Tree (c-MST) problem (Gavish, 1982).

Considering a complete, undirected graph $G = (V, E)$, $V = \{1, 2, ..., n\}$ be the set of nodes and $E = \{(i, j) \mid i, j \in V\}$ be the set of edges. For a subset of nodes $S(\subseteq V)$, define $E(S) = \{(i, j) \mid i, j \in S\}$ be the edges whose endpoints are in S. Let c_{ij} be the fixed cost related to edge (i, j), and d_I the demand at each node $i \in V$, where by convention the demand of the root node $d_1 = 0$. $d(S)$ the sum of the demands of the nodes of S. The subtree capacity is denoted κ. The centralized network design problem can be formulated as follows:

$$\textbf{CND:} \quad \min \ z(x) = \sum_{i=1}^{n-1} \sum_{j=2}^{n} c_{ij} x_{ij} \tag{14}$$

$$\text{s. t.} \ \sum_{i=1}^{n-1} \sum_{j=2}^{n} x_{ij} = n - 1 \tag{15}$$

$$\sum_{i \in S} \sum_{j \in S, j>1} x_{ij} \leq \mid S \mid -\lambda(S), S \subseteq V \setminus \{1\}, \mid S \mid \geq 2 \tag{16}$$

$$\sum_{i \in U} \sum_{j \in U, j>1} x_{ij} \leq \mid U \mid -1, U \subset V, \mid U \mid \geq 2, \{1\} \in U \tag{17}$$

$$x_{ij} = 0 \text{ or } 1, \ i = 1, 2, ..., n - 1, \ j = 2, 3, ..., n. \tag{18}$$

The parameter $\lambda(S)$ in (16) refers to the bin packing number of the set S, namely, the number of bins of size k needed to pack the nodes of items of size d_i for all $i \in S$. Up to now, all heuristic algorithms for these problems are only focused on how to deal with the constraints to make the problem simpler to solve. On the approach of cutting plane algorithms (Gouveia, 1995 and Hall, 1996) or branch-bound algorithm (Malik and Yu, 1993), while the network topology of the problem is usually neglected. Recently, Zhou and Gen proposed a GA approach, using a degree-based permutation, explained in last section, to encode the candidate solution for the problem (Zhou and Gen, 1997c). Zhou and Gen tested the problem given by Gavish (Gavish, 1985). The example consists of 16 nodes, a unit traffic between each node and node 1 and with a capacity restriction $\kappa = 5$. The best solution is 8526. After 500 generation GAs simulation they got the optimal solution and its corresponding topology of a centralized tree. Figure 10 illustrates the result.

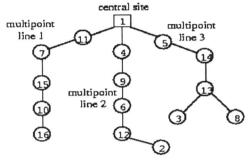

Figure 10. Optimal solution for the centralized network

7. Local area network design problem

In computer networking, local area networks (LANs) are commonly used as the communication infrastructure that meets the demands of the users in local environment. The computer networks typically consist of several LAN segments connected together via bridges. The use of transparent bridges requires loop-free paths between LANs segments. Therefore, the spanning tree topologies can be used as active LANs configurations when designing a computer network system.

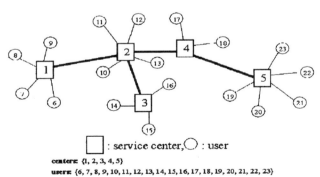

□ : service center, ○ : user

centers: {1, 2, 3, 4, 5}
users: {6, 7, 8, 9, 10, 11, 12, 13, 14, 15, 16, 17, 18, 19, 20, 21, 22, 23}

Figure 11. The sample LANs architecture

Considering the LANs with n service centers and m users. Figure 11 shows an example of the LANs with 5 centers and 23 terminals. Define the $n \times n$ service center topology matrix X_1, whose element x_{1ij} represents whether the centers i and j are connected. Assume that LANs are partitioned into n segments (service centers).

The users are distributed over those n service centers. The $n \times m$ clustering matrix X_2 specifies which user belongs to which center, whose element x_{2ij} means whether user j belongs to center i. Define the $n(n+m)$ matrix X called the spanning tree matrix $([X_1 \ X_2])$. The $M/M/1$ model is used to describe a single cluster (LAN segment) behavior. Then we can formulate the bicriteria LAN topology design problem as the following nonlinear 0-1 programming model:

$$\textbf{bc-LAN:} \quad \min \ \frac{1}{\Gamma}\left[\sum_{i=1}^{n}\frac{c_i(X)}{C_i - c_i(X)} + \sum_{i=1}^{n}\sum_{j=1}^{n}\beta_{ij} \cdot d_{ij}(X)\right] \quad (19)$$

$$\min \ \sum_{i=1}^{n-1}\sum_{j=i+1}^{n}w_{1ij}\cdot x_{1ij} + \sum_{i=1}^{n}\sum_{j=1}^{m}w_{2ij}\cdot x_{2ij} \quad (20)$$

$$\text{s. t.} \ \sum_{j=1}^{m}x_{2ij} \le g_i, \qquad i = 1, 2, \ldots, n \quad (21)$$

$$\sum_{j=1}^{n}x_{2ij} = 1, \qquad j = 1, 2, \ldots, m \quad (22)$$

$$c_i(X) < C_i, \qquad i = 1, 2, \ldots, n$$

where Γ is the total offered traffic, $c_i(X)$ is the total traffic at center i, $f_{ij}(X)$ is total traffic through link (k, l), C_i is the traffic capacity of center i, β_{ij} is the delay per bit due to the link between centers i and j, g_i is the maximum number which is capable of connecting to center i, w_{1ij} is the weight of the link between centers i and j, and w_{2ij} is the weight of the link between center i and user j.

A solution to the problem can be regarded as a spanning tree, where all users are terminals or leaf nodes and all centers are internal nodes. Gen, Ida, and Kim proposed a genetic algorithm to solve the bicriteria LANs topological design problem, minimizing the connecting cost and the average networks message delay (Gen, Ida, and Kim, 1998; Gen and Kim 1998). Prüfer number was used to encode all internal nodes and all leaf nodes are not included in the encoding. Genetic algorithms are used to search for Pareto optimal solutions and the TOPSIS method was used to calculate the compromise solution among Pareto solutions (Hwang and Yoon, 1981).

8. Shortest path problem

One of the most common problems encountered in analysis of networks is shortest path problem. The problem is to find a path between two designated nodes having minimum total length or cost. It is a fundamental problem that appears in many applications involving transportation, routing and communication. In many applications, however, there are several criteria associated with traversing each edge of a network. For example, cost and time measures are both important in transportation networks, economic and ecological factors for highway construction. As a result, there has been recent interest in solving bicriteria shortest path problem. It is to find paths that are efficient with respect to both criteria. There is usually no single path that gives the shortest path with respect to both criteria.

Instead, a set of Pareto optimal paths is preferred. Cheng and Gen proposed a compromise approach-based genetic algorithm to solve the bicriteria shortest path problem (Gen and Cheng, 2000). The compromise approach, contrary to generating approach, identifies solutions, which are closest to the ideal solution as determined by some measure of distance.

How to encode a path for a graph is a critical step. Special difficulty arises because (1) a path contains variable number of nodes, and (2) a random sequence of edges usually does not correspond to a path. To overcome such difficulties, Cheng and Gen adopted an indirect approach: encode some guiding information for constructing a path in a chromosome, but not a path itself (Gen and Cheng, 2000). A new encoding method, called *proposed priority-based encoding,* was introduced. In this method, the position of a gene was used to represent a node and the value of the gene was used to represent the priority of the node for constructing a path among candidates. The path corresponding to a given chromosome is generated by sequential node appending procedure with beginning from the specified node 1 and terminating at the specified node n. At each step, there are usually several nodes available for consideration, only the node with the highest priority is added into path. Recently, Cheng and Gen developed an evolution program using the priority-based encoding for solving a resource-constrained project scheduling problem (Cheng and Gen, 1998).

Consider the undirected graph shown in Figure 12. Suppose we want to find a path from node 1 to node 10. At the beginning, we try to find a node for the position next to node 1. Nodes 2 and 3 are eligible for the position, which can be easily fixed according to adjacent relation among nodes. The priorities of them are 3 and 4, respectively. The node 3 has the highest priority and is put into the path. The possible nodes next to node 3 are nodes 2, 5 and 6. Because node 6 has the largest priority value, it is put into the path. Then we form the set of nodes available for the next position and select the one with the highest priority among them. Repeat these steps until we obtain a complete path (1, 3, 6, 7, 8, 10). A detailed description of the path growth procedure is given in the next section.

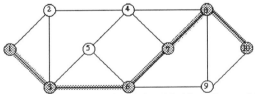

Figure 12. A simple undirected graph with 10 nodes and 16 edges

position: node ID	1	2	3	4	5	6	7	8	9	10
value: priority	7	3	4	6	2	5	8	10	1	9

Figure 13. An example of priority-based encoding

The compromise approach can be regarded as a kind of mathematical formulation of goal seeking behavior in terms of a distance function. It identifies solutions that are closet to the ideal solution as determined by following weighted L_p-norm:

$$r(z; p, w) = \left[\sum_{k=1}^{2} w_k^p \left| z_k - z_k^* \right|^p \right]^{1/p} \tag{23}$$

where $z^* = (z_1^*, z_1^*)$ is the ideal solution to the problem. The parameter p is used to reflect the emphasis of decision-makers. For the bicriteria shortest path problem, the ideal point can be easily obtained by solving two single criterion problems. For many complex problems, to obtain an ideal point is also a difficult task. To overcome the difficulty, a concept of proxy ideal point was suggested to replace the ideal point. The proxy ideal point is the ideal point corresponding to current generation but not to a given problem. In the other words, it is calculated in the explored partial solution space but not the whole solution space. The proxy ideal point is easy to obtain at each generation. Along with evolutionary process, the proxy ideal point will gradually approximate to the real ideal point.

9. Conclusion

With the development of modern society, the data communication has become very important part in human being's life. Actually, this trend will continue to the next century or even further future. Simultaneously, it also brings about many problems related with varieties of network designs to us. In this paper, we just gave out a brief review about our recent research works in this field. Different from many other conventional techniques, we developed the genetic algorithms to deal with all these network design problems. Our limited computational experience showed that the genetic algorithms approach is competitive to solve such kinds of networks design problems like multistage process planning problem, minimum spanning tree problem, centralized network design, local network design, and fixed charge transportation problem transportation problem. Especially, with the increase of problem scale, and some complicated constraints, the genetic algorithms showed their even great potential power to cope with all these network design problems. From this point of view, the genetic algorithms for them are not only means of algorithms or techniques, but also a kind of art in the sense that the problems were solved in coding space instead of the solutions space themselves. The paper therefore focused on such kind of state-of-the-art in the genetic algorithms approach on these network design problems.

Acknowledgement

This research work was supported by the International Scientific Research Program, the Grant-in-Aid for Scientific Research (No. 10044173: 1998.4— 2001.3) by the Ministry of Education, Science and Culture, the Japanese Government.

References

1. Awadh, B., Sepehri, N., and Hawaleshka, O. (1995) A Computer-aided Process Planning Model based on Genetic Algorithms, *Computers & Operations Research*, 22 841-856.
2. Bertsekas, D. and Gallager, R. (1992) *Data Networks*, 2nd ed., Prentice-Hall, New Jersey.
3. Chang, T. C. and Wysk, R. A. (1985) *An Introduction to Automated Process Planning Systems*, Prentice-Hall, Englewood Cliffs.

4. Cheng, R. (1997) *Study on Genetic Algorithm-based Optimal Scheduling Techniques*, PhD dissertation, Tokyo Institute of Technology, Japan.
5. Cheng, R. and Gen, M. (1998) An Evolution Program for the Resource-Constrained Project Scheduling Problem, *Computer Integrated Manufacturing*, 11(3) 274-287.
6. Cooper, L. (1963) Location-allocation problems, *Operations Research*, 11(3) 331-344.
7. Cormen, T.H., Leiserson, C.E., & Rivest, R.L. (1990) *Introduction to Algorithms*, The MIT Press.
8. Domschke, K. and Drex, A. (1984) An International Bibliography on Location and Layout Planning, Springer, Heidelberg.
9. Gavish, B. (1982) Topological Design of Centralized Computer Networks-Formulation and Algorithms, *Networks*, **12** 355-377.
10. Gavish, B. (1985) Augmented Lagrangian-based Algorithms for Centralized Network Design, *IEEE transaction on Commun.*, **COM-33** 1247-1257.
11. Gen, M. and Cheng, R. (1997) *Genetic Algorithms and Engineering Design*, John & Wiley Sons, New York.
12. Gen, M. and Cheng, R. (2000) *Genetic Algorithms and Engineering Optimization*, John & Wiley Sons, New York.
13. Gen, M., Cheng, W., and Wang, D. (1997) Genetic Algorithms for Solving Shortest Path Problems, *Proceedings of IEEE International Conference on Evolutionary Computation*, Indianapolis, Indiana, 401-406.
14. Gen, M., Ida, K. and Kim, J. R. (1998) A Spanning Tree-based Genetic Algorithm for Bicriteria Topological Network Design, *Proceedings of IEEE International Conference on Evolutionary Computation*, Anchorage, Alaska, 15-20.
15. Gen, M. and Kim, J. R. (1998) GA-based Optimal Network Design: A State-of-the-Art Survey, in Dagli, C. H., et al eds. *Intelligent Engineering Systems, Through Artifical Neural Networks*, 8 247-252, ASME Press, New York.
16. Gen, M. and Li, Y. (1998) Solving Multiobjective Transportation Problem by Spanning Tree-based Genetic Algorithm, in I. Parmee ed., *Adaptive Computing in Design and Manufacture*, Springer-Verlag, 98-108.
17. Gen, M. and Li Y. Z. (1999) Hybrid Genetic Algorithms for Transportation Problem in Logistics, *Journal of Logistics; Research & Applications* (forthcoming).
18. Goldberg, D.E. (1989) *Genetic Algorithms in Search, Optimisation and Machine Learning*, Addison-Wesley, Reading.
19. Gong, D., Gen, M., Yamazaki, G., & Xu, W. (1995) Hybrid Evolutionary Method for Obstacle Location-Allocation Problem, *Computers and Industrial Engineering*, 29(1-4) 525-530.
20. Gong, D., Gen, M., Yamazaki, G., & Xu, W. (1997) Hybrid Evolutionary Method for Capacitated Location-Allocation, *Engineering Design & Automation*, 3(2) 166-173.
21. Gottlieb, J. & Paulmann, L. (1998) Genetic Algorithms for the Fixed Charge Transportation Problem, *Proceedings of IEEE International Conference on Evolutionary Computation*, Anchorage, Alaska, 330-335.
22. Gouveia, L. (1995) A 2n Constraint Formulation for the Capacitated Spanning Tree Problem, *Operations Research*, **43**(1) 130-141.
23. Hall, L. (1996) Experience with a Cutting Plane Algorithm for the Capacitated Minimal Spanning Tree Problem, *INFORMS Journal on Computing*, 8(3) 219-234.
24. Holland, J. H. (1975) *Adaptation in Natural And Artificial Systems*, MIT Press, Cambridge, MA.
25. Hwang, C. and Yoon, K. (1981) *Multiple Attribute Decision-Making: Methods and Applications*, Springer-Verlag.
26. Jensen, A. P. and Barnes, J. W. (1980) *Network Flow Programming*, John Wiley, New York.
27. Katz, I. and Cooper, L. (1981) Facility Location in the Presence of Forbidden Regions; I. Formulation and the Case of the Euclidean Distance with one Forbidden Circle,

120

European Journal of Operational Research, **6** 166-173.

28. Kim, J. R., Gen, M. and Ida, K. (1999) Bicriteria Network Design using Spanning Tree-based Genetic Algorithm, *Artificial Life and Robotics*, **3** 65-72.

29. Kim, J. R., Gen, M., and Yamashiro, M. (1999) A Bi-level Hierarchical GA for Reliable Network Topology Design, *Proceedings of the 7th European Congress on Intelligent Techniques & Soft Computing*, Session CD-7, Aachen.

30. Kusiak, A. and Finke, G. (1988) Selection of Process Plans in Automated Manufacturing Systems, *IEEE Journal of Robotics and Automation*, **4**(4), 397-402.

31. Li, Y. Z. and Gen, M. (1997) Spanning Tree-based Genetic Algorithm for Bicriteria Transportation Problem with Fuzzy Coefficients, *Australian J. of Intelligent Inform. Processing Systems*, **4**(3/4) 220-229.

32. Li, Y. Z., Gen, M., and Ida, Y. (1998) Fixed Charge Transportation Problem by Spanning Tree-based Genetic Algorithm, *Beijing Mathematics*, **4**(2) 239-249.

33. Li, Y. Z. (1999) *Study on Hybridized Genetic Algorithm for Production Distribution Planning Problems*, PhD dissertation, Ashikaga Institute of Technology, Japan.

34. Malik, K. and Yu, G. (1993) A Branch and Bound Algorithm for the Capacitated Minimum Spanning Tree Problem, *Networks*, **23** 525-532.

35. McHugh, J. A. (1990) *Algorithmic Graph Theory*, Prentice Hall, New Jersey.

36. Multagh, B. and Niwattisyawong, S. (1984) An Efficient Method for the Muti-depot Location-Allocation Problem, *Journal of Operational Research Society*, **33** 629-634.

37. Narula, S. C. and Ho, C. A. (1980) Degree-constrained Minimum Spanning Tree. *Computers & Operations Research*, **7** 239-249.

38. Prüfer, H. (1918) Neuer beweis eines Satzes über Permutationen, *Arch. Math. Phys*, **27** 742-744.

39. Rosing, K. (1992) An Optimal Method for Solving (Generalized) Multi-Weber Problem, *European Operational Research*, **58**, 479-486.

40. Sancho, N. G. (1986) A Multi-objective Routing Problem, *Engineering Optimization*, **10**, 71-76.

41. Sniedovich, M. (1988) A Multi-objective Routing Problem Revisited, *Engineering Optimisation*, **13**, 99-108.

42. Sun, M., Aronson, J. E. Mckeown, P. G. and Drinka, D. (1998) A Tabu Search Heuristic Procedure for the Fixed Charge Transportation Problem, *European J. of Operational Research*, **106** 441-456.

43. Vignaux, G. A. and Michalewicz, Z. (1991) A Genetic Algorithm for the Linear Transportation Problem, *IEEE Transactions on Systems, Man, and Cybernetics*, **21** 445-452.

44. Walters, G. A. and Smith, D. K. (1995) Evolutionary Design Algorithm for Optimal Layout of Tree Networks, *Engineering Optimisation*, **24** 261-281.

45. Whatley, J. K. (1985) *SAS/OR User's Guide: Version 5. Netflow Procedure*, SAS Institute Inc., Cary, NC, 211-223.

46. Zhou, G. and Gen, M. (1997a) Evolutionary Computation on Multicriteria Production Process Planning Problem, in B. Porto editor, *Proceedings of the IEEE International Conference on Evolutionary Computation*, 419-424.

47. Zhou, G. and Gen, M. (1997b) A Note on Genetic Algorithm Approach to the Degree-Constrained Spanning Tree Problems, *Networks*, **30** 105-109.

48. Zhou, G. and Gen, M. (1997c) Approach to Degree-Constrained Minimum Spanning Tree Problem Using Genetic Algorithm, *Engineering Design and Automation*, **3**(2) 157-165.

49. Zhou, G. and Gen, M. (1999) A New Tree Encoding for the Degree-Constrained Spanning Tree Problem, *Soft Computing* (forthcoming).

50. Zhou, G. (1999) *Study on Constrained Spanning Tree Problems with Genetic Algorithms*, PhD dissertation, Ashikaga Institute of Technology, Japan.

Chapter 3

Design Representation Issues

Adaptive Techniques for Evolutionary Topological Optimum Design
H. Hamda, M. Schoenauer

Representation in Architectural Design Tools
U.-M. O'Reilly, P. Testa

Fitting of Constrained Feature Models to Poor 3D Data
C. Robertson, R.B. Fisher, N. Werghi, A.P. Ashbrook

Exploring Component-based Representations – The Secret of Creativity by Evolution?
P.J. Bentley

Adaptive Techniques for Evolutionary Topological Optimum Design

Hatem Hamda and Marc Schoenauer

CMAP – UMR CNRS 7641 – École Polytechnique – France

{Hatem.Hamda,Marc.Schoenauer}@polytechnique.fr

Abstract

This paper introduces some advances in Evolutionary Topological Optimum Design, thanks to extensive use of adaptive techniques. On the genotypic side, a variable length representation is used: the complexity of the representation of each individual is evolved by the algorithm rather than being prescribed by some fixed mesh of the design domain, resulting in self-adaptive complexity. On the phenotypic side, an original adaptive mechanism is proposed that maintains both feasible and infeasible individuals, thus exploring both sides of the boundary of the feasible region, where the optimum structure is known to lie. Not only does this improves the results of past work in on Evolutionary Topological Optimum Design on standard benchmark bidimensional cantilever problems, but it also allows to address three-dimensional problems who had up to now stayed beyond reach for evolutionary algorithms.

1 Introduction

Early works on adaptivity in the framework of Evolutionary Algorithms (EAs) mainly involved the on-line tuning of operator parameters, like the population-level tuning of operator probabilities for Genetic Algorithms [1] and mutation variance in the 1/5th rule for Evolution Strategies [2], or the individual-level self-adaptive mutation step-size in Evolution Strategies [3] or Genetic Algorithms [4]. Some useful definitions and a survey can be found in [5].

A few work addressed the issue of adaptive fitness, either by considering a prescribed series of fitness functions to gradually drive the population toward the optima of the target objective [6] or by adapting some penalty coefficient for CSP [7] or integer programming [8].

Very few work deal with adaptivity at the representation level, and they are concerned with binary representations [9, 10]. However, in the context of Design, the importance of adaptivity in representation has been recently highlighted: In the previous edition of this event, J. Gero's contribution [11] emphasized what could be called 'off-line adaptivity': a new representation is derived from a careful analysis of the results of a first EA that uses a rather raw representation. Using that new representation, much better results are obtained. And in this volume, P. Bentley's paper [12] praises what he calls component-based representation, by opposition to more standard parameter-based representations.

This paper focuses on adaptivity for EAs in the context of Topological Optimum Design (TOD), regarding both the representation and the fitness function.

By opposition to the binary representation used in most previous works, the Voronoï representation for the TOD problem is adaptive in the sense that it is a variable length representation: the number of 'genes' is evolved by the algorithm. Hence the complexity of the representation is self-adaptive (i.e. is adjusted by the evolution itself) at the individual level.

The problem at hand is a constrained problem: An original population-level adaptive penalty method is proposed, ensuring that some feasible and some infeasible individuals remain in the population. This leads to an efficient exploration of the neighborhood of the boundary of the feasible domain, where the solution is known to lie.

The paper is organized as follows. Section 2 presents the mechanical background and briefly reviews some previous works, discussing their limitations. Section 3 introduces the variable length Voronoi representation, together with its variation operators. Its advantages over the standard binary representation are discussed and enforced by mesh-dependency experimental results. Section 4 describes the original adaptive penalty method used to explore the neighborhood of the boundary of the feasible region. Comparative results involving different penalty methods demonstrate that, for the TOD problem at hand, the adaptive strategy outperforms all other methods. In section 5, experimental results on difficult cantilever benchmark problems are presented: the proposed algorithm finds good quality solutions for the 2D 10×1 cantilever, and is able to propose alternative original solutions to a 3D problem. Section 6 closes the paper with some discussion and proposal for further research directions.

2 Background

2.1 The mechanical problem

The general framework of this paper is the problem of finding the optimal shape of a structure (i.e. a repartition of material in a given *design domain*) such that the mechanical behavior of that structure meets some requirements (e.g. a bound on the maximal displacement under a prescribed loading). The optimality criterion is here the weight of the structure, but it could involve other technological costs.

The mechanical model used in this paper will be the standard two-dimensional (except in section 5.2) plane stress linear model, and only linear elastic materials will be considered (see e.g. [13]).

Throughout this paper, the most popular benchmark problem of Optimum Design will be used, that is the optimization of a cantilever plate: the design domain is rectangular, the plate is fixed on the left vertical part of its boundary (displacement is forced to 0), and the loading is made of a single force applied on the middle of its right vertical boundary. Figure 2.1 shows the design domain for the 2×1 cantilever plate problem.

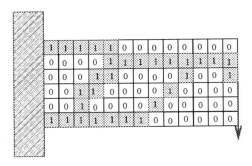

Figure 2.1: *The 2×1 cantilever plate test problem, and a bitarray representation of a structure derived from a regular mesh (here a 13×6 mesh).*

2.2   Previous works

This paper concentrates on TOD problems, in which no assumption is made on the topology of the structure, in contrast with the domain of shape optimization [14]. The most up-to-date approach to TOD is that of homogenization, introduced in [15], which deals with a continuous density of material in $[0, 1]$. This relaxed problem is known to have a unique solution in the case of linear elasticity and for one single case [16] – and the corresponding numerical method does converge to that non-physical solution [17], which is further forced to a feasible solution (with boolean density). This approach is insofar limited to the linear-elasticity/single-loading case, and cannot address loadings that apply on the (unknown) actual boundary of the shape (e.g. uniform pressure).

A possible approach to overcome these difficulties of TOD is to use stochastic optimization methods. Simulated Annealing has been used to find the optimal shape of the cross-section of a beam in [18]; and Evolutionary Algorithms have been used to solve the cantilever problem presented in Section 2.1 in [19, 20, 21]. The above limitations of the homogenization method have been successfully overcome by these works – in [21, 22] for instance, results of TOD in nonlinear elasticity, as well as the optimization of a bicycle (a 3-loading case) and that of an underwater dome (where the loading is applied on the unknown boundary) are presented: all are out of reach for the homogenization method.

2.3 Representations of structures for TOD

The most crucial step when constructing an EA is the choice of representation. All the works cited in previous section that address TOD problems with EAs use the same 'natural' binary representation, termed *bitarray* in [21]: it relies on a mesh of the design domain – the same mesh that is used to compute the mechanical behavior of the structure in order to give it a fitness (see section 2.4). Each element of the mesh is given value 1 if it contains material, 0 otherwise.

In spite of its successes in solving TOD problems [21], bitarray representation suffers from a strong limitation due to the dependency of its complexity on that of the underlying mesh. Indeed, the size of the individual (the number of bits used to encode a structure) is the size of the mesh. Unfortunately, according to both the theoretical results in [23] and the empirical considerations in [24] the critical population size required for convergence should be increased at least linearly with the size of the individuals. Hence it is clear that the bitarray approach will not scale up when using very fine meshes. This greatly limits the practical application of this approach to coarse (hence imprecise) 2D meshes, whereas Mechanical Engineers are interested in fine 3D meshes!

These considerations appeal for some more compact representations whose complexity does not depend on a fixed discretization. The ultimate step in the direction of complexity-free representation is to let the complexity itself evolve and be adjusted by the EA.

Two such alternative representations have been proposed in [25], and used to solve non destructive control problem in [26]. The rest of the paper will use one of these, namely the *Voronoï* representation (see section 3.1).

2.4 Fitness computation

The problem tackled in this paper is to find a structure of minimal weight such that its maximal displacement stays within a prescribed limit D_{lim} when some given pointwise force is applied on the loading point (see Figure 2.1). The computation of the maximal displacement is made using a Finite Element Analysis solver (kindly provided by F. Jouve [27]).

From mechanical considerations, all structures that do not connect the loading point and the fixed boundary are given an arbitrary high fitness value. Moreover, the material in the design domain that is not connected to the loading point – and has thus no effect on the mechanical behavior of the structure – is discarded (see [21, 22] for a detailed discussion on both these issues). In summary, for connected structures, the problem is to minimize the (connected) weight subject to the constraint $D_{Max} \leq D_{lim}$, where D_{Max} its maximal displacement computed by the FEM under the prescribed loading. The constraint handling method will be detailed in section 4.

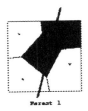
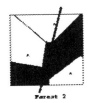

Figure 3.1: *The Voronoï representation, and its crossover operator: A random line is drawn across both diagrams, and the sites on one side are exchanged*

3 Adaptive representation

3.1 Voronoï representation

Voronoï diagrams: Consider a finite number of points $V_0, \ldots, V_N$ (the *Voronoï sites*) of a given subset of $\mathbf{R}^n$ (the design domain). To each site V_i is associated the set of all points of the design domain for which the closest Voronoï site is V_i, termed *Voronoï cell*. The *Voronoï diagram* is the partition of the design domain defined by the Voronoï cells. Each cell is a polyhedral subset of the design domain, and any partition of a domain of $\mathbf{R}^n$ into polyhedral subsets is the Voronoï diagram of at least one set of Voronoï sites (see [28] for a detailed introduction to Voronoï diagrams, and a general presentation of algorithmic geometry).

The genotype: Consider now a (variable length) list of Voronoï sites, each site being labeled 0 or 1. The corresponding Voronoï diagram represents a partition of the design domain into two subsets, if each Voronoï cell is labeled as its associated site. Example of Voronoï representations can be seen in Figure 3.1.

Decoding: Practically, the fitness of all structures will be evaluated using a fixed mesh. A partition described by Voronoï sites is easily mapped on any mesh: the subset (void or material) an element belongs to is determined from the label of the Voronoï cell in which the gravity center of that element lies.

Initialization: the initialization procedure for the Voronoï representation is a uniform choice of the number of Voronoï sites between 1 and a user-supplied maximum number, a uniform choice of the Voronoï sites in the structure, and a uniform choice of the boolean void/material label.

Variation operators: The variation operators for the Voronoï representation are problem-driven:

- The **crossover operator** exchanges Voronoï sites on a geometrical basis. In this respect it is similar to the specific bitarray crossover described in [29]. Figure 3.1 is an example of application of this operator.

- The **mutation operator** is chosen by a roulette wheel selection based on user-defined weights among the following operators:

  - the *displacement mutation* performs a self-adaptive Gaussian mutation on the coordinates of the sites, as defined in [3]: one standard deviation is

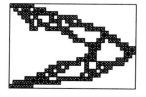 

(a) Bitarray result (weight = 0.267)   (b) Voronoï result (weight = 0.272)

Figure 3.2: Results for the 32 × 22 mesh after approx. 80000 FEM analyses.

attached to each coordinate, and undergoes log-normal mutation before being used as step-size for the mutation of the coordinate.

- the label *mutation* randomly flips the boolean attribute of one site.
- the *add* and *delete* mutations are specific variable-length operators that respectively randomly add or remove one Voronoï site on the list.

3.2 Evolutionary experimental conditions

Unless otherwise stated, the experiments presented further on have been performed using the following settings: Standard GA-like evolution (linear rank-based selection and generational replacement of all parents by all offspring) with populations size 80; At most 40 Voronoï sites per individual; Crossover rate is 0.6 and mutation rate per individual is 0.3; Weights among the different mutations are 1/2 for the displacement mutation and 1/6 for the 3 other mutations; All runs are allowed at most 2000 generations, and the algorithm stops after 300 generations without improvement; all plots are the result of 21 independent runs; All CPU times are given related to a Pentium II processor running at 300MHz under Linux.

3.3 Bitarrays vs Voronoï

The first experiments performed with Voronoï representation aimed at comparing it with the bitarray representation, in term of both quality of the solutions and computational cost. Figure 3.2 shows a typical example of a result obtained using the bitarray representation (taken from [21]) and the result on a very similar case obtained using the Voronoï representation (though for technical reasons, the exact mechanical conditions could not be reproduced). The problem is the 2D 2 × 1 cantilever plate discretized into a 32 × 22 mesh. Both trials were run for 100 000 evaluations, using a fixed penalty approach for the fitness function (see section 4).

The main difference between both final solutions is that the structure obtained using the Voronoï representation does not exhibit the small holes that appear in the solution of the bitarray representation. However, this difference is probably not due to the well known difficulty that binary GAs usually have to fine tune the very last bits: remember that the solution to the TOD problem lies in the relaxed space of structures

(a) - $D_{lim} = 20$, $D_{max} = 19.77$
weight = 0.21, 19 sites

(b) - $D_{lim} = 10$, $D_{max} = 9.959$
weight = 0.44, 13 sites

Figure 3.3: *Result for the Voronoï representation on the 10×20 mesh for the 1×2 cantilever plate problem (the structures are fixed at the bottom and the force applies at center top) for two different values of the constraint D_{lim}. CPU cost is around 2.6s per generation.*

made of infinitely many holes of infinitely small size (see section 2.2). Hence it seems that the bitarray approach tries to go toward that solution, limited only by the mesh it relies on. On the other hand, the Voronoï representation can only generate more regular shapes, finally giving a much more satisfactory solution, from a technological perspective.

Things are quite different for the case of the 1×2 cantilever problem, where the solution is known to be the perfect "V" shape: Figure 3.3 shows that the Voronoï representation is able to find such shape whereas the bitarray representation could hardly fine-tune the boundary of the structures (see [22]).

3.4 Mesh-dependency results

In order to acknowledge the non-dependency of the results of the Voronoï representation with respect to the complexity of the mesh, different regular meshes for the 2×1 cantilever plate were used under the same mechanical and evolutionary conditions.

Monitoring the evolution of the result in terms of number of fitness evaluations should show whether the refinement of the mesh does modify the convergence speed, i.e. the number of FEM analyses needed to reach a given solution (note that the computational time needed for one fitness computation increases with the size of the mesh, and that there is no way to avoid it).

Figure 3.4 shows the evolution of the fitness of the best individual (averaged over 21 runs) for the $10 \times 20, 20 \times 40$ and 40×80 meshes for two different values of the constraint on the maximal displacement ($D_{lim} = 10$ and $D_{lim} = 20$). Whereas Figure 3.4-a shows the expected perfect independence w.r.t. mesh size, the results of Figure 3.4-b first seemed to contradict our hypothesis. However, a closer look at the solutions gave the explanation:

The best solution obtained in a run on the 10×20 (weight 0.44, maximal displacement 0.997) was projected on the 20×40 mesh, it ended up with a weight of 0.43125 and a maximal displacement of 11.265! So the projection error due to the coarseness of the mesh was responsible for the difference in the plots – and as the relative difference is greater for light structures, this explains that it shows mainly on the first plot (see Figure 3.2). But note that even on Figure 3.4-b, the dynamic behavior of the EA is roughly the same for the three meshes – up to the difference in final solution. Moreover, the number of Voronoï sites in the best solutions for the three

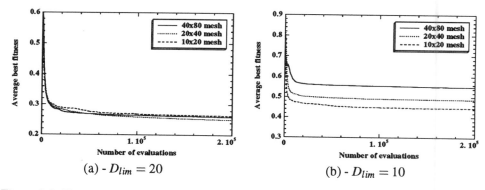

Figure 3.4: *Weight of optimal structure w.r.t. number of evaluations for the 10×20, 20×40 and 40×80 meshes of the 1×2 cantilever – see solutions Figure 3.3. The CPU costs are respectively about 2.6s, 7.7s and 80s per generation.*

series of runs is roughly the same: respectively 18.1, 19.9 and 21.6 for the 10×20, 20×40 and 40×80 meshes.

4 Adaptive penalty

The problem at hand is a constrained problem (see section 2.4). Constrained optimization has been recently a very active field in Evolutionary Computation, and many specific methods have been designed (see e.g. [30] for a survey).

Moreover, it is clear from mechanical considerations that the solution lies on the boundary of the feasible domain – at least for the continuous problem. Furthermore, specific methods exist to explore the boundary of the feasible domain when the constraint is know to be active at the optimum [31]. Unfortunately, for the TOD problem, first, the solution of the discretized problem does not lie exactly on that boundary, and second, that boundary is out of reach for direct sampling. On the other hand, the penalty method is straightforward to implement in any situation. Hence, as in all previous work [21], the constraint on the maximal displacement of the structure was handled by penalization.

4.1 Penalty methods

Introducing the positive penalty parameter α, the fitness function to minimize is (x^+ is the positive part of x)

$$Weight + \alpha(D_{max} - D_{lim})^+ \qquad (1)$$

However, adjusting α is not an easy task (see again [31]). Static penalty, where α is kept constant, can give very good results, but requires a very careful tuning. Dynamic penalty, where α is modified according to a user-defined schedule, as proposed in [32] or in the framework of TOD in [21], requires a lucky guess for the schedule.

Adaptive penalty parameters have been successfully used in the context of (discrete) Constraint Satisfaction Problems [7], where the objective is to find at least one feasible individual. In the context of parameter optimization, an adaptive scheme has been presented in [33]: the penalty parameter is updated according to the feasibility of the best individual in the population only. The new adaptive penalty method proposed here updates the penalty parameter based upon global statistics of feasibility in the population. Its main goal is to explore the neighborhood of the boundary of the feasible region by trying to keep in the population individuals that are on both sides of that boundary (the same idea lead to the Segregated GA [34], that used two different fixed penalty parameters to achieve the same goal).

4.2 Population-based adaptive penalty

The objective is to maintain in the population a minimum proportion of feasible individuals as well as a minimum proportion of infeasible individuals. Denote by $\Theta^k_{feasible}$ the proportion of feasible individuals at generation k, and by Θ_{inf} and Θ_{sup} two user-defined parameters. As small penalty parameters favor the infeasible individuals (and vice-versa), the following update rule for α is proposed to try to keep $\Theta^k_{feasible}$ in $[\Theta_{inf}, \Theta_{sup}]$:

$$\alpha_{k+1} = \begin{cases} \beta \cdot \alpha_k & \text{if } \Theta^k_{feasible} < \Theta_{inf} \\ (1/\beta) \cdot \alpha_k & \text{if } \Theta^k_{feasible} > \Theta_{sup} \\ \alpha_k & \text{otherwise} \end{cases} \qquad (2)$$

with $\beta > 1$. User-defined parameters of this method are Θ_{inf}, Θ_{sup}, β and the initial value α_0. Note the limit case $\Theta_{inf} = \Theta_{sup}$ was also tried, but was discarded after the first experimental trials.

The robust values $\beta = 1.1, \Theta_{inf} = 0.4$, and $\Theta_{sup} = 0.8$ were used in all experiments presented here.

Note that the variations of α are non monotonous, and hence there is no a priori guarantee that the best individual in the population is feasible. It can even happen that the population contains no feasible individual – though in that case the steady increase of α will rapidly disadvantage infeasible individuals.

4.3 Comparative results

This section summarizes and discusses the comparative results obtained for different approaches of penalty function.

Initially, several numerical tests were performed on the 10×20 regular mesh using static penalty, and confirmed that for small values of α (e.g. 0.01) the optimal structures are very light, but violate the constraint, while for large values of α (e.g. 10000) all structures visited during evolution are feasible, but the resulting structures are heavy. A good intermediate value (30) was chosen, giving reasonable results in most cases. It will also be used as the starting value for dynamic and adaptive methods.

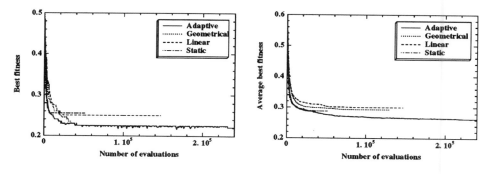

Figure 4.1: *Best and averaged fitnesses (out of 21 runs) for different penalty approaches on the 10×20 cantilever problem with $D_{lim} = 20$.*

Two dynamic penalty methods were also included in the comparison, namely a linear increase of α from α_0 to $10 \times \alpha_0$ and the geometrical increase defined in [21] where α is multiplied by a factor $\beta = 0.01$ every 10 generations.

Figure 4.1 shows plots of the best and average fitnesses for the four methods. All final best individuals of all runs were feasible.

First, the plot for the adaptive method clearly shows that the best fitness in the population sometimes correspond to an infeasible individual, which explains the rough aspect of the plot for the best fitness.

With respect to comparison, Figure 4.1 is a rather typical situation: the adaptive strategy clearly outperforms all other methods on average, and equals or outperforms the best result of any of the other method. On the other hand, the geometrical dynamic approach can give results that are as good as the best ones from the adaptive approach, but shows a much larger variance over the 21 runs. Hence it performs poorly on average. The static penalty, when the penalty parameter is carefully tuned, can give rapidly quite good results, but proves unable to improve any more after a rather small number of generations.

5   Innovative results

Needless to say, all results of [21, 22] demonstrating the flexibility of Evolutionary TOD compared to the homogenization method can be reproduced or even improved by the approach presented here. But the rest of the paper concentrates on purely original results as far as Evolutionary TOD is concerned.

5.1   The 10×1 cantilever

The problem of the 10×1 cantilever (discretized using a 100×10 regular mesh) raises an additional difficulty: most of initial random structures do not connect the fixed boundary and the point where the loading is applied. Hence an alternate initialization procedure was used, where the average weight of random structures can be tuned (see [35] for details). Furthermore, the maximal number of sites for each individuals was

Figure 5.1: *Optimal structure on the* 100×10 *mesh for* 10×1 *cantilever plate.* $D_{lim} = 12$, *number of cells = 105. weight = 0.479. CPU time = 14s/gen.*

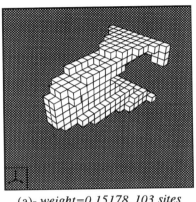

(a)- *weight=0.15178, 103 sites*   (b) - *weight=0.166, 109 sites*

Figure 5.2: *Two results for the symmetrical three-dimensional problem using a* $16 \times 7 \times 10$ *mesh for half of the structure, with same constraint (CPU time = 6mn/gen).*

increased (to 120), and the best results were obtained with a population size of 120. Figure 5.1 shows the most significant result for $D_{lim} = 12$.

5.2 Three-dimensional problem

This section introduces the first results of 3D TOD obtained using Evolutionary Computation (as far as we are aware of).

The design domain is a quadrangle subset of $\mathbf{R}^3$, and the problem is symmetrical: only half of the domain is discretized, according to a $16 \times 7 \times 10$ mesh. Its left face is fixed, and the loading is applied on the middle of the right face.

Here again the higher complexity of the problem lead to modify the settings: the population size was increased to 120 and the maximum number of Voronoï sites was increased to 120.

Figure 5.2 demonstrates that the algorithm was able to find some good solutions in ... a few days of CPU time (3D FEM analyses are far more costly than 2D for the same mesh size). Moreover, it also stress the ability of EAs to find multiple quasi-optimal solutions to the same problem, some of them quite original indeed when compared to the results of the homogenization method on the same problem.

6   Discussion and conclusion

The power of adaptivity in Evolutionary Computation is now widely acknowledged, and the results presented in this paper witness for that at two levels: the representation and the fitness.

Many issues remain open for the adaptive penalty method proposed in this paper. The main question is that of the performance of this method when the optimum is not close to the boundary of the feasible region. Automatic computation of the parameters (e.g. the initial guess for the penalty parameter α) from initial sampling of random individuals is also desirable. And of course that method has to be compared to other evolutionary constraint handling methods.

But the main new direction regarding the fitness that could bring huge improvement for Mechanical Engineers might be the use of multi-objective techniques rather that constraint handling methods. There is now a large body of work in the area of multi-objective Evolutionary Computation (see e.g. K. Deb's contribution [36]). And indeed, even the problem of the cantilever plate actually amounts to minimize both the weight and the maximal displacement under several loadings – which is a multi-objective problem. Moreover, modal optimization problems also are multi-objective, as one usually wants the part that is being optimized against bad vibrations to have some minimal stiffness in some prescribed loading situations.

On the representation side, the adaptive Voronoï representation was able to discover simple structures using few Voronoï sites (section 3.3) and more complex structures using many more sites (section 5). Allowing more sites to the simple problems did not significantly modify the results, while limiting the maximum number of sites for the complex problems forbid the emergence of good solutions. As the upper bounds fixed for the number of sites were never hit, one hypothesis, that further experiments will try to check, is that the main effect of the maximum number of sites takes place during initialization.

But though the effects of self-adaptive complexity clearly appear, a close look at the final solutions shows that many Voronoï sites could be removed without any modification of the phenotype (even on the simple structures of Figure 3.3). On the one hand, removing such useless sites could focus the fine tuning on the really useful sites. And on the other hand, if only crucial sites remain in the structures, it might be possible to deduce new and more efficient representations for Evolutionary TOD, along the lines of [11]. Further work will try to remove such useless sites, either by increasing the rate of the delete mutation or by deterministically looking for the useless sites.

Finally, let us quote that adaptivity is also important in the mind of the engineer/programmer: only by getting rid of his past habits will he be able to achieve what will look like artificial creativity, and will only be the reflection of his own.

References

[1] L. Davis. Adapting operator probabilities in genetic algorithms. In J. D. Schaffer, editor, *Proceedings of the 3rd International Conference on Genetic Algorithms*, pages 61–69.

Morgan Kaufmann, 1989.

[2] I. Rechenberg. *Evolutionstrategie: Optimierung Technisher Systeme nach Prinzipien des Biologischen Evolution.* Fromman-Hozlboog Verlag, Stuttgart, 1973.

[3] H.-P. Schwefel. *Numerical Optimization of Computer Models.* John Wiley & Sons, New-York, 1981. 1995 − 2nd edition.

[4] T. Bäck and M. Schütz. Intelligent mutation rate control in canonical GAs. In Z. W. Ras and M. Michalewicz, editors, *Foundation of Intelligent Systems 9th International Symposium, ISMIS '96*, pages 158–167. Springer Verlag, 1996.

[5] R. Hinterding, Z. Michalewicz, and A. E. Eiben. Adaptation in evolutionary computation: A survey. In T. Bäck, Z. Michalewicz, and X. Yao, editors, *Proceedings of the Fourth IEEE International Conference on Evolutionary Computation*, pages 65–69. IEEE Press, 1997.

[6] M. Schoenauer and S. Xanthakis. Constrained GA optimization. In S. Forrest, editor, *Proceedings of the 5th International Conference on Genetic Algorithms*, pages 573–580. Morgan Kaufmann, 1993.

[7] A.E. Eiben and Z. Ruttkay. Self-adaptivity for constraint satisfaction: Learning penalty functions. In T. Fukuda, editor, *Proceedings of the Third IEEE International Conference on Evolutionary Computation*, pages 258–261. IEEE Service Center, 1996.

[8] J. C. Bean and A. B. Hadj-Alouane. A dual genetic algorithm for bounded integer programs. Technical Report TR 92-53, Department of Industrial and Operations Engineering, The University of Michigan, 1992.

[9] C.G. Schaefer. The argot strategy: Adaptive representation genetic optimizer technique. In J. J. Grefenstette, editor, *Proceedings of the 2nd International Conference on Genetic Algorithms*, pages 50–55, 1987.

[10] D. E. Goldberg, B. Korb, and K. Deb. Messy genetic algorithms: Motivations, analysis and first results. *Complex Systems*, 3:493–530, 1989.

[11] J.S. Gero. Adaptive systems in designing: New analogies from genetics and developmental biology. In I. Parmee, editor, *Adaptive Computing in Design and Manufacture*, pages 3–12. Springer Verlag, 1998.

[12] P. Bentley. Exploring component-based representations. In *This volume*, 2000.

[13] P. G. Ciarlet. *Mathematical Elasticity, Vol I : Three-Dimensional Elasticity.* North-Holland, Amsterdam, 1978.

[14] J. Cea. Problems of shape optimum design. In E. J. Haug and J. Cea, editors, *Optimization of distributed parameter structures - Vol. II*, volume 50, pages 1005–1088. NATO Series, Series E, 1981.

[15] M. Bendsoe and N. Kikushi. Generating optimal topologies in structural design using a homogenization method. *Computer Methods in Applied Mechanics and Engineering*, 71:197–224, 1988.

[16] G. Allaire and R. V. Kohn. Optimal design for minimum weight and compliance in plane stress using extremal microstructures. *European Journal of Mechanics, A/Solids*, 12(6):839–878, 1993.

[17] G. Allaire, E. Bonnetier, G. Francfort, and F. Jouve. Shape optimiztion by the homogenization method. *Nümerische Mathematik*, 76:27–68, 1997.

[18] G. Anagnostou, E. Ronquist, and A. Patera. A computational procedure for part design. *Computer Methods in Applied Mechanics and Engineering*, 97:33–48, 1992.

[19] E. Jensen. *Topological Structural Design using Genetic Algorithms*. PhD thesis, Purdue University, November 1992.

[20] C. D. Chapman, K. Saitou, and M. J. Jakiela. Genetic algorithms as an approach to configuration and topology design. *Journal of Mechanical Design*, 116:1005–1012, 1994.

[21] C. Kane and M. Schoenauer. Topological optimum design using genetic algorithms. *Control and Cybernetics*, 25(5):1059–1088, 1996.

[22] C. Kane. *Algorithmes génétiques et Optimisation topologique*. PhD thesis, Université de Paris VI, July 1996.

[23] R. Cerf. An asymptotic theory of genetic algorithms. In J.-M. Alliot, E. Lutton, E. Ronald, M. Schoenauer, and D. Snyers, editors, *Artificial Evolution*, volume 1063 of *LNCS*, pages 37–53. Springer Verlag, 1996.

[24] D. E. Goldberg, K. Deb, and J. H. Clark. Genetic algorithms, noise and the sizing of populations. *Complex Systems*, 6:333–362, 1992.

[25] M. Schoenauer. Representations for evolutionary optimization and identification in structural mechanics. In J. Périaux and G. Winter, editors, *Genetic Algorithms in Engineering and Computer Sciences*, pages 443–464. John Wiley, 1995.

[26] M. Schoenauer, L. Kallel, and F. Jouve. Mechanical inclusions identification by evolutionary computation. *European Journal of Finite Elements*, 5(5-6):619–648, 1996.

[27] F. Jouve. *Modélisation mathématique de l'œil en élasticité non-linéaire*, volume RMA 26. Masson Paris, 1993.

[28] F. P. Preparata and M. I. Shamos. *Computational Geometry: an introduction*. Springer Verlag, 1985.

[29] C. Kane and M. Schoenauer. Genetic operators for two-dimensional shape optimization. In J.-M. Alliot, E. Lutton, E. Ronald, M. Schoenauer, and D. Snyers, editors, *Artificial Evolution*, number 1063 in LNCS. Springer Verlag, Septembre 1995.

[30] Z. Michalewicz and M. Schoenauer. Evolutionary Algorithms for Constrained Parameter Optimization Problems. *Evolutionary Computation*, 4(1):1–32, 1996.

[31] M. Schoenauer and Z. Michalewicz. Boundary operators for constrained parameter optimization problems. In Th. Bäck, editor, *Proceedings of the 7th International Conference on Genetic Algorithms*, pages 322–329. Morgan Kaufmann, 1997.

[32] J.A. Joines and C.R. Houck. On the use of non-stationary penalty functions to solve nonlinear constrained optimization problems with GAs. In Z. Michalewicz, J. D. Schaffer, H.-P. Schwefel, D. B. Fogel, and H. Kitano, editors, *Proceedings of the First IEEE International Conference on Evolutionary Computation*, pages 579–584. IEEE Press, 1994.

[33] A. B. Hadj-Alouane and J. C. Bean. A genetic algorithm for the multiple-choice integer program. Technical Report TR 92-50, Department of Industrial and Operations Engineering, The University of Michigan, 1992.

[34] R. G. Leriche, C. Knopf-Lenoir, and R. T. Haftka. A segragated genetic algorithm for constrained structural optimization. In L. J. Eshelman, editor, *Proceedings of the 6th International Conference on Genetic Algorithms*, pages 558–565, 1995.

[35] L. Kallel and M. Schoenauer. Alternative random initialization in genetic algorithms. In Th. Bäck, editor, *Proceedings of the 7th International Conference on Genetic Algorithms*, pages 268–275. Morgan Kaufmann, 1997.

[36] K. Deb. Multi-objective evolutionary algorithms: Past, present and future. In *this volume*, 2000.

Representation in Architectural Design Tools

Una-May O'Reilly
Artificial Intelligence Lab, MIT
unamay@ai.mit.edu

Peter Testa
Dept. of Architecture, MIT
ptesta@mit.edu

Abstract

This paper will outline the software development tools of the Emergent Design Group at Massachusetts Institute of Technology. The tools are based on the principles of artificial life and evolutionary algorithms. In view of our experiences we discuss a central issue of ALife and EA-based tools: representation of a design.

1 Introduction

The central mission of the Emergent Design Group (EDG, [1]) at M.I.T. is to research architectural morphology with a specific focus on the emergent and adaptive properties of architectural form. One avenue we pursue is the development of interactive software tools based on the principles of artificial life (ALife, e.g. [2, 3]) and evolutionary algorithms (EAs, e.g. [4]). Another avenue is the development of prototype building systems. We seek to apply both technologies to real world problems.

Through the development of our EA-based and ALife-based tools we seek to generate, discover, simulate, analyze, display, and predict behaviour of complex spatial and material systems. Some questions we study include: How can emergence, the process by which a limited set of carefully chosen and constrained building blocks, introduce complexity in the developmental process of form? How can architectural forms co-evolve within a dynamic environment? How can we design open spatial systems in which the distributed whole influences its components? What makes this approach to architectural design particularly appealing is that it allows designers and users to creatively explore a vastly expanded design space. It results in adaptive systems in which forms and structures respond to changing conditions. It is time based and can engage complex and divergent design parameters. Furthermore, it results in a continuously improving solution.

The software tools we design are intended to be experimental or investigative platforms. They focus on the conceptual design stage when form is a primary issue and much of the design scenario is still ill-defined. The conceptual stage is early in the overall process of architecture design. The architect ideally engages a computer-based tool at this point as an inspirational, creative design partner. The computer visual medium supplements hand sketches and the software tool acts as a means of expressing ideas addressing considerations such as architectural program, site or material. The tool functions not as one complete, linear investigative progression but as a set of non-linear, short investigations that permit the flexibility to examine certain issues in isolation and then combine and integrate the findings of separate investigations.

We stress that our tools should ideally seamlessly integrate into the complex, overall architecture design process that includes the construction of physical models and structural elements using computer-aided manufacturing processes (CAD/CAM). For example, an architect can transfer a three dimensional form design from our tool into FormZ that unfolds it into two dimensional format for laser cutting then assembly into a three dimensional counterpart physical model. Alternatively, digital files of a grown or evolved form can be directly used to produce a three dimensional model using 3D printing technology. This technology is useful in providing the designer with immediate feedback and is an effective tool in communicating design intentions and formal properties to manufacturers for full scale prototyping.

The group's membership is interdisciplinary. The two leaders, Testa and O'Reilly are architect and computer scientist respectively. We are aided by undergraduates and graduate students who major in Architecture, Computer Science or Mathematics. Our strategy is to rapidly build distinct tools that each focus primarily on a few issues that we recognize as challenging. We leave the non-essential components of the tool quite primitive. We put our tools quickly into the hands of their users - both students and mature architects, so that we can draw on user feedback and our observations of the tool's performance.

We now implement our tools within existing CAD/CAM tools through their applications-programmer interface (API). This has the benefit that we do not have to include or build from scratch rendering and drawing facilities that have already been provided. It also serves our ambition of seamless process integration because after running the tool (or, perhaps, during) we ensure that the outcome is modifiable in the CAD/CAM tool environment.

Our experiences have made it clear that evolution-based and ALife-based concepts have many powerful elements that make them ideal for incorporation into design software. This submission starts with brief descriptions of four software tools we have created over the last three years. We then describe the internal representation of a design in each of our tools. Representation is a challenge inherent in EA-based and ALife-based tool design. We draw observations from our experience in order to offer a set of criteria that an internal representation should fulfill in order to facilitate a powerful tool.

2 Tools of the Emergent Design Group

2.1 GermZ: Genetically Recombinant Modeling

GermZ is an Artificial Life system that includes genetic programming. It functions as a generative snowboard design tool. It is programmed in C++ as an application in Alias—Wavefront Studio. GermZ features a computational ecology of bacteria that develop from a "genetic" definition to form a colony that populate and migrate down the side of a mountain slope. The user uses either a file or the mouse to place a colony of bacteria on a modeled mountain slope which is enhanced with attractors and repellors to influence bacteria flow and the subsequent course design. The lifetime of a bacterium consists of moving on the slope leaving a trail wherever it collides with the ground. The trails of a colony (population) collectively define a surface interpretable as a snowboard course. The life-path of a bacterium depends on three forces: 1) the mountain terrain, 2) its interaction with other bacteria, attractors or repellers, and, 3) gravity. The terrain (e.g. model of rocks or topographical info) repels the bacterium away from the ground, bacteria repel each other and gravity forces a bacterium towards the ground. When ground and a bacterium collide some momentum of the bacterium is absorbed and eventually it will stop.

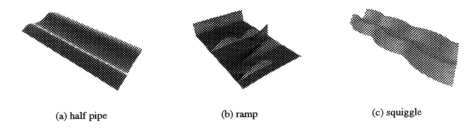

(a) half pipe   (b) ramp   (c) squiggle

Figure 2.1: Three GermZ bacteria colonies shaped by environmental influences.

(a)   (b)   (c)

Figure 2.2: Three different GermZ offspring of half pipe, ramp and squiggle bacteria colonies.

Attractors (e.g. food for the bacteria) can exist on the terrain and they pull bacteria towards them with a force that decreases at $1/r^2$ where r is the distance between a bacterium and the attractor. An attractor can be "eaten" by a bacterium and thus no longer exist. Repellors never cease existing and they, conversely, push bacteria away from themselves.

At present, homogenous colonies of bacteria have been specified and then subjected to mutation and crossover. This resulted in a "half pipe" bacteria colony evolving to one called "ramp" which spiked its motion resulting in randomly distributed moguls. "Squiggle" bacteria emerge from deliberately placed attractors and repellors. Half-pipe, ramp and squiggle are shown in Figure 2.1. In Figure 2.2 three different offspring of mating among the half-pipe, ramp and squiggle colonies are shown.

2.2 MoSS: Morphogenetic Surface Structure

The MoSS tool [5] provides architects with a model of surface growth in three dimensions. It incorporates a specialized implementation of 3D Lindenmayer systems (L-systems, [6]). In addition, just as growth is influenced by tunable factors, an architect can use MoSS to model surface geometry within a three dimensional shaping environment. Figure 2.3 shows two different surfaces generated in MoSS as examples.

MoSS surface structure generation is accomplished via a growth environment consisting of a volumetric boundary, attractors and repellors. The attractors and repellors disrupt the joints of surface plates and bend them away from each other leaving gaps. MoSS currently has three filters that address, in various ways, this distortion of surface geometry. All filters attempt to move vertices on the surface so that they match up to their neighbors and form a closed surface.

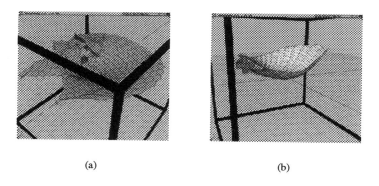

(a)   (b)

Figure 2.3: MoSS surfaces inside Alias Wavefront. Black lines are bounding box.

MoSS also fits into a larger design process; its output can be used as input towards the construction of physical models and structural elements using computer aided manufacturing processes (CAD/CAM). Three dimensional MoSS surfaces are unfolded using CAD/CAM software (Autodessys FormZ) into two dimensional format. These patterns are cut using a flat bed laser cutter. The laser cutter alternately cuts and scores the surface allowing for reassembly of the complex curvature of any MoSS generated surface. The 3D models in Figure 2.4 are two examples.

MoSS is a plug-in to Alias—Wavefront Studio written in C++. It interfaces with A—W to draw its forms. The plug-in aspect implies MoSS is integrated with an existing CAD tool.

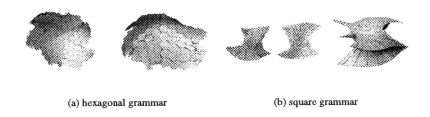

(a) hexagonal grammar   (b) square grammar

Figure 2.4: Laser cut 3D models of MoSS surfaces.

2.3 Emergent Design Java Toolbox

The Emergent Design Java toolbox [7] incorporates concepts of artificial life that allow architects to realize conceptual experiments in which the elements of an architectural scenario are endowed with agency and dynamic, spatial interaction. Elements of the scenario combine and interact spatially within a site and over time a global design emerges. Spatial areas or elements that are distinct are represented with different colors. The visualization includes element movement which conveys the influence of spatial proximity.

Figure 2.5 shows two different applications that are on the WWW at [8]. These case studies and others are detailed in [7].

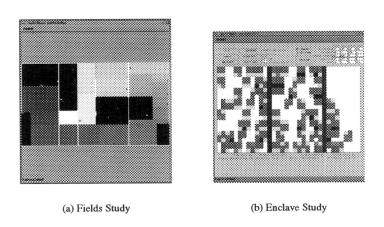

(a) Fields Study   (b) Enclave Study

Figure 2.5: Emergent Design Java Toolbox applications.

Investigations follow a practical and uncomplicated process. The architect defines the goals of an investigation concerning how spatial elements may be configured within a bounded area (i.e. site). Then, initial conditions (e.g. initial quantities of elements, conditions of site) and the elements and the relationships among them are identified. How elements influence each other and under what conditions they interact is stated. This style of thinking is very common to architects e.g., the architect says"this is what happens when *a* comes into contact with *b* or, when the corner of any element of class *e* touches the boundary".

After the toolbox (available at [9]) is customized it is run and simulations are evaluated. Multiple runs can be gathered and analyzed for general properties or one particular run may be scrutinized because it shows something unanticipated or interesting. Often, in view of the results, the architect revises the initial conditions to try out different relationships and actions. If the outcomes are satisfactory, they are now used to take the next step in the larger process of solution definition.

2.4 RGP: The Rule Genetic Programmer

The rule genetic programmer uses genetic programming (GP, [10]) to explore different arrangements of functional spaces in 3D combination. It is implemented in Java and uses the Java3D package. All applications use the same GP "language" (i.e. primitives and interpreter) but implement their own design evaluation mechanism (i.e. fitness function).

In the rhizome application the design team needed to design a high density, multi-programmatic building (i.e. a rhizome) within an existing urban area. Their objectives raised issues such as: how should the volume of the building be apportioned among architectural programs and how should different programs be placed in a relation to each other within the building?

The desired complex three dimensional spatial relationships of a rhizome were codified in the form of a preference table where each program had an affinity rating for every other program, including itself. This matrix is used by RGP's fitness function to evaluate the quality of a rhizome. The matrix defines a class of rhizomes rather than one optimal one. RGP uses it to evolve and discover designs that meet the preferences in different high quality ways. Examples of evolved rhizomes and an axonometric of the final proposal are shown in Figure 2.6.

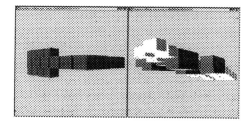

(a) RGA-evolved rhizomes   (b) Rhizome Axonometric

Figure 2.6: Rule Genetic Algorithm Rhizome Application

3   EA-based and ALife-based Representations

3.1   GermZ

In GermZ each bacterium is, again, data plus executable structure. The genotypic representation is 1) a finite state machine (fsm) with probabilistic transitions that also depend on mountain conditions, 2) a mass, and, 3) a personal field strength. The fsm records a drive schedule for each state which, given the bacterium's current velocity, specifies its next move and whether there is a state change. The fsm is variable length.

In GermZ the genotype and the finite state machine interpreter culminate in behavior that is the tool's central artifact. That is, the traces of bacteria movement constitute the primary elements of the snowboard course, rather than the bacteria themselves. Thus, there is no phenotype, per se, and this absence is unimportant. The purpose of the executable structure lies in the process it generates over simulation time.

All genotypic information can be altered by crossover or mutation. Two parent bacteria generate an offspring bacterium which inherits its mass and field strength randomly from either parent. The fsms of the two parents are treated as hierarchical graphs and a sub-graph from each is combined via crossover to generate the fsm of the offspring. Mutation takes place infrequently at a background rate and with a rate proportional to the length of the fsm during reproduction. Because of GermZ representation and interpreter, GermZ has a compelling mix of ALife and EA aspects. Bacteria both, "live" and evolve.

3.1.1   *MoSS*

MoSS is an extended 3D Cartesian space L-system. An L-system is a set of re-write rules that are serially applied to an initial axiom and interpreted geometrically. This geometry can be thought of as "agent" based. That is, an agent moves through space with a position vector and three orthogonal axis vectors. Its momentary orientation is determined on the basis of the re-write rule that might, for example, direct it to move ahead, upwards or to the side by a certain degree.

For the 3D space of MoSS we define α, β, γ, alpha, beta and gamma as three angular degrees for specifying movement in the roll, pitch and yaw (X, Y, Z axes) directions. The symbols $+$ and $-$ indicate the agent is to roll the alpha specified degrees in either the positive or negative direction along the axis of movement. Likewise the symbols $\wedge$ and $\&$ indicate positive and negative pitch rotation and the symbols $<$ and $>$ indicate positive and negative yaw rotation. Movement forward in MoSS without drawing a line segment, is denoted by the symbol f in addition to the typical use of the symbol F for drawing a line segment. As in all L-systems, MoSS interprets the symbols [and] to represent the action of pushing and popping the agent state in space. Thus the agent's position can be "memorized", it can perform movements from that point and then revert back to that point to perform other movements.

In MoSS we focus on grammars that define surfaces. For example, given $alpha = beta = gamma = 90\circ$, a square is defined by:

$$S \to F > F > F > F$$

And, a cube is defined by making six calls to S to form each face followed by a command to return the agent to its original position, then a bracketed section to form the top of the box and, finally, placement of the box bottom:

$C \rightarrow Sf \wedge Sf \wedge Sf \wedge Sf \wedge [< \wedge S]f > f \wedge S$

A $2X2X2$ block of cubes (see Figure 3.1) is created with two additional productions that, first, create 4 cubes in a $2X2$ square, then, second, stack two $2X2$ squares on top of each other :

$B2 \rightarrow CfC > fCf > fC$
$B \rightarrow B2 \wedge f\&B2$

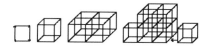

Figure 3.1: MoSS, L-System Cube Series

Replication or infinite sequences are defined by recursive productions or recursively coupled productions. For more details, see [5] and [11].

L-systems, like the representation of GermZ, are executable structures. Like GermZ, the phenotype is unimportant in MoSS. Instead, the process of execution yields the artifact of interest i.e., surface structure. L-systems have a wide range of powerful expression that emulates growth and derives natural morphogenetic structures. This makes them particularly attractive to architects. Given experience with students using MoSS, the concepts of grammatical definition and re-writing are simple for users to grasp but to design grammars that have some predictability is very difficult. The intrinsic non-linearity of L-systems presents this disadvantage so it is virtually impossible to eliminate.

3.2 RGP

In RGP the primitive set is composed of conditions and actions that we term rules. Each design is a sequence of rules that unfold into a decision tree. If a rule is an action, a volume is placed in the space and its program is designated. If a rule is a condition, the condition is a predicate that tests some property of the space such as proximity, placement, volumetric limits, or placement corresponding to the preference table. Depending on the predicate being true or false, a branch to another rule is followed. The conditions can be compounded predicates using OR, AND, NOT.

The execution of the rule tree is dependent upon state information stored in the interpreter. For each run of the rule tree, the interpreter has a designated location and orientation. The rule tree specifies the direction to grow relative to this position and orientation. After growing, the position of the interpreter is set to the location of the new growth and the orientation of the interpreter is updated to the direction of the new growth relative to the previous position. Each execution of the rule tree can yield only one new growth. If the current line of growth fails to generate a new area, the algorithm attempts to grow from the next most recent line of growth.

This sequential nature of the decision tree interpretation in RGP has a strong influence on the way in which its designs are generated and their nature. While a sufficiently rich range of designs have been investigated with the two RGA applications we have conducted, there are other appealing ways to execute the designation of sub-volumes within a volume. For example, one might start within the volume and carve out the sub-volumes in a multicentric fashion rather than incrementally add them. This might be done in parallel with attention on conflict resolution strategies. This, however would require totally new design of RGP's language and interpreter.

In addition, the condition predicates in RGP are general rather than application specific. Application aspects are intended to be expressed via the fitness function. One could foresee applications in which the predicates are the logical way to express application specific states. Admitting new predicates to the RGP at the application level becomes technically difficult because there must be support within the representation of a design to allow these states to be queried.

3.3   Emergent Design Java Toolbox

The EDJ toolbox is implemented as a collection of Java classes. For clarity, we will italicize Java object classes in the descriptions that follow. The two dimensional region of architectural inquiry or environment of the simulation is termed a *site*. A *site* is defined generally so that it can be conceived in a variety of ways: for example, as a large-scale map of states or counties to investigate transcontinental transportation routes, or, as a dining room to study the layout of furniture.

The *site* can be refined in terms of its sub-parts and in terms of the architectural elements that are imposed on it. We term a sub-part, in general, a *piece* and an architectural element, in general, a *zone*. A *zone* is an object that is imposed on the *site*, rather than a component of it. The bulk of an application consists of the applet creating *zones* and enabling their interaction among each other (which results in zone placement, removal or movement) based on different rules, preferences, and relationships. A *zone* can represent a table, room, building, farm, city, or whatever else the architect wishes to place down on the *site*. A zone can be modified: one can change its color, translate, scale, and skew it, or change its type (e.g. "farm"). In terms of its display properties, a *zone* can be transparent (either partially or fully), and can lie on top of or below other *zones*.

The *site* consists of *pieces*. By choosing the dimensions of a *piece*, the level of granularity for regarding the simulation is chosen. A *piece* has its private state so that it can store information about various local properties of the *site*, as well as information about the uppermost *zone* on it.

We have implemented a particular means of property specifications for a *site* which we call a *siteFunction*. A *siteFunction* is a general means of representing various qualities or quantities that change across the Site and which may be dependent on time. It could, for example, be used to model sun cover, wind, pollution, rain, crime, income levels, etc. A siteFunction is a function of four variables (x, y, z (3D space which a piece defines), and t (time)). One can create all sorts of interesting relationships between functions and other functions, as well as between zones and sitefunctions or vice versa. By creating feedback loops (where the output of, say, a function is used as input to another function, or to a Zone, whose output could be in turn given as input to the object from which it received input), one can very easily set the stage for emergent behavior between human creations (Zones) and the environment (SiteFunctions).

The foundation software is intentionally very flexible, and can be used to model almost anything of a spatial nature. By defining a group of *zones* and *siteFunctions*, and defining ways for them to interact with each other locally, one can set up the circumstances that lead to extremely complex, unexpected global behavior.

The purpose of its class abstractions in the toolbox is to enable the implementation of different applications that, despite their differences, still use a base of common software. The software is conveniently described in terms of the objects and their behavior we conceived as integral to any emergent design application. This generality has been proven by student exercises which have worked at different design scales (e.g. housing aggregation at a site level, patterns of an atomic housing unit and programmatic layout of a single building).

3.4 Criteria for an Ideal Representation

None of our internal representations are completely ideal. However, each has brought us to a better understanding of what constitutes an ideal representation. First, the representation must allow the tool to serve a variety of applications. From our perspective, whatever the tool, its set of envisaged applications are similar in terms the general class of design outcomes (i.e., spatial morphology and form) but disparate in terms of considerations which motivate the investigation such as architecture program or site conditions. For example, RGA produces 3D volumes with internal space designations. It can accommodate diverse applications that consider different issues because its language supports general predicates for state tests and actions that designate volume usage. MoSS can generate different surfaces with the same grammar because its extended L-system interpreter takes boundary, attractors and repellors into account. The EDJ toolbox classes were chosen to support scenarios with the fundamental commonality of agents, movements in space and local interactions with general architecture abstractions such as the site, its constituents (zones) and elements imposed on the site (pieces).

Next, a tool should be expected to require application specific tailoring but only at well defined interfaces such as refinement of the evaluation criteria or definition of the site, rather than through modification of its representation. MoSS and GermZ use environments, in which the interpreter acts, to accommodate properties integral to site conditions. In RGP, the fitness function is the interface by which application specific desire outcomes can be defined. In the EDJ toolbox application specific tailoring is at the software engineering level. The specific classes and class hierarchy of the toolbox should encourage sub-class specialization which will provide inheritance and data encapsulation.

Next, the final representation of the design, whether phenotype or interpretive artifact, should exploit the viewing, interpretative and editing facilities of the enclosing CAD/CAM tool to as large an extent as possible. Architects conduct a complex, non-linear process as they design. They use many different tools and combine the outcomes of these tools in various ways. Any EA-based or ALife-based tool must seamlessly integrate into their existing design process. We have found that implementing our tools within an existing tool's API provides such continuity. A tool that generates a design (i.e. user-level, external representation) that can immediately be manipulated by other tools (e.g. by its elements being selectable and modifiable or by being transferable to form flattening software) is extremely valuable, if not essential. This approach coincidentally provides tool design advantages. Only the key paradigm or inspiration driving the tool needs to be implemented.

Next, we have also found that an ideal internal representation should also be intuitive so an architect can specify an initial design if it is not entirely generated by the tool itself. If the interpretive process mapping the genotype to the phenotype is complicated, it falls on the internal representation to support user expression of designs. That is, the power of executable structures typically implies the user can not fashion the external representation and expect the tool to determine its corresponding internal representation. Therefore, it is vital that the language of internal representation be accessible to the user in terms of ease of expression.

Finally, our users have repeatedly emphasized that an effective tool is one which facilitates a process we term "interruption, intervention, and resumption" (IIR). Without the ability to influence the discovery process of the tool, an architect feels at its mercy and robbed of creative control.

IIR can be facilitated *directly* or *indirectly*. To date, if at all, we have only offered our users indirect control over the process of design generation. In both GermZ and MoSS the user changes the placement of attractors and repellors prior to a run to obtain a new design. It is fair to say that is only an implementation detail that prevents the attractors and repellors or environment from being modifiable during a run. This would facilitate IIR, however, it would not solve the whole problem. The indirect nature of manipulating what influences design generation rather than manipulating a design directly in order to manifest a change would remain problematic. "Nudging" quickly becomes tedious and is often non-intuitive. Direct IIR is more desirable. However, precisely because EA-based and ALife based tools use representations that have genotypic and phenotypic levels and interpretation to provide mapping between the genotype and phenotype, it remains a "holy grail". For a tool to allow direct IIR, the altered phenotype must be sent through the interpretation process backwards to obtain the genotype that would specify it. However, this reversible process is not simple or always possible. Interpretation is not a one to one mapping so there could be many genotypes that generate a phenotype. Which one should be chosen? Alternatively, a change the designer imposes may not even be expressible by the language of the genotype and its interpreter.

For some applications, one may be willing to balance the difficulties of non-intuitivity and awkward usage of indirect IIR as a consequence of representation with the purpose and benefits of using the paradigms. For others, it is simply not acceptable. We think this is an important problem which will determine the ultimate benefit an EA-based or ALife-based tool can deliver.

4 Summary

In conclusion, we have provided brief descriptions of four tools of the Emergent Design Group. The particular representations schemes of each tool have been presented and lead to several criteria for EA-based or ALife-based representations. For example, a representation must tread the fine line between generality and the ability to be application specific. Furthermore, we have observed that EA-based or ALife-based tools represent a design with a genotype to phenotype mapping that, while the essence of the paradigm, presents major difficulties in accommodating interruption, intervention and resumption of a design tool. In an ongoing project called Agency, we are investigating the issues of direct IIR. One option we propose is to provide a small, closed set of direct manipulations within the GUI for which the genotype changes are known so a manipulation can be reverse mapped back to the genotype. Another option is to record the manipulations and simply suffix them onto the genotype. They would not be genetically alterable but they would be applied after the genotype had been mapped to phenotype. The ultimate success of the tools rests on devising solutions to these difficulties.

References

[1] Emergent Design Group: URL web.mit.edu/arch/edg/.

[2] C. G. Langton, editor. *Artificial Life*. Santa Fe Institute Studies in the Sciences of Complexity, Proc. Vol. VI. Addison-Wesley, Reading, MA, 1989.

[3] C. G. Langton, C. Taylor, J. D. Farmer, and S. Rasmussen, editors. *Artificial Life II*. Santa Fe Institute Studies in the Sciences of Complexity. Addison-Wesley, Reading, MA, 1992.

[4] Thomas Bäck. *Evolutionary Algorithms in Theory and Practice: Evolution Strategies, Evolutionary Programming, Genetic Algorithms*. Oxford University Press, Oxford, UK, 1996.

[5] P. Testa, U.M. O'Reilly, M. Kangas, and A. Kilian. Moss: Morphogenetic surface structure, a software tool for design exploration. In *Proceedings of Greenwich 2000 Digital Creativity Symposium*, pages 71–80, London, UK, 2000. University of Greenwich.

[6] P. Prusinkiewicz and J. Hahn. *Lindenmayer Systems, Fractals and Plants*. L.N. in Biomathematics, No. 79. Springer Verlag, 1989.

[7] P. Testa, U.M. O'Reilly, D. Weiser, and I. Ross. Emergent design studio: Interactive model of design and computation. In Submission to Environment and Planning B: Planning and Design.

[8] Fields investigation: URL web.mit.edu/arch/eds/studios/cware/studentwork.htm.

[9] Emergent Design Java Toolbox available from URL web.mit.edu/arch/eds/toolbox/about.htm.

[10] J. R. Koza. *Genetic Programming: On the Programming of Computers by Means of Natural Selection*. MIT Press, Cambridge, MA, 1992.

[11] MoSS documentation at URL web.mit.edu/arch/edg/projects/moss/moss.htm.

Acknowledgements: GermZ software was developed by Brian Clarkson and Hyung-Jin Kim. Emergent Design Java Toolbox was developed by Ian Ross. MoSS was developed by Markus Kangas and Axel Kilian. Devyn Weiser participated in all projects as a design consultant.

Fitting of Constrained Feature Models to Poor 3D Data

C. Robertson, R. B. Fisher, N. Werghi, A. P. Ashbrook

Machine Vision Unit, Institute for Perception, Action and Behaviour
Division of Informatics, University of Edinburgh,
Edinburgh, EH1 2QL, UK
craigr@dai.ed.ac.uk

Abstract

In this work we have addressed the question of whether it is possible to extract parametric models of features from poor quality 3D data. In doing so we have examined the applicability of an evolutionary strategy to the problem of fitting constrained parametric models. In the first phase, a background surface is fitted and removed leaving points of discontinuity associated with the feature. Then the discontinuities are classified, using the RANSAC algorithm, into drilled hole artifacts or blade edges suggesting drilled slots. This information, as well as the set of discontinuity points is passed to the Genocop III algorithm, proposed by Michalewicz [3], for optimization using a priori geometric constraints. Results, example times for convergence and comparisons with known ground truths are given.

Keywords: Model fitting, Ransac, evolutionary algorithms.

1 Introduction

Shape analysis of objects from range data (captured three dimensional co-ordinates of surface points) is a key problem in computer vision with several important applications in manufacturing, such as assembly, quality control and reverse-engineering. In this work we have examined whether it is possible to extract parametric models of features from poor quality 3D data. This kind of reconstruction problem is generally formulated as a nonlinear programming problem (NLP), which tries to optimally fit the data to candidate shape descriptions. The NLP optimises a function subject to several constraining equations and inequalities. Especially with nonlinear constraints, it is notoriously difficult to optimise and there is no known method to guarantee a satisfactory solution. Traditional search techniques, such as gradient descent, are unsatisfactory for the solution of NLPs, due to the local nature of their search methods and the reliance on smooth derivatives in the search-space. In previous work [8, 9] we examined the applicability of evolutionary strategies to the problem of fitting constrained lines, planes and degenerate second order surface to both synthetic and acquired object range data. In this paper we examine the applicability of fitting parametric models to poor quality range data from small complex artifacts such as drilled holes and machined slots of differing cross-section. The GenocopIII algorithm developed by Michalewicz [3], Ch.7 has been used and extended in this paper by adding a complex model fitting function. It is an evolutionary algorithm which is specialised to handle constrained function optimisation and particular to it is the handling of non-linear constraints. It uses real-valued genes and includes methods to deal with linear, non-linear, domain and inequality constraints. We have used a special-ised fitness function (described in section 3.5), which is applied to the problem of fitting parametric 3-dimensional model chromosomes to range data while simultaneously applying several necessary constraints. The constraints applied are of two types : domain (the restriction on the parameter size) and relational (relationships between surfaces that are known *a priori*).

Since this problem has a specific context it is important to illustrate it. Our group is researching the reverse engineering of machined parts. These parts are often complex and possess many surfaces which may have known geometric relationships. Segmentation and parameterisation of the captured 3-dimensional range data is a difficult multi-part task involving the following elements:

1. *Data collection.* This is performed using a moving-bed, orthogonal laser stripe ranger which provides data at up to 0.5mm steps in the X-Y plane. Noise on the data is around 0.15 *mm* standard deviation.

2. *Data registration.* This is performed using a variant of the **iterated closest point** algorithm [1].

3. *Segmentation.* There are many ways of segmenting the 3d dataset, most are based upon changes in local surface curvature followed by some form of least-squares optimisation, for example [4].

4. *Exploitation of constraints.* Constraints may be applied to exploit know-ledge about surface relationships.

The formulation of constraints and the application of constraint-based cor-rection and optimisation of surface fitting has been achieved previously [2] with notable success using the several constraint application strategies. There are, however, some associated problems with this approach: complex formu-lation of the constraint function; heavy reliance on the global convexity of the solution space; reliance on accurate initial estimate of solution.

The 'processing pipeline' that is required for this approach also can lead to a build-up of problems that must be solved in the constraint application stage. Previously [8, 9] it was demonstrated that simultaneous fitting and constraint management using all the available data could be achieved by a

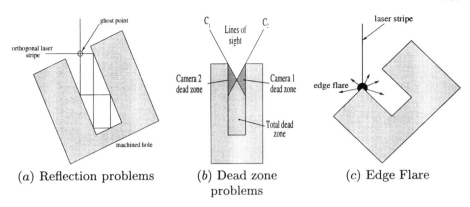

(a) Reflection problems   (b) Dead zone   (c) Edge Flare
                              problems

Figure 1: Laser ranger problems

single evolutionary algorithm using careful chromosome management and good generation of starting conditions.

In this paper we address the fitting of complex models of artifacts such as slots and drilled holes which are often found in machined mechanical parts. There are, however, several problems unique to this fitting, stemming from the size of the artifacts and the nature of the machined surfaces from which they are made.

2 Method

2.1 Data Quality and Sampling Issues

The sampling density of the range data is 0.5mm in the X and Y dimensions. This is a good density for the fitting of large convex surfaces, often of around 100 cm^2 area, where the samples can easily be accumulated to build a very accurate model. However, when gathering data from drilled holes or slots it is the points of discontinuity that are the important points. The parametric fitting of the supporting surface is relatively straightforward and leaves the complex mass of data representing the artifacts (see figure 2). Reflections from the inside walls of the artifact also cause problems because the machined surface is highly reflective and often has a shape that amplifies this. This causes ghost points that although consistent between two viewing cameras, are anomalous, as shown in figure 1(a). Camera dead-zones also cause data to be lost from the bottom of both drilled holes and machined slots when a camera can no longer track the laser stripe, figure 1(b). There is a further complicating factor in finding artifacts which is due to laser specularities around points of rapid discontinuity such as the edges of holes or slots. These specularities (edge flare) are due to the finite width of the laser which becomes split at the discontinuity. Although local this can mean that the edges of such artifacts are very difficult to accurately ascertain, as shown in figure 1(c).

Data (plotted as XYZ points) which shows many of these problems are

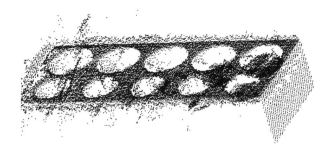

Figure 2: Typical poor data

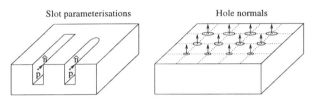

Figure 3: Artifact Models

shown in figure 2.

2.2   Primitives and Models

Our feature models have only three basic primitives:

1. *Straight edges*, where planes in the data intersect.

2. *Curves edges*, where machined holes or curved slots end.

3. *Planes*, which are often the supporting surfaces of the machined artifacts as well as the bases of slots and holes.

The three models we have experimented with are :

- *Orthogonal drilled hole* : Circular surface discontinuity and a plane defining the bottom of the hole.

- *Square ended slot* : Represented by three or four lines of discontinuity defining its edges and a plane defining the bottom.

- *Cylindrical ended slot* : Represented by two lines of discontinuity defining its edges, one or two semi-circular endings and a plane defining the bottom.

Although there are many other possible models we have used these to demonstrate the principle. Further to the basic models, our optimization uses geometric relationships between the models such as colinearity of drilled holes, slots with the same axis normal and distance relationships.

2.3 Full Models

For reconstruction, definitions of full models are required, which consistin of primitive collections and their relationships to each other . In our case we have built models only by the application of constraints at the optimization stage. For example, since a slotted part is characterized by planes intersecting at various angles, we use a model of this only using those angular constraints, adding distance constraints when they are known. Examples of models used in this work are given below:

- *Cylindrical slotted part.* This part has three machined slots at different depths. The model of this consists of one unit vector normal, describing the slot direction, three 3D points describing the positions of the slots and one parameter specifying the slot widths. The parameterisation has *a priori* constraints that: the slots are the same width; the slots are all lying in the same direction; the slot widths also define the estimate of the diameter of the machine tool used.

- *Square slotted part.* This part also has three machined slots at different depths. The model of this consists of one unit vector normal, describing the slot direction, three 3D points describing the positions of the slots and one parameter specifying the slot widths. The parameterisation has *a priori* constraints that: the slots are the same width; the slots are all lying in the same direction; the end planes are lying along the same direction.

- *Drilled hole part 1.* This part consists of two sets of five holes drilled at different depths and radii. Each set of holes is constrained to be co-linear and hole radii of each set constrained to be equal. One set has a radius of seven millimetres, the other is nine millimetres. Hole spacing is a free variable. See figure 3(a).

- *Drilled hole part 2.* This part consists of three sets of holes drilled at different depths and radii. Each set of holes is constrained to be co-linear with the hole radii of each set constrained to be equal. The radii are : (5x)5mm, (4x)3mm, (4x)4mm. Number of holes at that radius are shown in brackets. Each set of holes is constrained to be co-linear and radii of each set constrained to be equal. See figure 3(b).

- *Compound slot part.* This part consists of four slots, two square-ended, two round ended machined into the same block. The square slots have one square closed end, one open end. This is repeated with the same

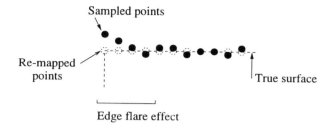

Figure 4: Discontinuity re-mapping to reduce edge flare

for the round ended slots. The parameterisation has *a priori* constraints that: the slots are the same width; the slots are all lying in the same direction; the end planes are lying along the same direction.

These models are encoded implicitly inside the optimization function as explained in section 3.2.

3 Algorithm

Our algorithm consists of three serial processes :

1. Extraction of primitives using the *Ransac* algorithm [11].

2. Heuristic identification of models present.

3. Model fit optimisation using the *Genocop III* algorithm [3] using geometric constraints.

3.1 Primitive Extraction

Supporting surfaces were extracted from the dense range data using a *Ransac* plane finder [11]. Sub-pixel positions of discontinuities on these surfaces were then found by applying a local gradient operator and re-projecting the discontinuities back to the supporting surface. This has the effect of reducing the edge flare effects quite substantially as shown in figure 4.

Once a set of discontinuities has been found, a composite Ransac line and circle finder was used to extract likely primitives together with their parameterisations. Example images of the range data, segmentation and primitive extraction are shown in figure 5(a-d).

Once primitives have been extracted, when appropriate, the point sets and parameterisations are passed to an evolutionary algorithm for further optimisation using geometric constraints. The details of this algorithm are discussed below.

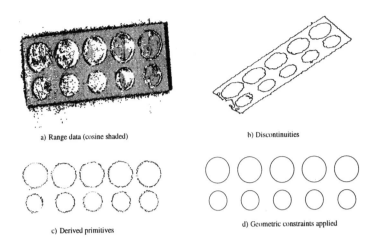

a) Range data (cosine shaded)   b) Discontinuities

c) Derived primitives   d) Geometric constraints applied

Figure 5: Primitive Extraction Process

3.2 Model Formulation

In standard GAs binary encoding forms the chromosomes in the solution, however in an evolution program each gene is a floating point number. Genes are then concatenated into a single chromosome.

In the case of **lines** on the support plane, we have used the 4 gene parametric representation : $A :< \hat{n}, d >$ where $\hat{n}$ is the unit normal describing the line and d is the constant defining its minimum distance from the origin. In the case of **circles**,also in the support plane, we have used the 4 gene parametric representation : $A :< \mathbf{P}, r >$ where $\mathbf{P}$ is the centre point and r is the radius. The normal is inherent in the supporting plane direction.

A full chromosome, G, describing a given artifact set, is represented by a set of concatenated part-chromosomes: $G = \{A_j\}$. This representation of a set of artifacts as a chromosome is much shorter than, say, a set of general quadrics as explored in [8]. This cuts down the complexity of constraints and makes it amenable to straighforward manipulation without the need to employ constraints on the forms themselves, only on sets of forms taken as systems. Since many of the parameter constraints are implicit, the parameters can be optimised under just a handful of constaints (mostly on their domains) which is relatively cheap computationally.

3.3 Genocop II and III

Calculus based methods assume that the objective function, $f(\mathbf{x})$, and all constraints are continuously differentiable functions of $\mathbf{x}$. The general approach is to transform the non-linear problem (NLP) into a sequence of sub-problems and then solve those, requiring an explicit computation of the objective function. Some of these methods become ill-conditioned and fail.

Genocop II uses a sequential quadratic penalty function and is formulated as the optimisation of the function:

$F(\mathbf{x}, r) = f(\mathbf{x}) + \frac{1}{2r}\overline{C}^T\overline{C}$, where $r > 0$, f is the user defined evaluation function and $\overline{C}$ is the vector of active constraints, $c_1, \ldots, c_l$.

Attia has provided solutions to the instability of this approach [10]. The set of all constraints, C, is divided into the linear constraints, L, the non-linear equations, N_e and the non-linear inequalities, N_i. A set of active constraints, A is then built from N_e and the violated constraints from N_i (a constraint is said to be violated if it is more than some tolerance δ from its correct value), which are called V. The structure of Genocop II is outlined in [3]. Inside its main loop, Genocop I optimizes the function $F(\mathbf{x}, r) = f(\mathbf{x}) + \frac{1}{2r}\overline{A}^T\overline{A}$. Several mutation operators take an initially identical population and introduce diversity to it. At convergence, the best individual, $\mathbf{x}^*$, is saved and the the value of the penalty parameter is decreased.

Most of the essential elements of Genocop III are the same as those of Genocop II. However, in this algorithm two populations are kept, a reference set $\overline{R}$ and a search set $\overline{S}$. The reference population is a set of fully feasible individuals which satisfy all the constraints whereas the search population may not. At each iteration, the search population are allowed to move around the solution space and are repaired back onto the constraint manifold. If the search point is $\overline{S}$ and the reference point is $\overline{R}$, then a random point $\overline{Z}$ is created from the segment between $\overline{S}$ and $\overline{R}$ by generating a value, $a \in [0, 1]$ then :

$$\overline{Z} = a\overline{S} + (1 - a)\overline{R} \tag{1}$$

Once a feasible $\overline{Z}$ is found, if it is better than $\overline{R}$, then that reference point is replaced with some probability. As the iterations progress, the set of reference points converge to the maximum or minimum on the search space.

3.4 Evaluation Function and Point Assignment for the Fitting

In our application of Genocop III the evaluation function is novel in that it is a multi-objective optimization. For each point, x_i, the true geometric distance to the theoretical primitive is computed and this is used as the least-squares error for that point relative to that primitive.

$$e_i = min_p\{dist(\mathbf{x}_i, S_p)\} \tag{2}$$

where e_i is the error for the point $\mathbf{x}_i$, p is the index of M theoretical primitives, i is the index of N points, S_p is the parameterised primitive and $dist$ is the distance to that primitive.

The evaluation function to be minimised is then the sum of these minima :

$$E = \sum_{i=0}^{i=N} e_i \tag{3}$$

3.5 Starting Conditions and Relational Constraints

In virtually all cases, domain constraints on individual genes are used to narrow the search space for that gene. These are represented as one permanent part of the sequential quadratic penalty function matrix [3] used in the evaluation function. A good example of where domain constraints can reduce the search space is in the case of the three parameters describing a unit normal. Each of these parameters can never be outside the range $[-1, +1]$ so these make good domain constraints.

In-chromosome relational constraints are straightforward to formulate when a parametric form is used. For example, consider two planes:

$P_1 = < n_{x1}, n_{y1}, n_{z1}, d_1 >$ and

$P_2 = < n_{x2}, n_{y2}, n_{z2}, d_2 >$

which are known *a priori* to be orientated orthogonally. In this case, the chromosome would have the form :

$G = \{n_{x1}, n_{y1}, n_{z1}, d_1, n_{x2}, n_{y2}, n_{z2}, d_2\}$

and the orthogonality constraint would then appear as a non-linear inequality of the form: $(n_{x1} * n_{x2}) + (n_{y1} * n_{y2}) + (n_{z1} * n_{z2}) \leq \epsilon$ where ϵ is the constraint tolerance value.

In order to perform the optimisation, Genocop requires a starting position on the constraint manifold. This is the seed for the search points which are then mutated around it. It is also used to produce the set of reference points as described earlier in section 3.4. In our case this means designing a chromosome which is both close to being a concatenation of the individual least-squares results for the part-chromosomes as well as fulfilling the domain and relational constraints. Starting conditions for increasingly complex solutions with increasingly complex constraints have previously been found to be difficult [8] for the general quadric. However, when a parametric representation is used, start conditions become very simple to generate, even when many constraints are used.

4 Results

4.1 Drilled Holes

There are many examples of drilled holes to be found in the machined objects which we regularly use to test our algorithms. Often these holes are filled before scanning in order to reduce the problems discussed earlier and because the supporting surfaces are often the most important ones to parameterize.

4.1.1 Primitive Extraction

Primitives were extracted using the RANSAC algorithm as discussed earlier. The exact locations of the holes are not known beforehand but in some cases we know the inter-hole distance and can use this as a method of assessing positional accuracy. Mean error in radius estimation for 25 drilled holes was found to be 0.165mm (standard deviation 0.128 mm) using holes ranging from

5mm to 10mm radius. On our calibrated drilled block this gave an average inter-hole positional error of 0.235mm, over estimates of 24 difference positions.

4.1.2 Optimization

We used a calibration block drilled with 10 holes (example data from which is shown in figure 2) where the inter-hole distances are known, to apply geometric constraints. This gave us an improvement in positional accuracy, positional errors became 0.134mm on average (standard deviation 0.120 mm), with the same set of 24 observations. Radius estimates for the calibration block improved slightly, with the average error in radius estimate now being 0.163mm.

4.2 Slots

4.2.1 Primitive Extraction

Slot lengths were found to have the highest degree of variation, with an average error of 2.061 mm (standard deviation 1.565 mm) over 72 examples. Slot angles were taken using pairs of slot sides. Mean slot angle error was 0.850^o (standard deviation 0.769^o) over some 36 samples. Slot depths were harder to collect although of those that were possible, the angular error - that is the difference in surface normal between the slot bottom and the supporting surface - was found to be under 5° (4.89^o) although no claims are made for this result as its stability is poor. Slot widths could not be properly ascertained since the lines were not necessarily found to be parallel. Variations in error were between 1 and 3mm.

4.2.2 Optimization

After optimization the slot results improved significantly in terms of angular error, since it was used as a geometric constraint and was based on the very accurate estimate of the supporting surface normal, and width variations. The width of all slots found in each of the slotted objects was constrained to be the same and the chromosome representing it consisted of the normal direction of the slot, 3 point chromosomes defining the positions and one width which was applicable to all of the slots. This degree of constraint produced slot widths that had a consistent error which is clearly systematic. Over 8 sets of data (that is 8 sets of three slots) the slot direction normal error was 0° with a width error of 0.5mm (standard deviation 0.04mm). The removal of this error is discussed in section 5.

4.3 Example timings

For primitive extraction the average time for processing is under 1 minute on a 200MHz UltraSparc (average load) when analysing 8000 line hypotheses per dataset and 3000 hole hypotheses. This number of hypotheses is relatively generous however. When the Genocop III algorithm is used the average time for a run is between 5 minutes and 10 minutes depending upon the initialisation used (random or directed). Random initialisation is often achievable but not desirable as the position of the early reference solutions in the search space are very far away from the desirable solutions. This can effectively double processing time, or more, depending on the number of parameters that are used.

5 Conclusions

In this paper we have posed the question of whether it is possible to extract and optimize parametric models of machined artifacts from poor quality range data. We have shown that this is not only possible but that it may be done in reasonable amount of time. In previous work [8, 9] we have shown that optimal surface fitting under geometric constraints for large and well formed surfaces was feasible with an evolutionary algorithm. In this paper we have demonstrated that the method is equally applicable to model fitting for poor data. The preprocessing of the data using ransac to extract primitives has also greatly improved the speed of the algorithm both in terms of segmentation and approximation.

Good results have been derived from data that is very difficult to analyse under normal circumstances.

Since the data in these examples suffers from particularly acute sampling problems it is necessary to do further modelling of the error characteristics in order to remove the systematic errors that we have discovered. These almost certainly come from either the quantisation of the sampling space or from the surface characteristics of the data. In order to improve this aspect of the approach more work needs to be done to design a better model of where real discontinuities occur relative to our data discontinuity points.

We intend to implement a constraints-based model desription language that will be used in all phases of the optimization, from primitive finding through to the final optimization using genocop. This will have the added bonus that the *a priori* geometric constraints can be used in all of the subparts of the scheme instead of just the final optimization. Finally, we envisage a full processing system that will be able to extract (i.e. segment), fit and finally optimize very complex models consisting of many surfaces and artifacts such as those described. High level descriptions will be pieced together by the user from a central library and the processing pipeline will create the CAD model, complete with optimized parameterisations.

Acknowledgements

The work presented in this paper was funded by a UK EPSRC grant GR/H86905.

References

[1] D. Eggert, A. W. Fitzgibbon, R. B. Fisher. "Simultaneous registration of multiple range views for use in reverse engineering", Proc. Int. Conf. on

Pat. Recog., pp 243–247, Vienna, Aug. 1996.

[2] N. Werghi, R.B. Fisher, A. Ashbrook, C.Robertson "Modelling Objects Having Quadric Surfaces Incorporating Geometric Constraints." Proc. ECCV'98, pp.185-201, Freiburg, Germany, June 1998.

[3] Z. Michalewicz, *Genetic Algorithms + Data Structures = Evolution Programs*, Third Edition, Springer, 1996.

[4] A. Hoover, G. Jean-Baptiste, X. Jiang, P. J. Flynn, H. Bunke, D. Goldgof, K. Bowyer, D. Eggert, A. Fitzgibbon, R. Fisher, "An Experimental Comparison of Range Segmentation Algorithms", IEEE Trans. Pat. Anal. and Mach. Intel., Vol 18(7), pp673–689, July 1996

[5] M. Gen and R. Cheng, "A Survey of Penalty techniques in Genetic Algorithms", In *Proceedings of the IEEE International Conference on Evolutionary Computation 1996*, 1996.

[6] Z. Michalewicz and N. Attia, "Evolutionary Optimization of Constrained Problems", in , *Proceedings of the 3rd Annual Conference on Evolutionary Programming*, San Diego, CA, 1994, pages 98-108. World Scientific.

[7] Z. Michalewicz, D. Dasgupta, R. G. Le Riche and M. Schoenauer, "Evolutionary algorithms for constrained engineering problems", special issue on *Genetic Algorithms and Industrial Engineering*, ed. M. Gen, G.S.Wasserman and A. E. Smith, *International Journal of Computers and Industrial Engineering*, 1996.

[8] C. Robertson, D. Corne, R. B. Fisher, N.Werghi, A.Ashbrook, "Investigating Evolutionary Optimisation of Constrained Functions to Capture Shape Descriptions from Range Data", Proc. 3rd On-line World Conference on Soft Computing (WSC3) also in *Advances in Soft Computing - Engineering Design and Manufacturing*, eds. Roy, Furuhashi and Chawdhry, Springer-Verlag, 1998.

[9] C. Robertson, R. B. Fisher, N. Werghi, A. Ashbrook. "An Evolutionary Approach to Fitting Constrained Degenerate Second Order Surfaces", to be published in the Proceedings of EvoIASP'99, Sweden (28th May 1999), eds. Poli, Voigt, Cagnoni, Corne, Smith and Fogarty, Springer-Verlag Berlin, 1999.

[10] N. F. Attia, *New Methods of Constrained Optimization Using Penalty Functions*, Ph.D Thesis, Essex University, United Kingdom, 1985.

[11] R. C. Bolles and M. A. Fischler, "Random Sample Consensus: A Paradigm for Model Fitting with Applications to Image Analysis and Automated Cartography", Technical Note 213, Artificial Intelligence Center, SRI International, Menlo Park, California, 1980.

Exploring Component-based Representations - The Secret of Creativity by Evolution?

Peter J. Bentley

Department of Computer Science, University College London

e-mail: P.Bentley@cs.ucl.ac.uk

Abstract. This paper investigates one of the newest and most exciting methods in computer science to date: employing computers as creative problem solvers by using evolution to explore for new solutions. The paper introduces and discusses the new understanding that explorative evolution relies upon a representation based on components rather than a parameterisation of a known solution. Evolution explores how the components can be arranged, how many are needed, and the type or function of each. The extra freedom provided by this simple idea is remarkable. By using evolutionary computation for exploration instead of optimisation, this technique enables us to expand the capabilities of computers. The paper describes how the approach has already shown impressive results in the creation of novel designs and architecture, fraud detection, composition of music, and creation of art. A framework for explorative evolution is provided, with discussion of the significance and difficulties posed by each element. The paper ends with an example of creative problem solving for a simple application - showing how evolution can shape pieces of paper to make them fall slowly through the air, by spiraling down like sycamore seeds.

1.   Introduction

Similar to the field of neural computation, evolutionary computation comprises a set of techniques inspired by natural processes. Instead of being modelled on the workings of brains, evolutionary algorithms are based on natural evolution. Problems are solved by using populations of solutions that 'reproduce', and 'die' according to how well they satisfy the problem, resulting in the emergence of good solutions after a number of generations.

Despite the first evolution-based programs dating back to the 1960's, this subset of computer science has only in the last ten years become recognised as a field of research in its own right. Today, hundreds of researchers world-wide, numerous departments and over twenty international conferences a year devote their efforts to understanding and applying evolution to problems.

The steady increase of interest in evolution-based approaches seems to be due to the fact that evolution provides a quick and easy way to solve difficult problems. It is now commonplace to evolve solutions for scheduling, machine learning, anomaly detection, ordering problems, data mining, control, game playing, reliable and fault tolerant systems, design optimisation and many other problems [1]. Evolutionary biologists now use evolutionary algorithms to increase understanding of natural evolution [2].

Evolution allows us to overcome many of the problems associated with searching for difficult and complex real-world solutions. But traditional implementations of evolutionary search suffer from the same fundamental drawbacks as all conventional search algorithms. They rely on a good parameterisation to permit them to find a good solution. If we are optimising a propeller blade, but the parameterisation does not permit the width of the blade to vary, then the computer will never be able to find solutions with different widths. If there is no parameter for something, then the computer cannot modify it. Evolution, like all search algorithms, is limited and constrained by the representation it can modify.

However, recent work has removed many of these limits and constraints. We can now use evolution to explore further than ever before. Evolution can search for good search spaces, even as it searches within a space. It can alter the dimensionality of a space, the parameterisation, the representation of solutions, and at the same time, find good solutions in this ever-changing hyperspace. Evolution can be used to modify so much of the problem that the very concept of search space begins to become unhelpful. The advantages of such freedom of search are plain: the evolved solutions now resemble inventions rather than improvements [3]. And because so many of the constraints of representation are thrown away, our computers begin to seem creative in the way they arrive at these surprising results.

2.   Explorative Evolution

Most of the advances in explorative evolution have grown up on my home turf, evolutionary design [1]. And it is easy to see why this has happened. Design problems such as architecture, engineering design, and aesthetic design are horribly complex, with huge numbers of (often conflicting) objectives, many constraints and often thousands of parameters [4]. But the most difficult aspects of these design problems are *people*. Designs are usually used by people - we live in architecture, we use and interact with the things around us. We like and dislike things almost at random, and as fashions change, so our preferences and requirements change. This means that a design specification will usually be a moving target [5]. It will have many unknowns, and the few things that are true one week will not necessarily be true the next week.

Designers take such things for granted. They know that their designs will be revised and modified many times until clients are satisfied (or until time or costs prevent further changes). And architecture is perhaps the most extreme type of design in this respect. There are so many different rules, opinions, preferences and materials that for every new building, that there will be an infinite number of possible design solutions. Exploring these solutions forms part of the difficult job of being an architect. Consequently, it is no coincidence that the first forays into explorative and generative evolutionary design were made by architects. Some of the earliest work was performed by Prof. John Frazer, who spent many years developing evolutionary architecture systems with his students [6]. He showed how evolution could generate many surprising and inspirational architectural forms, and how novel and useful structures could be evolved. His methods often involved the use of components such as cellular automata, which were evolved and sometimes wrapped in surfaces to generate smooth exteriors. In Australia, the work of Prof. John Gero and colleagues such as Dr. Michael Rosenman (both architects) also investigated the use of evolution to generate new architectural forms. This work concentrates on the generation of new floor-plans for buildings, showing over many years of research how explorative evolution can create novel floor-plans that satisfy many fuzzy constraints and objectives [7]. They even show how evolution can learn to create buildings in the style of well-known architects [8].

Designers and architects still remain at the forefront of this area of research.

Today increasing numbers are creating evolutionary architecture systems capable of generating novel forms, structures, buildings and even towns. For example, Paul Coates of the University of East London has shown how evolution can generate coherent plans for housing estates and buildings, as well as innovative building exteriors [9]. Prof. Celestino Soddu of Italy uses evolution to generate everything from novel table-lamps to castles to three-dimensional Picasso sculptures [10]. The work of Dr. Ian Parmee at the University of Plymouth has revealed a variety of methods for handling the complex and fluid nature of engineering design problems over the years [4].

Artists are also keen researchers into explorative evolutionary systems. The use of evolution to generate form, judged entirely on aesthetics, was first shown by Prof. Richard Dawkins in the mid 1980's [11]. His work inspired the well-known work of William Latham and Stephen Todd, who had considerable success in evolving artistic images and animations [12]. The work of Karl Sims was also inspired by Dawkins, and also showed the astonishing complex and aesthetically pleasing images that evolution could generate [13]. Evolutionary art systems have now become very popular, with many programs available and some now being incorporated into art and CAD packages [14].

All of these examples of evolutionary systems used evolution as an explorer, not as an optimiser. Normally guided by a human, the software is used to investigate many possible solutions, to provide inspiration and to give a feel for the range of useful solutions. Whilst producing impressive results, such software always received some criticism that the presence of a human to guide the direction of search by evolution was the key to the success of these systems. My own work in this area, and the work of others such as Adrian Thompson and John Koza, provided some of the first convincing demonstrations that evolution was capable of innovation without human guidance. I showed how evolution would find surprising and creative solutions to design problems, even when only software guidance was provided. By telling the computer the desired function in the form of a set of evaluation routines, but not anything about the design itself, I removed the human from the loop and showed conclusively the power of explorative evolution [15],[16]. Adrian also demonstrated the power of evolution, this time to generate novel electronic circuit designs which were tested using field programmable gate arrays [17]. In a similar vein, John Koza has been demonstrating the use of genetic programming to find novel computer programs for some years, i.e. to use computers to program themselves [18]. Work such as this began the recent change in thinking about evolution. No longer were we content to regard an evolutionary algorithm as 'another optimiser'. Our evolutionary programs were now independently solving problems for us, and finding creative solutions that surprised us.

Since then, research in this area has expanded rapidly. Adrian Thompson spends his time trying to develop ways to analyse how his evolved circuits work [19]. Julian Miller has expanded upon Thompson's work, showing how evolution can create circuits that seem to work on entirely new principles [20]. John Koza has recently announced the completion of a 1000-processor Beowulf computer which will be used for the sole purpose of 'using genetic programming as an automated "invention machine" for creating new and useful patentable inventions'. His group has recently demonstrated the use of evolution to generate many different types of analogue circuit, many of which mirror or outperform our best human-created circuit designs [21]. Jordan Pollack has shown how evolution can generate highly novel structures such as bridges and cranes [22] and has now begun to use these methods for the evolution of robot bodies. Karl Sims used evolution to create the bodies and brains of 'virtual creatures' capable of swimming, walking, jumping and competing in virtual environments [23]. Husbands et al [24] used evolution to generate novel propeller-like forms. Many research departments around the world use evolution to generate new and highly complex neural networks for various applications. My own work in this area continues, as I use evolution to compose music that cannot be distinguished from human compositions [25]. (For a more comprehensive review of work in this area, see [26]).

3. Exploring the Explorer

The research described here aims to investigate explorative evolutionary computation. To date there has been no significant research aimed at understanding the difference between exploration and optimisation. There has been little increase in our understanding of how evolution can innovate since we began demonstrating this ability. This project aims to discover answers to these fundamental questions and to explore, characterise and explain how we can use these answers to further increase the capabilities of explorative evolutionary computation.

The first question to be tackled is the most fundamental: what is evolution doing when it explores, that it is not doing when it optimises?

Traditional views of an evolutionary algorithm regard this search technique as an optimiser. A better term is actually 'satisficer' [27]. Evolution never tries to find globally optimal solutions. It merely propagates improvements through the population. In doing so, evolution walks a mysterious and winding path through the search space. Sometimes this path may reach a dead-end (premature convergence). Sometimes the path may go around in circles. And occasionally the path may lead to a global optimal - but there is no guarantee of this. The wanderings of evolution are like those of an explorer.

But exactly what does evolution explore? This is determined by the representations it uses. With a fixed-length parameterisation, explorations do resemble optimisation. For example, if there are only three parameters defining the solution and a single fitness function, evolution will behave in much the same way as an optimiser - it will find the best values for those three parameters such that the solutions are as fit as possible. But with a different kind of representation, the behaviour of evolution changes. When the parameters do not define the solution directly, when they define a set of components from which the solution is constructed, the idea of optimisation becomes inappropriate. Now evolution *explores* new ways of constructing the solution by changing the relationships between components. It can vary the dimensionality of the space by adding or removing elements. It can explore alternatives instead of optimising a single option. When we examine the representations, objectives and goals of optimisation and exploration, the difference between these approaches becomes clearer:

Optimisation: A knowledge-rich encoding of the problem is used, i.e. a solution is parameterised using the fewest possible parameters (and thus both minimising the search space, but also minimising potential for exploration). Evolution is used only to find the best parameter values within that parameterisation. Objective functions are normally predefined and fixed. The goal is to find a global or near global optimum with minimal computation.

Exploration: A knowledge-lean representation is used, and instead of a parameterisation of a solution, a set of low-level components is defined. Solutions are then constructed using the components, allowing exploration (sometimes at the expense of size of search space and ability to locate optima). Objective functions may vary at any time. The goal is to identify new and interesting solutions - normally more than one is desirable - and these solutions must be good. However, finding global optima may be undesirable, impractical or even impossible.

This difference between optimisation and exploration is rarely considered, let alone defined. And yet when the literature in this area is examined, it becomes evident that every system that performs exploration uses this component-based representation. Researchers who evolve architecture use evolution to manipulate components such as walls or bricks. Researchers who evolve electronic circuits use evolution to modify components such as logic gates, transistors and capacitors. Artists use evolution to create art out of primitive shapes: swirls, spheres, tori, or out of mathematical functions such as cosine, arctangent and factorial. Likewise, computer scientists using GP evolve programs from components contained within the function and terminal sets. My own work used cuboid blocks to construct designs, and currently uses fuzzy logic functions to construct fuzzy rules and musical notes to construct compositions. Every explorative evolutionary system relies on some kind of component-based representation. Figure 1 illustrates this idea, showing how components allow increased freedom for evolution.

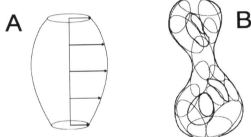

Figure 1 Optimisation of fixed parameterisation versus component-based exploration. Shape A uses minimal parameters and is knowledge-rich (the height of the shape and the swept 2D outline is assumed by this representation). Shape B is constructed by wrapping an envelope around a collection of 3D ellipses. By varying the relative positions and sizes of the ellipses, vastly more innovative and creative forms can be generated.

Evolutionary algorithms, and in particular genetic algorithms and genetic programming, naturally lend themselves to this kind of exploration. It is standard practice to use evolution to manipulate coded versions of solutions (genotypes) and to map these onto actual solutions (phenotypes). This use of genotypes and phenotypes means that the distinction between, and use of, components of solutions and complete solutions is natural to this field.

4. A Framework for Explorative Evolution

From this novel intuitive understanding, a framework for explorative evolutionary systems can be constructed, containing the following five components:
1. An evolutionary algorithm.
2. A genetic representation.

3. An embryogeny using components.
4. A phenotype representation.
5. Fitness function(s).

To summarise, an explorative evolutionary system requires some kind of evolutionary algorithm to generate new solutions. The algorithm modifies genotypes defined by the genetic representation, which must be designed to minimise disruption caused by the genetic operators. An embryogeny (or mapping process) must decode the genotype, and using some kind of components, must construct the phenotype. The phenotype representation must be designed such that it permits quick and efficient evaluation by the fitness function(s). It is likely that the evolutionary algorithm, the genetic representation and to some extent the embryogeny, will be generic and suitable for reuse for most problems without modification. The phenotype representation and fitness functions must be specific to the current application of the system. The following sections explore the elements of the framework for explorative evolutionary systems in more detail.

Evolutionary Algorithm: The evolutionary algorithm forms the core of any evolutionary system. There are four main EAs in use today: the genetic algorithm, genetic programming, evolutionary strategies and evolutionary programming. Only the GA and GP are commonly used for explorative purposes. The reason for this can be found in the way these algorithms work. The genetic algorithm maintains genotypes and phenotypes, with a mapping between the two. As described earlier, this distinction has helped to encourage some GA researchers to use component-based genotype representations that map onto the phenotype representations, thus allowing explorative evolution to begin. In the same way, genetic programming also makes use of genotypes (this time with tree-structures) that are mapped onto phenotypes such as programs, images or circuits. GP has the advantage that its genetic representation *requires* the use of smaller components (in the function and terminal sets), so all applications of GP demonstrate the explorative power of evolution. This explains why the first notion of "invention machine" came from John Koza, the inventor of GP - his algorithm ensures that explorative evolution will always take place. In contrast, algorithms such as evolutionary strategies and evolutionary programming make no distinction between genotype and phenotype. By directly modifying the solution and with no provision for mapping to new representations, these approaches make the use of components to construct solutions more difficult to implement - but not impossible.

Within any evolutionary algorithm there are other issues that must be tackled. Handling multiple objectives, multimodality, noise, premature convergence, fuzzy or changing fitness functions must all be considered. Solutions to all of these problems, using ideas such as Pareto optimality, region identification, speciation, variable or directed mutation rates and steady-state GAs are now emerging in evolutionary computation [26], [4], [28]. These issues, although important, are not the most significant consideration for explorative evolution. Indeed, even the choice of evolutionary algorithm (or indeed any other search algorithm) is secondary to the representations, for it is the representations that permit evolution to explore.

Genotype Representation: The genotype representation defines the search space of the algorithm. A poor representation may enumerate the space such that very dissimilar solutions are close to each other, making search for better solutions harder. For explorative evolutionary computation, where genes will represent (directly or indirectly) a variable number of components, the search space is typically of variable dimensionality, thus making its design even harder [7].

There are also other problems. Because of the use of components to represent solutions, the likelihood of epistasis dramatically increases. Not all component-based representations will have this effect (e.g. a voxel representation allows both exploration and zero epistasis). However, most components are inherently linked to their companions for the solution to work as a whole. A circuit relies on the links between its components, a melody relies on links between notes, a house relies on links between walls, a program relies on links between commands. These linkages all mean that corresponding genes become epistatically linked, resulting in potentially serious problems for evolution. With polygeny so prevalent in these problems, great care must be taken in the design of the genotype representation and corresponding genetic operators to minimise the disruption of inheritance.

Practitioners of GP have long been aware of these problems, with many solutions now in existence. Modifications can be made to the genetic representation to increase functionality and decrease disruption, e.g. ADFs, ADIs, ADLs, etc. [21]. Genetic operators that enforce typing help ensure that genetic trees are not shuffled too drastically during the production of offspring [29].

GAs do not require the use of tree-structured genotypes, so genetic representation-based problems are often less prevalent. GAs can be used to evolve variable-length genotypes and structured genotypes, typically with operators designed to perform crossover only at points of similarity between two parent genotypes, for example [30]. Advanced GAs designed to minimise damage caused by disruption of epistatic links between genes have also been demonstrated [31]. In addition, GAs do not suffer from the classic GP problem of bloat, where genotypes tend to increase in size, with redundant genetic material becoming ever greater in solutions. It is clear that the creation of suitable genetic representations and corresponding operators is a considerable problem in its own right. Furthermore, recent research seems to indicate that significant benefits may be gained from using less complex genetic representations and operators, instead making use of embryogenies of greater complexity [32].

Embryogeny: An embryogeny is a special kind of mapping process from genotype from phenotype. Within the process, the genotype is now regarded as a set of 'growing instructions' – a recipe which defines how the phenotype will develop. Polygeny is common, phenotypic traits being produced by multiple genes acting in combination. My own research in this area has revealed some of the potential of these advanced mapping processes. Advantages include reduction of the search space, better enumeration of search space, the evolution of more complex solutions, and adaptability [26].

Embryogenies are widely used for explorative evolutionary systems, for they provide the mechanism for constructing whole solutions from components. Three types of embryogeny are used today, the first and most common being *external*, where a programmer writes the software that performs the mapping, and the process cannot be evolved [26]. More recently, *explicit* embryogenies have become popular, with every step of the growth process explicitly held as part of the genotype, and evolved [26]. Examples include Cellular Encoding (used by Koza and team for the evolution of analogue circuits [21], Lindenmayer Systems (used by Coates for the evolution of architectural forms [9] and shape grammars (used by

Gero for the evolution of floor plans [8]. Despite the considerable success of these embryogenies, they often require complex additions to genetic representations and operators to allow evolution to work. The third type of growth process, the *implicit* embryogeny, has shown the most exciting results and greatest potential in recent work. Instead of evolving the mapping as a set of explicit steps in the genotype, an implicit embryogeny uses a set of rules, typically encoded as binary strings in a GA genotype. For each solution, a 'seed' component is created, and then the rules are iteratively applied. Over many iterations, with rules activating and suppressing each other, the growth, position, and type of new components are built up, finally resulting in the development of a complete solution. This emergent growth process shows remarkable properties of scalability, with the genotype describing solutions of increasing complexity without any increase in the number of rules needed - the rule-directed growth process is simply allowed to run for more time. This is in contrast to all other approaches, which require significantly larger genotypes to define the increased growth of more complex solutions [32].

The process of mapping from genotypes to phenotypes is clearly of importance to the investigation of explorative evolution. Issues of scalability, evolvability, and biases induced in search have yet to be considered by researchers in any great depth. Increased understanding in this area would benefit both computer science and developmental biology.

Phenotype Representations: Once constructed by the embryogeny, the resulting solution is defined by the phenotype representation. Typically this representation is application-specific - if we are evolving circuits, the representation might define networks of connected components, if we are evolving buildings, the representation might define exterior shapes and/or interior walls, floors and stairs. An important criterion is evaluation - typically the phenotype representation will be designed to allow direct evaluation by fitness functions, without intermediate transformations or calculations. A poor choice will detrimentally affect processing times and solution accuracies.

The distinction between genotype, embryogeny and phenotype representations is often blurred in this field. Some GP practitioners regard all three to be the same. Others, such as Jakobi's work on evolving neural networks [33] or Taura's work on evolutionary configuration design [34], use different and distinct representations for each stage. It is still unclear whether the component growing process of the embryogeny should use the same representation as the phenotype - should the phenotype be represented by blocks or should the blocks be merged into a single, whole description of the solution? For example, should an evolved musical composition be represented by a series of notes or by a single, complex waveform in the phenotype? The answer is likely to depend on the fitness functions.

Fitness Functions: The fitness functions must provide an evaluation score for every solution. For explorative evolutionary computation, this is often a little harder. Typically the use of components rather than a parameterised solution means that early results can be chaotic to say the least. A design of a car may resemble a shoe; a melody may sound like a burglar alarm. Before evolution has had time to improve these initially random solutions, they can be nothing at all like the desired result. And yet the fitness functions must always be able to provide a fitness score that makes sense. The task is made even harder when unknowns or approximations must be incorporated into the evaluation, or when constraints and objectives are varied to aid exploration. Potential solutions include the use of custom-designed modular functions [16] and fuzzy logic [4] to cope with such problems.

Many explorative systems use human input to help guide evolution. Artists can completely take over the role of a fitness function [12], [13], and more recent work has investigated the use of these techniques for evolving photo-realistic images of faces for the identification of criminals [35]. These applications raise numerous human-computer-interfacing issues, i.e., will an explorative system detrimentally affect the style of an artist or the memory of a crime victim? These software tools have been shown to aid imagination and creativity, but how best to let the user inform evolution of his/her preferences and how best for the computer to report the structure and contents of the space being explored? Clearly, further research is required to address these issues.

5. Example Creative Application

To demonstrate the potential of explorative evolution, this section of the paper describes a simple application - the generation of shapes that, when constructed out of paper, fall as slowly as possible to the ground. A simple explorative evolutionary system was developed, following the framework provided above. A 'simple GA' was used as the evolutionary algorithm [36]. The genotype representation comprised a flat chromosome of 16 binary coded genes. A very basic mapping process or embryogeny was used to derive 8 (x, y) parameter values from each chromosome, join each vector to its successor by an edge, join the last vector to the fist with an edge, and fill the resulting shape. This process was performed by executing PostScript instructions output by the system, printing the shapes, and using a scalpel to cut out the shapes. The resulting paper phenotypes ('represented' by reality) were then tested by releasing them from a height of 150mm three times in succession. Each phenotype was allocated their fitness score by calculating: 1/(time1+time2+time3), ensuring that evolution would attempt to generate phenotypes with increased 'falling times'.

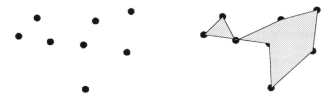

Figure 2 The *paper fall* application uses eight vertices as its components (left). These are extracted from the genotype and transformed into real paper shapes by the use of a 'join the dots and fill the shape' embryogeny (and someone to print and cut out the shape).

This application illustrates the use of an extremely basic component-based embryogeny representation. As figure 2 illustrates, the components are simply eight vertices with (x, y) positions. Despite these components having no size and no type, the unconstrained freedom of position of each vertex relative to all other vertices means that this component-based representation allows the definition of a vast number of different shapes. Clearly this is a knowledge-lean representation (no information about which shapes are best is provided). It should also be clear that the representation allows evolution to explore the solution space in an unconstrained manner, and that there will be many millions of good and bad solutions for this problem.

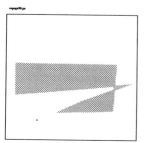 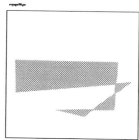

Figure 3 Two paper shapes evolved to fall slowly through the air. Both are members of the final generation, and both use a smaller 'arm' or flap to cause the shape to rotate as it falls, like a sycamore seed. (Not shown at scale used for testing.)

Predictably, the use of real-life testing is time-consuming and laborious (especially if you make the mistake of performing the experiment yourself instead of enlisting the services of a student). Because of this, population sizes of 10 individuals were used in an evolutionary run of 10 generations. Despite these excessively low values, evolution was able to make significant improvements on the time taken for the shapes to reach the ground. Times taken for initially random shapes to fall 150mm varied from 0.7 seconds to 1.8 seconds. By the tenth generation, all shapes took, on average, more than 2 seconds to fall the same distance. Figure 3 shows two of the solutions in the final population. Convergence has begun to occur, with most shapes using the same technique of having a smaller flap which causes the shapes to rotate as they fall. The main benefit of this solution appears to be the way rotation stabilises the motion of the shape, preventing it from slipping sideways in the air and plummeting to the ground at tremendous speed. Even with the very limited resources evolution was given, a 'creative' solution to the problem was found.

6.   Conclusions

Creativity is a very hard concept to define, but this paper does not suggest that the creativity of people depends on the exploration of component-based representations. However, evolution is very different from the human brain. It is apparent that all of the evolutionary systems that claim to produce 'creative' or innovative results do rely on the use of component-based representations with evolution. It is also clear that such unconstrained and knowledge-lean representations do provide far greater freedom for evolution to innovate, compared to knowledge-rich representations based on a parameterisation of a solution.

With this insight into enabling creativity by evolution, we can create a framework for explorative evolutionary systems, comprising: evolutionary algorithm, genotype, embryogeny and phenotype representation and fitness functions. Although looking familiar, the framework employs a component-based embryogeny to map genotypes into phenotypes, which raises more unusual issues for the genotype and phenotype representations and the fitness functions.

This framework was illustrated by the *paper fall* application, which used binary strings for the genotype representation, a component-based embryogeny using eight vertices to form shapes, and real phenotypes constructed from paper. Even using such simple techniques, evolution was able to generate some creative solutions in the form of 'sycamore seed' shapes, demonstrating the power of explorative evolution. Although it may not be the only reason for the success of evolution for these problems, it seems that exploring components is one of the secrets of creativity by evolution.

References

1. Bentley, P. J. (Contributing Editor), 1999a. *Evolutionary Design by Computers*. Morgan Kaufman Publishers Inc., San Francisco, CA.
2. Dawkins, R., 1996. *Climbing Mount Improbable* Penguin Books, Ltd..
3. Bentley, P. J. and Corne, D. (Eds), 1999. Proceedings of the *AISB'99 Symposium on Creative Evolutionary Systems (CES)*. Published by AISB, Sussex, UK. ISBN 1 902956 03 6.
4. Parmee, I., 1999. Exploring the Design Potential of Evolutionary Search, Exploration and Optimization. In Bentley, P. J. (Ed.) *Evolutionary Design by Computers*. Morgan Kaufman Publishers Inc., San Francisco, CA.
5. French, M., 1999. The Interplay of Evolution and Insight in Design. In Bentley, P. J. (Ed.) *Evolutionary Design by Computers*. Morgan Kaufman Publishers Inc.
6. Frazer, J., 1995. *An Evolutionary Architecture*. Architectural Association, London.
7. Gero, J. S. & Kazakov, V., 1996. An exploration-based evolutionary model of generative design process. *Microcomputers In Civil Engineering* 11, 209-216.
8. Schnier, T. and Gero, J. S., 1996. Learning genetic representations as alternative to hand-coded shape grammars, in J. S. Gero and F. Sudweeks (eds), Artificial Intelligence in Design'96, Kluwer, Dordrecht, pp.39-57
9. Coates, P., (1997) Using Genetic Programming and L-Systems to explore 3D design worlds. *CAADFutures'97*, R. Junge (ed), Kluwer Academic Publishers, Munich.
10. Soddu, C., 1995 Recreating the city's identity with a morphogenetic urban design. 17th International Conference on Making Cities Livable, Freiburb-im-Breisgau, Germany, Sept. 5-9 1995.
11. Dawkins, R., 1986. *The Blind Watchmaker*. Longman Scientific & Technical Pub.
12. Todd and Latham, 1999. The Mutation and Growth of Art by Computers. In Bentley, P. J. (Ed.) *Evolutionary Design by Computers*. Morgan Kaufman Publishers Inc.
13. Sims, K., 1991. Artificial Evolution for Computer Graphics. *Computer Graphics*, **25**, No.4, 319-328.
14. Rowbottom, A., 1999. Evolutionary Art and Form. In Bentley, P. J. (Ed.) *Evolutionary Design by Computers*. Morgan Kaufman Publishers Inc., San Francisco, CA.
15. Bentley, P. & Wakefield, J., 1997. Conceptual Evolutionary Design by GAs. *Engineering Design and Automation Jnl* 3:2, John Wiley & Sons, Inc, 119-131.
16. Bentley, P. J. & Wakefield, J. P., 1997b. Generic Evolutionary Design. Chawdhry, P.K., Roy, R., & Pant, R.K. (eds) *Soft Computing in Engineering Design and Manufacturing*. Springer Verlag London Limited, Part 6, 289-298.
17. Thompson, A., 1995. Evolving Fault Tolerant Systems. *Genetic Algorithms in Engineering Systems: Innovations and Applications*, IEE Conf. Pub. 414, pp. 524-529.
18. Koza, J., 1992. *Genetic Programming: On the Programming of Computers by Means of Natural Selection*. MIT Press.
19. Thompson, A. & Layzell, P., 1999. Analysis of Unconventional Evolved Electronics.

172

Communications of the ACM, April 1999 - Volume 42, Number 4, pp71-79.

20. Miller, J., Kalganova, T., Lipnitskaya, N., Job, D., 1999. The genetic algorithm as a discovery engine: strange circuits and new principles. To appear in Bentley and Corne (eds), *Creative Evolutionary Systems*, Morgan Kaufman Pub.

21. Koza, J.R. Bennett III, F. R., Andre, D. & Keane, M. A., 1999. *Genetic Programming III: Darwinian Invention and Problem Solving*. Morgan Kaufmann Pub.

22. Funes, P. and Pollack, J., 1999. The Evolution of Buildable Objects. In Bentley, P. J. (Ed.) *Evolutionary Design by Computers*. Morgan Kaufman Publishers Inc.

23. Sims, K., 1999. Evolving Three-Dimensional Morphology and Behaviour. In Bentley, P. J. (Ed.) *Evolutionary Design by Computers*. Morgan Kaufman Publishers Inc.

24. Husbands, P., Jermy, G., McIlhagga, M., & Ives, R., 1996. Two Applications of Genetic Algorithms to Component Design. In *Selected Papers from AISB Workshop on Evolutionary Computing*. Fogarty, T. (ed.), Springer-Verlag, Lecture Notes in Computer Science, pp. 50-61.

25. Bentley, P. J., 1999b. Is Evolution Creative? In P. J. Bentley and D. Corne (Eds) Proceedings of the *AISB'99 Symposium on Creative Evolutionary Systems (CES)*. Published by The Society for the Study of Artificial Intelligence and Simulation of Behaviour (AISB), pp. 28-34.

26. Bentley, P. J., 1999c. An Introduction to Evolutionary Design by Computers. Chapter 1 in Bentley, P. J. (Ed.). *Evolutionary Design by Computers*. Morgan Kaufman Publishers Inc., San Francisco, CA, 1-73.

27. Harvey, I., 1997. Cognition is not Computation: Evolution is not Optimisation. In *Artificial Neural Networks - ICANN97*, Gerstner, Germond, Hasler, and Nicoud (eds).

28. Vavak, F., & Fogarty, T., 1996. Comparison of Steady State and Generational GAs for Use in Nonstationary Environments. *Proceedings of the IEEE 3rd International Conference on Evolutionary Computation ICEC'96*, published by IEEE.

29. Page, J., Poli, R. and Langdon, W., 1999. Smooth Uniform Crossover with Smooth Point Mutation in Genetic Programming: A Preliminary Study, In R. Poli, P. Nordin, W. B. Langdon and T. Fogarty (Eds.), *Proceedings of the Second European Workshop on Genetic Programming - EuroGP'99*, Goteborg, May 26-27, 1999, Springer-Verlag.

30. Bentley, P. J. & Wakefield, J. P., 1996. Hierarchical Crossover in Genetic Algorithms. In *Proceedings of the 1st On-line Workshop on Soft Computing (WSC1)*, (pp. 37-42), Nagoya University, Japan.

31. Goldberg, D., 1999. The Race, the Hurdle, and the Sweet Spot: Lessons from Genetic Algorithms for the Automation of Design Innovation and Creativity. In Bentley, P. J. (Ed.) *Evolutionary Design by Computers*. Morgan Kaufman Publishers Inc.

32. Bentley, P. J. and Kumar, S., 1999. Three Ways to Grow Designs: A Comparison of Embryogenies for an Evolutionary Design Problem. In *Genetic and Evolutionary Computation Conference (GECCO '99)*, pp.35-43.

33. Jakobi, N., 1996. Harnessing Morphogenesis. In *Proceedings of the international Conference on information Processing in Cell and Tissue*.

34. Taura, T. and Nagasaka,, 1999. Adaptive growth type representation for 3D configuration design. In Bentley, P.J. (Guest Ed.) *First Special Issue on Evolutionary Design, Artificial Intelligence for Engineering Design, Analysis and Manufacturing* (AIEDAM) v13:3, Cambridge University Press, 171-184.

35. Hancock, P. and Frowd, C., 1999. Evolutionary Generation of faces. In Bentley, P. J. & Corne, D. W. (Eds) *Proceedings of the AISB'99 Symposium on Creative Evolutionary Systems (CES)*. Published by AISB, Sussex, UK. ISBN 1 902956 03 6.

36. Goldberg, D. E., 1989. *Genetic Algorithms in Search, Optimization & Machine Learning*. Addison-Wesley.

Chapter 4

Manufacturing Applications

Evolutionary Design of Facilities Considering Production Uncertainty
A.E. Smith, B.A. Norman

A Fuzzy Clustering Evolution Strategy and its Application to Optimisation of Robot Manipulator Movement
J.C.W. Sullivan, B. Carse, A.G. Pipe

An Industry-based Development of the Learning Classifier System Technique
W.N.L. Browne, K.M. Holford, C.J. Moore

A Genetic Programming-based Hierarchical Clustering Procedure for the Solution of the Cell-formation Problem
C. Dimopoulos, N. Mort

Evolutionary Design of Facilities Considering Production Uncertainty

Alice E. Smith
Department of Industrial and Systems Engineering
Auburn University
Auburn University, AL 36849-5346 USA
aesmith@eng.auburn.edu

Bryan A. Norman
Department of Industrial Engineering
University of Pittsburgh
Pittsburgh, PA 15261 USA
banorman@engrng.pitt.edu

Abstract. This paper considers the design of facilities while explicitly incorporating production uncertainty in the form of product quantity expected values and standard deviations. Block layout designs can be optimized directly for robustness over a prespecified range of uncertainty. This formulation is solved using a random keys genetic algorithm meta-heuristic with a flexible bay construct of the departments and total facility area.

1. Introduction

Facility design problems generally involve the partition of a planar region into departments (work centers or cells) along with an aisle structure and a material handling system to link the departments. The primary objective of the design problem is to minimize the costs associated with production and materials movement over the lifetime of the facility. Such problems occur in many organizations, including manufacturing cell layout, hospital layout [8], construction site management [34], and environmental management [4]. For U.S. manufacturers, between 20% to 50% of total operating expenses are spent on material handling and an appropriate facilities design can reduce these costs by at least 10% to 30% [19]. Tompkins [31] wrote, "Since 1955, approximately 8 percent of the U.S. GNP has been spent annually on new facilities. In addition, existing facilities must be continually modified...These issues represent more than $250 billion per year attributed to the design of facility systems, layouts, handling systems, and facilities locations...". Altering facility designs due to incorrect decisions, forecasts, or assumptions usually involves considerable cost, time, and disruption of activities. On the other hand, good design decisions can reap economic and operational benefits for a long time period.

The design problem primarily studied in the literature has been "block layout" which specifies the placement of the departments, without regard for aisle structure and material handling system, machine placement within departments or input/output locations. Block layout is usually a precursor to these subsequent steps, termed "detailed layout."

1.1 Formulation of the Unequal Area Block Layout Problem

A block layout where departments may have different areas and/or different shapes was first formulated by Armour and Buffa [1] as follows. There is a rectangular region, A, with fixed dimensions H and W, and a collection of n required departments, each of specified area a_j and dimensions (if rectangular) of h_j and w_j, whose total area = A = H×W. There is a material flow F(j,k) associated with each pair of departments (j,k) which generally includes a traffic volume in addition to a unit cost to transport that volume. There may also be fixed costs between departments j and k. The mathematical objective is to partition A into n subregions representing each of the n departments, of appropriate area, in order to:

$$\min \sum_{\substack{j=1}}^{n} \sum_{\substack{k=1 \\ j \neq k}}^{n} F(j,k)d(j,k,\Pi) \tag{1}$$

where d(j,k,Π) is the shortest rectilinear distance between centroids of department j and department k in the layout Π. They approached this problem by requiring all departments comprise contiguous rectangular unit blocks, and then applying departmental adjacent pairwise exchange. Other unit block approaches include Bazaraa [2]), Hassan et al. [12], Meller and Bozer [18], and Ziai and Sule [35]. Drawbacks of the unit block approach are that departmental shapes can be impractical and each department must be discretized artificially into blocks.

A general linear programming approach was formulated by Montreuil and others [20, 21] to avoid departmental overlap and minimize interdepartmental flow costs, when first given a "design skeleton". Drawbacks of the mathematical programming approach are: (a) inability to enforce exact departmental areas, (b) rapid increase in the number of variables and constraints as the problem size increases, and (c) dependence on an initial layout skeleton. Related approaches (e.g., [14, 33]) have the disadvantage that the final layout is a central cluster of departments that may be very different in shape than the normal rectangular bounding area H×W. Foulds and others [9-11] pioneered a graph theory approach, however it suffers from a rapid increase in search space size as the design problem size increases and the difficulty in translating the graph to a block layout that resembles a physical facility.

Another formulation for unequal area block layout is slicing trees, where the departments and the bounding facility are required to be rectangular and the layout is represented by alternating vertical and horizontal slices [6, 28, 29]. A related formulation is the flexible bay structure (Figure 1) [30]. This structure first allows slices in a single direction, creating bays, which are then subdivided into departments by perpendicular slices.

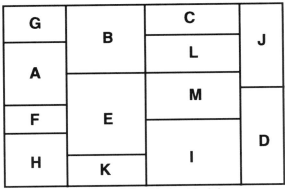

Figure 1. Typical flexible bay block layout.

For both the unit and non-unit block approaches mentioned above, the objective of minimizing distance between centroids (equation 1) causes departments to be shaped as unrealistically elongated rectangles if an aspect ratio[1] or a minimum side length constraint on each department is not imposed.

1.2 Stochastic Plant Layout

The research described in section 1.1 is predicated on the idea that all flows and costs are known with certainty, however the idea of multiple uncertain scenarios motivated the research on stochastic plant layout. Shore and Tompkins [27] studied four possible scenarios based on product demand. They optimized each scenario separately using the unit block approach to unequal area problems, and then selected the layout that had the lowest penalty when considering the likelihood of each scenario. This was called the most "flexible" layout. Rosenblatt and Lee [26] first developed the idea of "robustness". They considered a single-period problem where product demand (and thus production) could be represented as a three-point random variable. Robustness is defined as the number of times a layout falls within a prespecified percentage of optimal when enumerating the optimal layouts for each production scenario. This approach was demonstrated on a small quadratic assignment problem (QAP)[2] problem. Rosenblatt and Kropp [25] studied the stochastic QAP formulation using an expected value objective function, which depends on the projected demand/cost scenarios assumed along with their probability of occurrence. Kouvelis and Kiran [15] studied layout using stochastic product demand, routings, and process plans. They assigned a probability to each production scenario, developed dominance conditions, and developed exact methods to solve small- to medium-sized instances of the QAP formulation. Kouvelis et al. [16] also used the QAP formulation to study layout based on robustness of the layout as defined by

---

[1] Aspect ratio is the ratio of the longer side to the shorter side.

[2] The QAP problem restricts all departments to be interchangeable, that is, of identical size and shape.

Rosenblatt and Kropp [25]. They consider alternative plant states by not only including multiple periods, but by also including different product manufacturing scenarios. Taking a different approach, Cheng et al. [5] considered material flow as a random variable by assuming it to be a fuzzy number. They demonstrated their approach using a genetic algorithm optimization routine and a slicing tree construct for unequal area block layout. Layouts were ranked by converting the fuzzy measure to a mean measure using the procedure of Lee and Li [17].

A common theme of the work cited in this section is that uncertainty is handled by identifying a few mutually exclusive production scenarios then assigning a probability mass to each or assigning a continuous distribution to demand. A second common theme is the implicit assumption of risk neutrality, that is, decisions on optimal layouts depend on expected values. Furthermore, this work has been used mostly in the QAP formulation of design.

2. Formulation Using Expected Value and Variances of Products

The basic idea of this paper is to characterize uncertainty by assuming an expected value and a variance for the forecasted amount of each product to be handled in the facility. The optimization problem, then, involves the minimization of the area under the total material handling cost curve (Figure 2) for a prespecified range of uncertainty subject to constraints on departmental shapes given a fixed total rectangular area A with fixed H and W, and fixed departmental areas, a_j.

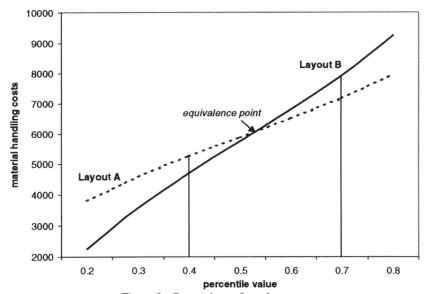

Figure 2. Comparison of two layouts.

To more formally state the objective function, there are N independent products each with an expected demand or production volume ($\bar{v}$) and a standard deviation (σ). The expected value and standard deviation are volume per unit of time (e.g., day, week, or month). Invoking the central limit theorem of sums [13], the probability distribution of the total material handling costs is normal, even when only a few products are involved. For each product, it must be known which departments will be included in the product manufacture, assembly, or handling. With this formulation, the variability of the forecasts of each product can be considered separately. An established product might have low variability of forecast while a new or future product may have high variability. The product volumes, variability, and routings along with unit material handling costs and departmental areas and shape constraints are the required information prior to the design phase.

This formulation enables a direct mathematical construction of a *robustness metric* for competing designs. The robustness measure examines the cost of the design over a user-specified range of statistical percentiles and weights the percentiles by their likelihood. This can be seen in Figure 2 where $p_L=0.4$ and $p_U=0.7$ for designs A and B. While design A performs better at higher p values and design B performs better at lower p values, it is not clear which layout is superior over the specified range of 0.4 to 0.7. Integration, however, will yield this answer directly. The smaller area under the curve will be the more cost-effective design for that range of uncertainty. For a prespecified range of p_L to p_U, the optimization problem is:

$$\min R(\Pi) = \int_{p_L}^{p_U} \left[E(\Pi) + z_p s(\Pi)\right] dp \tag{2}$$

$$\text{s.t.} \quad r_j \le R_j \ \forall j$$

where

$$E(\Pi) = \sum_{i=1}^{N} \overline{v}_i \left(\sum_{j=1}^{n} \sum_{\substack{k=1 \\ j \ne k}}^{n} \delta_{ijk} \left(\left| c_{xj} - c_{xk} \right| + \left| c_{yj} - c_{yk} \right| \right) \right)$$

$$s(\Pi) = \sqrt{ \sum_{i=1}^{N} \sigma_i^2 \left(\sum_{j=1}^{n} \sum_{\substack{k=1 \\ j \ne k}}^{n} \delta_{ijk} \left(\left| c_{xj} - c_{xk} \right| + \left| c_{yj} - c_{yk} \right| \right) \right)^2 }$$

where r_j aspect ratio of dept j

R_j maximum allowable aspect ratio of dept j

$\overline{v}_i$ expected volume for product i per unit time, where $i = \{1, 2, ... , N\}$

σ_i^2 variance of volume of product i per unit time

$\delta_{ijk} = \begin{cases} 1 \text{ if product } i \text{ is transported from dept } j \text{ to dept } k \\ 0 \text{ if product } i \text{ is not transported from dept } j \text{ to dept } k \end{cases}$

c_{xj} x-coordinate of dept j centroid

C_{yj} $\quad y$ – coordinate of dept j centroid

Z_p $\quad$ standard normal z value for percentile p

Because the relationship between z and α is defined by the cumulative Normal probability function, Φ, equation 2 cannot be calculated directly. However, it can be rearranged as shown below, so that the objective function during optimization can be calculated in a closed-form manner. Taking equation 2 and using the identity, $z_p=\Phi^{-1}(p)$, yields

$$R(\Pi) = E(\Pi)(p_U - p_L) + s(\Pi)\int_{p_L}^{p_U} \Phi^{-1}(p)dp$$

letting $p = \Phi(x)$, $x=\Phi^{-1}(p)$ and therefore:

$$R(\Pi) = E(\Pi)(p_U - p_L) + s(\Pi)\int_{\Phi^{-1}(p_L)}^{\Phi^{-1}(p_U)} x\frac{1}{\sqrt{2\pi}}e^{-\left(\frac{x^2}{2}\right)}dx$$

Carrying out the integration yields the final objective function:

$$\min R(\Pi) = \min\left(E(\Pi)(p_U - p_L) + \frac{s(\Pi)}{\sqrt{2\pi}}\left[\frac{1}{e^{\frac{(\Phi^{-1}(p_L))^2}{2}}} - \frac{1}{e^{\frac{(\Phi^{-1}(p_U))^2}{2}}}\right]\right) \quad (3)$$

where the inverse cumulative normal calculation, Φ^{-1}, is performed numerically for p_L and p_U at the onset of the optimization.

3. Optimization Methodology

While the formulation shown in the previous section could be optimized by any method, a genetic algorithm (GA) meta-heuristic was chosen. The representation in this paper uses the random keys (RK) encoding of Norman and Bean [3, 22-24]. This encoding assigns a random $U(0,1)$ variate, or random key, to each department in the layout and these random keys are sorted to determine the department sequence. Consider the thirteen department example of Figure 1. The chromosome of random keys given below, when sorted in ascending order, would create the sequence depicted in Figure 1.

A	B	C	D	E	F	G	H	I	J	K	L	M
.16	.28	.49	.93	.37	.19	.07	.24	.74	.81	.43	.55	.66

The random keys encoding eliminates the need for special-purpose crossover and mutation operators or repair mechanisms to maintain encoding integrity for permutations because crossover or mutation always results in a set of random keys which can be sorted to determine a feasible permutation. Bay divisions are encoded by adding an integer to each random key. The integer indicates the bay number for the department. Consider the chromosome presented below which would decode to the layout shown in Figure 1.

A	B	C	D	E	F	G	H	I	J	K	L	M
1.16	2.28	3.49	4.93	2.37	1.19	1.07	1.24	3.74	4.81	2.43	3.55	3.66

Parents are selected using tournament selection with a tournament size of 2, and uniform crossover is used. Children replace parents, except for the best solution in the population, which is retained, unaltered, from generation to generation. The random keys of children are resorted after crossover. Mutation operates either on the bay structure (with probability of 0.20) or on departmental sequence (with probability of 0.80). For the latter, a departmental sequence random key is randomly perturbed by ± 20%, truncating at either 0 or 1. For the former, half of the bay break mutations randomly change a department's bay integer to another existing bay integer while the other half randomly change a department's bay integer to a randomly chosen integer from 1 to n. The problem of infeasibility due to violation of the aspect ratio constraints is handled using the adaptive penalty approach of Coit et al. [7]. The penalty imposed on infeasible layouts is a function of both the infeasibility of the layout and the relative fitness of the best feasible solution $(R(\Pi)_{feas})$ and best unpenalized solution $(R(\Pi)_{all})$ yet found. The objective function used is:

$$R(\Pi)_p = R(\Pi) + (R(\Pi)_{feas} - R(\Pi)_{all})(n_i)^3 \qquad (4)$$

where n_i is the number of infeasible departments in the layout, $R(\Pi)$ is the unpenalized objective function value from equation 2 and $R(\Pi)_p$ is the penalized objective function value.

4. Computational Results

A test problem was developed based on the 10-department problem of van Camp [32]. The department areas and the facility area were taken directly from van Camp [32], a maximum aspect ratio of 5 was used for all departments and a set of products with their routings and forecasted means and standard deviations were added as given in Appendix A. Four p ranges were used: (0.40-0.60), (0.25-0.50), (0.50-0.75), (0.25-0.60). The first range is symmetric about the 50th percentile and therefore is equivalent to optimizing the expected value of material handling costs only. The second range is a somewhat pessimistic view of the forecast while the third range is a somewhat optimistic view of the forecast. The last range is more pessimistic than optimistic. From experimentation, the GA parameters selected were a population size of 50 and a crossover probability of 1.00. The probability that a solution is selected for mutation was 0.50, and if selected, the probability of changing a department's random key was 0.10. Table 1 presents results. The block layouts associated with these designs are given in Appendix B.

Table 1. Results of the van Camp problem [32] with aspect ratio = 5.

p range	R(Π)	E(Π)	s(Π)
0.40 – 0.60	1130.43	5652.17	3223.34
0.25 – 0.50	1100.37	5754.17	4165.54
0.50 – 0.75	1720.92	5876.97	2469.25
0.25 – 0.60	1643.59	5754.17	4165.54

First, it can be seen that as the p range changes, the relative contribution of expected value (mean costs) and variance (standard deviation of costs) changes. By remembering that these are the mean and standard deviation of costs over the entire facility, it is apparent why this happens. At a symmetric p range, the variance is irrelevant and the costs depend only on expected values. On the optimistic side of the forecast, the standard deviation is reduced although this may result in a slight increase in mean costs. This is because it becomes more important to place departments nearer each other that have either a high expected flow value or a high variance (or both). This serves to reduce the overall variance of the layout because distance is part of that calculation. On the pessimistic side, this logic is reversed. Departments for which flows have high variance can be placed further apart.

This can be demonstrated with a simple example. Consider there is a flow between departments i and j of mean = 100 and standard deviation = 0 and a flow between departments k and l of mean = 100 and standard deviation = 50. For $p = 0.50$, these two flows are identical and it is equally important to locate i close to j and k close to l. If $p > 0.50$, the flows between k and l will be > 100 so it is more important to locate k close to l than i close to j. This will reduce the overall variance of the layout. If $p < 0.50$, the flows between k and l will be < 100 so it is more important to locate i close to j than k close to l. This will increase the overall variance of the layout. This effect can be seen with the layouts from Appendix B. Product 1, which routes through departments 3, 5, and 10, has a low mean (10) and a very high standard deviation (100). Layouts optimized for $p < 0.5$ push departments 3 and 5 apart, while layouts optimized for $p > 0.5$ group departments 3, 5, and 10 together. The routing distances for Product 1 are 29.025, 39.225, and 19.505, working from top to bottom, which confirms the strong effect of p range on the distance that Product 1 routes through the facility.

Visual comparisons are made in Figure 3. The three distinct optimal layouts from Table 1 are graphed over the percentile range of 0.2 to 0.8. It is quite clear where each block layout dominates the others, where the equivalence points are, and the relative cost differences among the layouts for each value of p.

5. Conclusions

This paper presented a new manner to examine uncertainty in design of facilities. Uncertainty is characterized by forecasted product volume mean and standard deviation. This formulation allows uncertainty to be considered on a continuous scale without the need to provide probability mass or distribution assumptions. Furthermore, it does not depend on an implicit (and possibly, unrecognized) assumption of risk neutrality on the part of the designer. The robustness metric weights the quality of different layouts by considering the likelihood of different product volumes and the cost of producing those volumes when using the layout under consideration. While this paper considered only independent products, dependence could be handled by a straightforward extension to consider covariance.

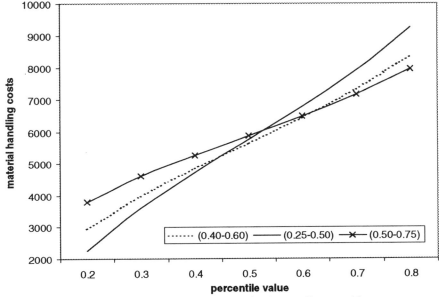

Figure 3. Comparison of three layouts for the van Camp problem.

Acknowledgment

Part of this research has been supported by the U.S. National Science Foundation grants DMI 9502134 and DMI 9908322.

References

[1] Armour, G C, Buffa, E S, 1963. A heuristic algorithm and simulation approach to relative allocation of facilities. *Management Science* 9:294-309.
[2] Bazaraa, M S, 1975. Computerized layout design: A branch and bound approach. *AIIE Transactions* 7:432-438.
[3] Bean, J C, 1994. Genetics and random keys for sequencing and optimization. *ORSA Journal on Computing* 6:154-160.
[4] Bos, J, 1993. Zoning in forest management: a quadratic assignment problem solved by simulated annealing. *Journal of Environmental Management* 37:127-145.
[5] Cheng, R, Gen, M, Tozawa, T, 1996. Genetic search for facility layout design under interflows uncertainty. *Japanese Journal of Fuzzy Theory and Systems* 8:267-281.
[6] Cohoon, J P, Paris, W D, 1987. Genetic placement. *IEEE Transactions on Computer-*

184

Aided Design 6:956-964.

[7] Coit, D W, Smith, A E, Tate, D M, 1996. Adaptive penalty methods for genetic optimization of constrained combinatorial problems. *INFORMS Journal on Computing* 8:173-182.

[8] Elshafei, A N, 1977. Hospital layout as a quadratic assignment problem. *Operational Research Quarterly* 28:167-179.

[9] Foulds, L R, Gibbons, P B, Giffin, J W, 1985. Facilities layout adjacency determination: An experimental comparison of three graph theoretic heuristics. *Operations Research* 33:1091-1106.

[10] Foulds, L R, Robinson, D F, 1976. A strategy for solving the plant layout problem. *Operational Research Quarterly* 27:845-855.

[11] Foulds, L R, Robinson, D F, 1978. Graph theoretic heuristics for the plant layout problem. *International Journal of Production Research* 16:27-37.

[12] Hassan, M M D, Hogg, G L, Smith, D R, 1986. Shape: A construction algorithm for area placement evaluation. *International Journal of Production Research* 24:1283-1295.

[13] Hogg, R V, Ledolter, J, 1992. *Applied Statistics for Engineers and Physical Scientists 2nd Edition.* Macmillan Publishing Co, New York, 155.

[14] Imam, M H, Mir, M, 1993. Automated layout of facilities of unequal areas. *Computers and Industrial Engineering* 24:355-366.

[15] Kouvelis, P, Kiran, A S, 1991. Single and multiple period layout models for automated manufacturing systems. *European Journal of Operational Research* 52:300-314.

[16] Kouvelis, P, Kurawarwala, A A, Gutierrez, G J, 1992. Algorithms for robust single and multiple period layout planning for manufacturing systems. *European Journal of Operational Research* 63:287-303.

[17] Lee, E S, Li, R L, 1988. Comparison of fuzzy numbers based on the probability measure of fuzzy events. *Operations Research* 15:887-896.

[18] Meller, R D, Bozer, Y A, 1996. A new simulated annealing algorithm for the facility layout problem. *International Journal of Production Research* 34:1675-1692.

[19] Meller, R D, Gau, K-Y, 1996. The facility layout problem: recent and emerging trends and perspectives. *Journal of Manufacturing Systems* 15:351-366.

[20] Montreuil, B, Ratliff, H D, 1989. Utilizing cut trees as design skeletons for facility layout. *IIE Transactions* 21:136-143.

[21] Montreuil, B, Venkatadri, U, Ratliff, H D, 1993. Generating a layout from a design skeleton. *IIE Transactions* 25:3-15.

[22] Norman, B A, Bean, J C, 1997. A random keys genetic algorithm for job shop scheduling. *Engineering Design and Automation* 3:145-156.

[23] Norman, B A, Bean, J C, 1999, Random keys genetic algorithm for complex scheduling problems. *Naval Research Logistics* 46:199-211.

[24] Norman, B A, Bean, J C, in print. Operation scheduling for parallel machine tools. *IIE Transactions.*

[25] Rosenblatt, M J, Kropp, D H, 1992. The single period stochastic plant layout problem. *IIE Transactions* 24:169-176.

[26] Rosenblatt, M J, Lee, H L, 1987. A robustness approach to facilities design. *International Journal of Production Research* 25:479-486.

[27] Shore, R H, Tompkins, J A, 1980. Flexible facilities design. *AIIE Transactions* 12:200-205.

[28] Tam, K Y, 1992. Genetic algorithms, function optimization, and facility layout design. *European Journal of Operational Research* 63:322-346.

[29] Tam, K Y, 1992. A simulated annealing algorithm for allocating space to manufacturing cells. *International Journal of Production Research* 30:63-87.

[30] Tate, D M, Smith, A E, 1995. Unequal-area facility layout by genetic search. *IIE*

Transactions 27:465-472.

[31] Tompkins, J A, 1997. Facilities planning: A vision for the 21st century. *IIE Solutions* August:18-19.

[32] van Camp, D J, Carter, M W, Vannelli, A, 1991. A nonlinear optimization approach for solving facility layout problems. *European Journal of Operational Research* 57:174-189.

[33] Welgama, P S, Gibson, P R, 1996. An integrated methodology for automating the determination of layout and materials handling system. *International Journal of Production Research* 34:2247-2264.

[34] Yeh, I-C, 1995. Construction-site layout using annealed neural network. *Journal of Computing in Civil Engineering* 9:201-208.

[35] Ziai, M R, Sule, D R, 1991. Computerized facility layout design. *Computers and Industrial Engineering* 21:385-389.

Appendix A - Test Problem Data

(All flows are bidirectional.)

Product 1:
Mean = 10, Standard Deviation = 100
Routing: Departments 3, 5, 10

Product 2:
Mean = 100, Standard Deviation = 10

Routing: Departments 1, 5, 8, 7
Product 3:
Mean = 50, Standard Deviation = 50
Routing: Departments 2, 9, 6, 4

Product 4:
Mean = 50, Standard Deviation = 25
Routing: Departments 2, 9, 5, 8, 7

Appendix B - Best Block Layouts

van Camp (0.40-0.60)

van Camp (0.25-0.50), (0.25-0.60)

van Camp (0.50-0.75)

A Fuzzy Clustering Evolution Strategy and its Application to Optimisation of Robot Manipulator Movement

J.C.W. Sullivan, B. Carse, and A.G. Pipe

Intelligent Autonomous Systems Engineering Lab
Faculty of Engineering, University of the West of England
Coldharbour Lane, Frenchay, Bristol BS16 1QY, United Kingdom
email : John.Sullivan@uwe.ac.uk, Brian.Carse@uwe.ac.uk,
Anthony.Pipe@uwe.ac.uk

Abstract. A new approach to constrained multi-modal function optimisation is presented based on a hybrid of the fuzzy k-means clustering algorithm and a multi-parental version of the evolution strategy paradigm. The Fuzzy Clustering Evolution Strategy (FCES) is described and experimental results are presented on a robot manipulator movement optimisation task. The task is framed as a constrained optimisation problem and Behavioural Memory constraint handling is applied.

1 Introduction

Since the earliest applications of evolutionary computation to numerical optimisation problems a recurrent theme has been the problem of reliably finding the global optimum in a multi-modal fitness landscape. Selection, acting on a finite population, will tend to cause stable convergence on a single peak and the consequent loss of genetic diversity prevents further exploration except as the result of random mutation. According to Wright's [24] *shifting balance hypothesis*, small sub-populations would range more widely over the adaptive landscape, because sampling error in small populations increases the importance of mutation and *genetic drift* relative to natural selection.

In section 2 we review previous attempts to encourage species-formation in evolutionary algorithms with particular emphasis on methods which use explicit clustering techniques. Our Fuzzy Clustering Evolution Strategy (FCES), which uses a fuzzy k-means clustering algorithm and specialised recombination operators is presented in section 4. Finally, in section 5 we demonstrate the effectiveness of this algorithm on the problem of determining the optimal joint angle trajectories for reaching movements of a redundant planar manipulator. We pose this as a constrained optimisation problem and use an adaptation of the Behavioural Memory [20] constraint handling method.

2 Species Formation and Clustering

One of the earliest Evolutionary Computation approaches to this problem, known as *crowding* has been used in *generation-gap* or *steady-state* models [5]. With crowding, each child will replace an individual which is most similar to itself (using a genotypic similarity metric) from a small random sample of size CF where CF is known as the *crowding factor*. This tends to maintain diversity in the population and can be viewed as promoting "niche" formation. Mahfoud [16] developed this idea further as the *deterministic crowding* method, having observed that stochastic errors in the replacement of population members caused loss of genetic diversity.

Further developments of the local selection approach are the Restricted Tournament Selection (RTS) technique of Harik [11] and the Adaptive Restricted Tournament Selection (ARTS) method of Roy and Parmee [18]. The basic idea of these methods is that selection is based on competition between similar (in terms of some similarity metric) individuals. In RTS, close individuals are identified from a randomly selected "window". In ARTS, on the other hand, a clustering algorithm (shared near neighbour [14]) is used in each generation to define the "neighbourhood" of each offspring. If the offspring has higher fitness than the closest individual in its neighbourhood, it replaces that individual in the population. The algorithm requires two parameters to control the "tightness" of clustering but otherwise requires no prior knowledge of the fitness landscape. In this respect ARTS was an improvement on Harik's [11] RTS algorithm in which *a priori* knowledge of the number of peaks is required in order to define the window size.

The concept of fitness sharing was introduced by Holland [13], and used by Goldberg and Richardson [10] to divide the population into different sub-populations according to similarity measures. This is achieved by reducing an individual's fitness according to the proximity of other individuals in its neighbourhood. The method requires the choice of a parameter σ_{share} which acts as a threshold of the distance metric in determining the number of population members within each niche. Deb and Goldberg [6] added a *mating restriction scheme* to the sharing GA. Mating partners were chosen by sampling the parent population for members within a prescribed distance σ_{mating}.

A major disadvantage of the fitness sharing approach is the assumption that the number of peaks is known in advance and that those peaks are uniformly distributed throughout the search space. Yin and Germay [26] introduced a GA which used a clustering algorithm in order to assign individuals to different niches together with fitness sharing in order to allow the formation of stable sub-populations or species. Clustering, using an adaptive K-means algorithm, was applied after the reproduction and crossover processes had generated a new offspring population. The fittest members were used as seed points for the initial cluster centroids. Using cluster analysis eliminates the requirement of knowing the number of peaks in the fitness landscape. However the algorithm requires two supplementary parameters : minimal distance (d_{min}) and maximal distance (d_{max}). As in [6] a restrictive mating scheme produced a considerable improvement in performance.

Hocaoğlu and Sanderson [12] used a minimal representation size cluster (MRSC) analysis to form an optimal (in terms of a representation size measure) hard partition of the population. A modified K-means clustering algorithm was used together with a multi-population GA. The resulting algorithm, MRSC_GA, was shown to identify all the optima in two multimodal test functions, and to generate multiple alternative solutions in a mobile robot path-planning application. In their algorithm, a number of sub-population GA's are run in parallel for several generations (the cluster interval). Clustering is then applied to the merged pool of individuals and the population is redistributed into sub-populations or species. The optimal number of species is determined by the MRSC criterion. As the species evolve the cluster interval increases. Some cross-species interaction is allowed by means of a cross-species recombination operator. This is essentially similar to the *island model* parallel GA except that the number of sub-populations is not fixed but varies adaptively at each cluster interval.

3 Clustering and Global Optimisation

Cluster analysis has, for some time, played an important role in pattern recognition and image processing. As basic tools for unsupervised data analysis and classification, clustering methods are important in data compression, statistics, neural networks and the natural sciences [7].

More recently, clustering algorithms have been incorporated into strategies aimed at global optimisation [22,23]. The aim in such methods is to concentrate sampled points around the local minima so that they can be recognised by a cluster analysis technique. In this way, multiple determination of local minima can be avoided and more search effort can be spent on global exploration, increasing the probability of finding the global optimum. The method proposed for grouping points around optima was to "push" each point toward a local optimum by performing a few steps of a local optimisation algorithm. In this manner it was hoped to identify *all* of the local optima and hence to find the global optimum reliably. The FCES approach takes this overall outline but uses an Evolution Strategy in place of the local hill-climbing algorithm and fuzzy clustering to identify the data partitions.

Clustering or partitioning a data set into compact, well separated subsets, is an optimisation problem in its own right and most approaches are based on algorithms which iteratively minimise some cost functional, such as the "energy" functional: $E = \sum_{i=1}^{K} \sum_{X_j \in C_i} \| X_j - V_i \|^2$ where C_i is the ith cluster with centroid or *prototype* V_i and X_j is an element of the data set. Most procedures which have been proposed to minimise cost functionals of this type are modifications or extensions of the basic ISODATA (Iterative Self-Organizing Data Analysis Technique A) of Ball and Hall [1], or its sequential relative, the K-means algorithm [15]. One disadvantage of the ISODATA and K-means algorithms is that they always result in a "hard" K-partition of the data set, even in cases where there is no such structure in the data. In such cases, invalid inferences can be drawn as a result.

3.1 Fuzzy Clustering

Fuzzy clustering extends the concept of fuzzy sets [27] to clustering, in that each data item may be fractionally assigned to multiple clusters. The central idea is the membership function which measures the degree to which objects satisfy imprecisely defined properties.

Ruspini [19] introduced the concept of a *fuzzy partition* of a data set. This is a "family" of fuzzy sets $A_1, \ldots, A_K$ on X such that

$$\forall x \in X, \sum_{i=1}^{K} \mu_{ix} = 1$$

where μ_{ix} denotes the degree of membership of x in set A_i. Ruspini stated that the advantage of using a fuzzy set representation is that stray points or points isolated between clusters could be identified.

The fuzzy K-means algorithm [3] is based on the iterative minimization of the following objective function:

$$J_q = \sum_{j=1}^{N} \sum_{i=1}^{K} (\mu_{ij})^q d^2(X_j, V_i); \qquad K \leq N$$

where X_j is the jth m-dimensional feature vector, V_i is the centroid of the ith cluster, $d^2(X_j, V_i)$ is a distance metric between X_j and V_i, N is the number of data points and K is the number of clusters.

The fuzzy membership μ_{ij} is calculated as

$$\mu_{ij} = \frac{\left(\frac{1}{d^2(X_j, V_i)} \right)^{\frac{1}{q-1}}}{\sum\limits_{i=1}^{K} \left(\frac{1}{d^2(X_j, V_i)} \right)^{\frac{1}{q-1}}}$$

The parameter q is a weighting exponent for μ_{ij} and controls the "fuzziness" of the resulting clusters. A value of 2 is often used on empirical grounds [25], and this results in considerable simplification of these expressions and reduction of computational effort.

4 Incorporating Fuzzy Clustering in the Evolution Strategy

The basis of this approach is that a clustering algorithm is used to form a partition of the parent (μ) population in a (μ, λ)-ES. The comma notation used here signifies extinctive selection, i.e. that each individual survives for only one generation, the parents for the next generation being selected deterministically from the offspring population (λ) [21].

In the absence of *a priori* knowledge of the fitness landscape some initial distribution of search points into the sub-populations is required. The fuzzy clustering approach allows the partition to be somewhat vague and ill-defined during the earlier stages of evolution, gradually becoming "harder" as the population migrates to higher fitness regions. It is important to realise that, in the FCES, selection is a global operator and the only function of the clustering algorithm is to promote localised gene pools for multi-parental recombination. Each parent's degree of membership of the cluster is used to "weight" its contribution to the genetic material of the offspring. The clustering algorithm operates on the parent population after the usual global (μ, λ) selection. Typically the parent population is less than half the size of the offspring population. The use of global selection allows a cluster (species) to become extinct or to migrate to better areas of the fitness landscape.

4.1 Recombination and Mutation Operators

Two new recombination operators are introduced here which rely on a fuzzy partition of the parent population in order to identify the underlying structure of the environmental manifold. Both operators are modifications of the multi-parent recombination operators proposed by Eiben et al. [8] and Beyer [2].

Intermediate recombination involves averaging the values at each locus on the genotype over a number of parents. Beyer [2] has shown that particularly rapid convergence is often obtained by letting the number of parents involved in this process (ρ) be as large as possible i.e. $\rho = \mu$. However such a strategy will tend to converge to a single point. In the operator proposed here the contribution from each parent is in proportion to its degree of membership of the randomly selected offspring cluster.

A *discrete* variant of this operator has also been implemented in which the value at each locus is taken by a random selection from the gene pool from a distribution which is weighted according to the degree of membership of each parent. The algorithm is essentially similar to a proportional selection scheme, except that the membership function for a randomly selected cluster is used instead of the fitness function. Particularly good results have been obtained with this recombination operator, and it was used to generate the experimental results presented in section 5.

Mutation operates on each offspring by the addition of normally distributed steps in each dimension. Step size adaptation is used [21] according to the *derandomized mutation* modification proposed by Ostermeier [17].

5 Manipulator Reaching Experiments

The experiments discussed in this section were conducted using a simulation of a three revolute joint (3R) planar manipulator and the tasks were concerned with the end-effector position (i.e. not orientation), giving one redundant degree of freedom. A sketch of a typical manipulator of this class is shown in figure 1. Also shown are the reference frames for the intrinsic $q_i, i \in [0, 2]$ and extrinsic x, y coordinate systems. The problem is formulated as a constrained optimisation problem, handled using an adaptation of the *behavioural memory* paradigm [20]. Here the constraint is that the reaching movement should terminate within a circular region in the workspace coordinates, centered on the target. The constraint function is equal to

$$c_{\text{reach}} = d^2(x_e, x_t) - \text{tol}$$

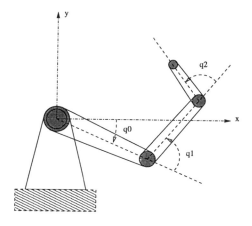

Fig. 1: 3R Planar Manipulator

where x_e is the end-effector position and x_t that of the target.

The initial optimization stage consists of using the constraint function as the cost function and consequently the minimizers constitute a set of manifolds, the *self-motion manifolds*, of dimension equal to the number of redundant degrees of freedom. In the case of a 3R planar manipulator the topology of this pre-image consists of either one or two closed rings depending on the position of the target in the workspace. This stage continues until a sufficient percentage of the population of search points is inside the feasible region or target area. This population then forms the basis for the second stage of learning in which the objective function is optimized and non-feasible points are rejected.

An important issue in optimal path planning is the form of representation or approximation used. In these experiments a B-spline representation has been used: the principal unknowns or object variables are the control points or "vertices" of the control polygons of B-spline curves representing the joint angle trajectory for each of the manipulator joints. This representation has the important advantages that each vertex exercises control only over a localised region of the function and that the curve is totally contained within the convex hull of the control points.

Fifth order B-splines with simple internal knots were used to ensure continuity of the first three derivatives i.e. C^3 continuity. The choice of B-spline order also has consequences for the efficiency of the learning algorithm since it affects the *epistasis* of the genetic coding. Davidor [4] has shown that representations with either too high or too low degree of epistasis are unsuitable for optimisation with genetic algorithms. B-splines allow a trade-off between local control and smoothness which can be adjusted to suit the demands of the task. For example, reducing the order of the B-spline basis functions reduces the support of each basis function and hence gives more local control of the joint angle trajectory. This is, however, at the cost of loss of continuity of the derivatives of the curve at the "knot" points and consequently of the smoothness of the movement. Increasing the B-spline order gives functions with wider support and therefore increased smoothness and more "linking" or epistasis between genes.

Constraints were applied to the end knots and vertices to give end-point interpolating curves with zero first, second and third derivatives at the ends. For this relatively simple task, eight B-spline functions were used for each joint, resulting in three unknowns per joint or a nine dimensional search space for the 3R manipulator. To illustrate the generation of a joint trajectory from this basis, figure 2 shows a typical joint angle curve with its associated scaled B-splines and the vertices of the control polygon which generated the curve.

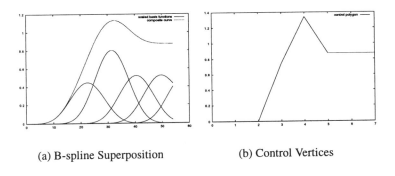

(a) B-spline Superposition   (b) Control Vertices

Fig. 2: B-spline Function Construction

5.1 Solving the Inverse Kinematics (IK) Problem

The problem formulation proposed above requires a preliminary stage (stage 0) in which the goal is to identify arm configurations which represent solutions to the IK problem at a point in the workspace corresponding to the target. There are an infinite number of possible solutions for each solution branch of the 3R planar manipulator and these lie on a 1-dimensional manifold. In the first series of experiments reported here, the target is a unit radius disc located at $x = 20, y = 20$ which is in the outer workspace annulus $\mathcal{W}_1$ so that there is only one solution branch. The objective therefore in this stage is to generate as many different arm configurations (which lie on this manifold) as possible in order to provide initial search points for the second stage of the Behavioural Memory process. The workspace of the 3R planar manipulator and the initial arm configuration used in the following experiments is shown in figure 3. Some idea of the structure of the target manifold can be gained from the scatter of the search points after convergence in the IK stage. In figure 4 the results from a run with 20 clusters in a parent population of size 200 is plotted after 80% of the population had converged inside the target area. Here the left-hand plot shows the population scatter resulting from 46 generations of evolution using the discrete fuzzy clustering recombination operator. Results from the intermediate variant of this operator are shown on the right. Note that the latter result was obtained after only 29

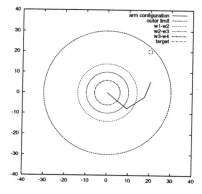

Fig. 3: Workspace of 3R Planar Manipulator

generations, but that the clusters are less evenly distributed around the manifold. The

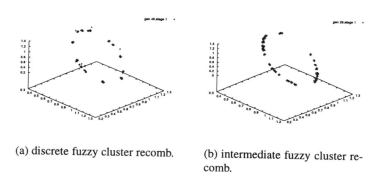

(a) discrete fuzzy cluster recomb.

(b) intermediate fuzzy cluster recomb.

Fig. 4: Population Scatter after IK Stage - Target Radius 0.1

inverse kinematics results shown above and the minimum jerk smoothing described in the following two sections are the outcome of selecting a target position in the $\mathcal{W}_1$ region (the outer annulus in figure 3). The number and topology of the pre-images of the kinematic mapping change at the boundaries of the workspace regions, these boundaries being known as *Jacobian surfaces*. A consequence of this is that, whereas the pre-image for positions in $\mathcal{W}_1$ is a single toroidal manifold, i.e. a single solution branch, there are two such manifolds for positions in $\mathcal{W}_2$. In figure 5 the results of an experiment with a target position at $x = 10.0, y = 5.0$ i.e. inside the $\mathcal{W}_2$ annulus are presented as a plot in joint space of the population after convergence of the inverse kinematics stage. Fuzzy clustering discrete recombination with 20 clusters was used in this experiment. Note that the two "S-shaped" sub-populations shown here are, in fact, separate "rings" or 1-dimensional manifolds which "wrap around" the generators for the third joint angle (q_2).

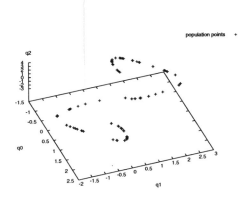

Fig. 5: Population after Convergence on Inverse Kinematics in $\mathcal{W}_2$

5.2 Experiments using an Intrinsic Minimum Jerk Cost Functional

Using joint variables as the coordinate frame for calculating the cost functional has the advantage that the forward kinematics does not need evaluating in order to compute the movement cost. The functional used in this stage (stage 1) is

$$J_{\text{intrinsic}} = \int_{t=0}^{\text{tmax}} \sum_{j=0}^{n} \frac{d^3\theta_j}{dt^3} dt.$$

Statistics from a series of runs of the FCES algorithm using this cost functional are presented in table 1.

Table 1: Minimum Jerk Reaching

pop. size		clusters			statistics of 20 runs					
μ	λ	init'l	conv'gd mean	final mean	optimum mean	s. dev	function evals mean	s.dev	generations mean	s. dev
20	100	10	2.3	2.3	3.7	1.56	5100.5	744.7	40	3.9
30	100	10	2.3	3.6	3.3	1.40	5840.3	424.1	49	2.6
40	100	10	2.1	4.9	4.2	1.73	6387.2	616.4	50	0
50	100	10	1.7	6.8	5.6	1.75	6384.8	541.0	50	0
30	100	5	1.2	2.55	3.47	1.36	6066.8	552.7	49	1.7
60	200	20	2.6	3.85	3.64	1.40	12859	1772.2	49	1.8

Apart from the numerical evaluation of the reduction of the cost functional, the results of this optimisation stage can be evaluated qualitatively according to the hypothesis

196

that application of the minimum jerk principle should result in smooth, straight paths for the end effector in the workspace. Figure 6 shows the paths traced by the "hand" before and after optimisation using the intrinsic formulation of the minimum jerk criterion. Each path drawn corresponds to the lowest cost individual in each of ten clusters. Note that fewer paths are shown after optimisation because some clusters became "extinct" during the process of evolution.

The other qualitative characteristic which has been predicted from the use of a minimum jerk cost functional is that the resultant velocity profile of the hand or end-effector should be a smooth, symmetrical bell-shaped curve [9]. As can be seen in figure 7 which shows a typical velocity profile after optimisation, these experiments confirm that hypothesis.

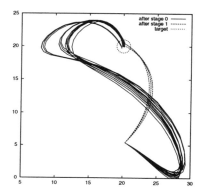

Fig. 6: Hand Paths Resulting from Optimisation using Intrinsic Min. Jerk

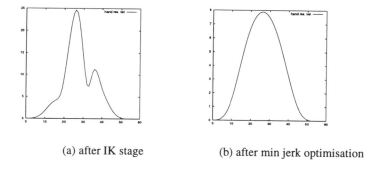

(a) after IK stage   (b) after min jerk optimisation

Fig. 7: Velocity Profiles - $(100, 300)$-ES 30 clusters

6 Conclusions

We have successfully applied the FCES method, with Behavioural Memory constraint handling, to the problem of generating optimally smooth reaching movements of a kinematically redundant robot manipulator. A constrained optimisation problem formulation was used, with the target accuracy requirement providing the single non-linear constraint. A minimum jerk cost functional was used to define the objective function. The results of these experiments showed that the FCES algorithm was able to identify a number of solutions to the inverse kinematics problem, revealing the topological structure of the self-motion manifold. In conjunction with a B-spline representation, the second stage of the behavioural memory procedure generated smooth goal-directed arm movements which exhibited many of the spatio-temporal characteristics of human arm movements.

References

1. G.H. Ball and D.J. Hall. A clustering technique for summarizing multivariate data. *Behavioural Science*, 12:153–155, 1967.
2. H.-G. Beyer. Towards a theory of evolution strategies : On the benefits of sex – the $(\mu/\mu, \lambda)$ theory. *Evolutionary Computation*, 3(1):81–111, 1995.
3. J.C. Bezdek. *Fuzzy Mathematics in Pattern Classification*. PhD thesis, Cornell University, 1973.
4. Y. Davidor. *Genetic Algorithms and Robotics, A Heuristic Strategy for Optimization*. World Scientific, Singapore, 1990.
5. K.A. De Jong. *An analysis of the behavior of a class of genetic adaptive systems*. PhD thesis, University of Michigan, 1975.
6. K. Deb and D.E. Goldberg. An investigation of niche and species formation in genetic function optimization. In J.D. Schaffer (ed), *Proc. 3rd Int. Conf. Genetic Algorithms*, pp 42–50, San Mateo, CA, 1989. Morgan Kaufmann.
7. R.O. Duda and P.E. Hart. *Pattern Classification and Scene Analysis*. Wiley-Interscience, New York, 1973.
8. A.E. Eiben, P-E. Raué, and Zs. Ruttkay. Genetic algorithms with multi-parent recombination. In Y. Davidor, H.-P. Schwefel, and R. Männer (eds), *Parallel Problem Solving from Nature – PPSN III*, pp 78–87, 1994. Springer, Berlin.
9. T. Flash and N. Hogan. The coordination of arm movements: An experimentally confirmed mathematical model. *J. Neuroscience*, 5(7):1688–1703, 1985.
10. D.E. Goldberg and J.J. Richardson. Genetic algorithms with sharing for multimodal function optimisation. In J. J. Grefenstette (ed), *Genetic Algorithms and their applications: Proc. 2nd Int. Conf. Genetic Algorithms*, pp 41–49, Hillsdale, New Jersey, 28-31 July 1987. Lawrence Erlbaum Associates.
11. G. Harik. Finding multi-modal solutions using Restricted Tournament Selection. In *Proc. 6th Int. Conf. Genetic Algorithms*, Pittsburgh, 1995.
12. C. Hocaoğlu and A.C. Sanderson. Multimodal function optimization using minimal representation size clustering and its applications to planning multipaths. *Evolutionary Computation*, 5(1):81–104, 1997.
13. J.H. Holland. *Adaptation in Natural and Artificial Systems*. University of Michigan, Ann Arbor, 1975.

14. R.A. Jarvis and E.A. Patrick. Clustering using a similarity measure based on shared near neighbours. *IEEE Trans. Computers*, 22(11), 1973.

15. J. MacQueen. Some methods for classification and analysis of multivariate observations. In *Proc. 5th Berkeley Symposium Math. Stat. Prob*, volume 1, pp 281–297, 1967.

16. S.W. Mahfoud. Crowding and preselection revisited. In R. Männer and B. Manderick (eds), *Parallel Problem Solving from Nature, 2nd Workshop, PPSN2*, pp 27–36. Elsevier, 1992.

17. A. Ostermeier, A. Gawelczyk, and N. Hansen. A derandomized approach to self-adaptation of evolution strategies. *Evolutionary Computation*, 2(4):369–380, 1995.

18. R. Roy and I.C. Parmee. Adaptive restricted tournament selection for the identification of multiple sub-optima in a multi-modal function. In *Proc. of the AISB Workshop on Evolutionary Computing, LNCS 1143*, pp 236–256, Brighton, 1996. Springer Verlag.

19. E. H. Ruspini. A new approach to clustering. *Inf. Control*, 15:22–32, 1969.

20. M. Schoenauer and S. Xanthakis. Constrained GA Optimization. In S. Forrest (ed), *Proc. 5th Int. Conf. Genetic Algorithms*, pp 573–580, San Mateo, CA, 1993. Morgan Kaufmann.

21. H-P. Schwefel. *Numerical Optimization of Computer Models*. Wiley, 1981.

22. A. Törn. Clustering methods in global optimization. In *Preprints of the 2nd IFAC Symposium on Stochastic Control*, pp 138–143, Vilnius, 1986.

23. A.A. Törn and A. Zilinskas. *Global Optimization*. Number 350 in Lecture Notes in Computer Science. Springer-Verlag, Berlin, 1989.

24. S. Wright. Evolution in Mendelian Populations. *Genetics*, 16:97–159, 1931.

25. X.L. Xie and G. Beni. A Validity Measure for Fuzzy Clustering. *IEEE Trans. Pattern Anal. Machine Intell.*, 13(8):841–847, 1991.

26. X. Yin and N. Germay. A fast genetic algorithm with sharing scheme using cluster analysis methods in multimodal function optimization. In *Proc. Int. Conf. Artificial Neural Networks and Genetic Algorithms*, pp 450–457, Berlin, 1993. Springer-Verlag.

27. L.A. Zadeh. Fuzzy sets. *Inf. Control*, 8:421–427, 1965.

An Industry-based Development of the Learning Classifier System Technique

WILLIAM N L BROWNE (corresponding author)
Department of Engineering, University of Leicester, Leicester, LE1 7RH, wnlb1@leicester.ac.uk

KAREN M HOLFORD
Division of Mechanical Engineering, Cardiff University, Queen's Building, PO Box 685, Cardiff, CF2 3TA, holford@cf.ac.uk

CAROLYNNE J MOORE
Division of Civil Engineering, Cardiff University, Queen's Building, PO Box 917, Cardiff, CF2 1XH, moorecj@cf.ac.uk

This paper describes the development of an Industrial Learning Classifier System for application in the steel industry. The real domain problem was the prediction and diagnosis of product quality issues in a Steel Hot Strip Mill. The properties of the data from this environment include multimodality (several optima), poor separation between fault levels and high dimensionality (many parameters). The method to develop the Learning Classifier System technique, based on deterministic simulated data, is presented. The advances made in the technique, which enhance its functionality in this type of industrial environment, are given. The novel methods developed are core to the Learning Classifier System technique and are not 'fixes' for given problems. They address the fitness measure, encoding alphabet, population scope, phases of training, genetic operators, life limits and removal of taxation schemes. These improvements allow the industrial LCS to function correctly in the simulated domain. Encouraging results from diagnosis of real data are presented; however, further work is needed for greater accuracy and to allow the prediction function to be used on-line. Learning Classifier Systems represent a potentially useful tool that combines the transparency of symbolic approaches (such as Decision Trees) with the learning ability of connectionist approaches (such as Artificial Neural Networks) to machine learning.

1.0 Introduction

The problem domain can be divided into the tasks of diagnosis (gaining knowledge) and prediction (determining future actions/events), for this application of knowledge learning in a Steel Strip Mill [1]. Following on from good knowledge, good predictions can be made [2]. The resource in this case is a large volume of past data containing the quality of steel strip as output and plant condition at time of rolling as input. This domain is not completely described, nor is the quality of the data known.

The lack of accurate mathematical models, voluminous data and unknown data quality resulted in machine learning techniques being considered. The results had to be transparent, so that knowledge could be gained and prediction accepted by engineers, operators and managers. Learning Classifier Systems (LCSs) were chosen as they met the above criteria when applied to theoretical domains. However, they require further development for industrial domains and so represent a fertile area of study.

The LCS concept, of a learning system in which a set of condition-action rules compete and co-operate for system control, was developed originally by Holland [3]. The rules plus the associated statistics, such as fitness measure, specificity and number of evaluations, are termed 'classifiers'. A classifier within the population will gain credit based on the system's performance in some environment when it is active. The cumulative credit is used to determine the probability of the classifier winning such system control competitions. If a LCS performs well under the control of some rule, that rule has an improved chance of winning further competitions for system control and therefore improving the likelihood of the overall system performing well. This 'fitness' also determines the influence of a particular classifier in an evolutionary process of rule discovery that seeks to improve the system's overall performance, replacing weak rules with plausibly better variations.

In an industry where the level of fault is an important classification, hierarchies that separate the severity of a fault are needed. Therefore, the LCS was initially tested on hierarchy formation in a single time step. However, this work does still benefit the rule chain functionality of the LCS as the methods of credit assignment and rule discovery [4] have to function correctly for both tasks.

The state of LCS technique prior to the start of the project was best summarised by Goldberg [5]: "LCS are a quagmire - a glorious, wondrous and inventing quagmire, but a quagmire nonetheless". The quagmire exists because of the development of computers and the consequent potential for machine learning paradigms - not because the LCS technique has fundamental flaws. Two decades of research have had many positive developments, but the current state showed that: "There's much to do in CFS [LCS] research!" [6].

The important characteristics of the domain are likely to be applicable to many other industrial domains. The type of data used to develop and test the LCS is described. The methodology of using deterministic simulated data in conjunction with real data is then explained. Following this the strategy for improving the LCS technique in a robust manner is presented. The multiple advances made in the LCS technique are described. These are supported by experimental results for both simulated and real data. Future work is suggested and then this work is concluded.

2.0 Domain

Understanding the characteristics of the data is critical in specifying the functionality that the LCS must provide. Data about the data (termed metadata), e.g. the number of fields (parameters), the number of cases (examples), can give useful insight. This can be analysed to determine the characteristics such as dimensionality, multimodality and separation. The methods within a LCS can be built based on this required functionality, as described below.

Dimensionality is the number of parameters considered. This is traditionally low in Learning Classifier System research, Smith and Dike [7] use eight for fighter manoeuvres simulation. High dimensionality, as likely in industrial domains, requires the LCS to be able to neglect superfluous parameters and quickly determine useful building blocks of information.

Multimodality relates to the number of optima in a domain above a threshold of interest. In this domain it is a function of the number of parameters that combine to cause a quality problem that requires investigation. The greater the parameter interaction, the harder the problem. The rule discovery is required to form links between parameters as well as keep useful parts of this information long enough to allow the links to form. This domain is expected to have high multimodality, due to its complex nature.

Lack of separation involves multiple quality levels that are close together in the search space. Consideration of the parameters grouped in areas related by properties shows that the levels of the same faults are likely to be connected and therefore not easily separated. The ability to form hierarchies that are stable, easy to adjust and well ordered is essential. Credit assignment and rule discovery are important, especially in the ability to develop all levels of the fault structure together.

Noise is a big problem in industrial data, especially as it is hard to measure and quantify, which leads to uncertainty in the data. Correct classification systems can produce incorrect results if sufficient noise is present to obscure the boundaries of a parameter's active range. Unfortunately, it is difficult to determine if incorrect results are due to an incorrect classification system or due to noise, if the effect of noise is unknown. Thus, to be able to develop a correct classifying system, the effect of uncertain noise must be countered. This does not mean that a LCS can ignore the effect of noise as it will be present in the real data. The functionality it must display to overcome noise includes clear rule boundaries, the ability to maintain small niches (noise has a greater effect at rule boundaries, which must be maintained in small niches) and the ability to clearly separate the levels of a hierarchy.

2.1 Description of Data

To achieve deterministic results that avoid the effects of noise in real data, simulated data was used. This also negates the need for a train and test set, as the data would be equivalent, speeding up the development process. Further assurance is that diagnosis can be used to verify the results with simulated data.

2.1.1 Simulated Data

The most commonly used test data is shown as examples of data in Table 1 and in hierarchy form in Table 2. In order to streamline the process of developing the LCS, only moderate dimensionality was used as high dimensionality needs large amounts of computation time to process the information. The data set includes a parameter (2) with less range than the maximum possible, in order to discover whether the LCS will extrapolate or generalise into the unseen values. Redundant parameters (3, 5, 7 and 8) test the appropriate use of wildcard and appropriate specificity levels. Obscure fault levels (level {3} and parameter 6) test the search abilities. The small niches also test the ability of credit assignment to maintain the niche once discovered; which is critical in noisy environments. The high modality generated in fault level three required the interaction of four parameters to be correctly identified. It also ensured that more than one correct hierarchy could be created. Integer, real and histogram parameters are included to prove the capability of the encoding [8].

Table 1 Simulated data examples.

Input								Output
130	19	4	233	1150	932	565	2	2
69	64	4	229	1366	852	720	1	0
105	2281	2	249	1210	970	598	1	1
103	334	2	206	1072	924	740	3	2
67	456	2	203	1252	824	701	1	0
49	579	3	226	1228	974	716	2	0
89	1731	2	200	1288	816	662	1	3
118	1254	3	242	1270	892	592	2	1

Table 2 Simulated data in an ideal homomorphic hierarchy.

Rank	Parameter - Upper Limit, Lower Limit. Resolution independent.								Output	Specificity
Type	Real	Int.	Real	Real	Hist.	Hist.	Real	Int		
No.	1	2	3	4	5	6	7	8		
1	99.9	2000+	5.5	239.9	1600.5	1000.5	750.5	3.0	[0]	7
	0	0	1.5	200.0	1000.0	800.0	550.0	1.0		
1	150.5	2000+	5.5	250.5	1600.5	1000.5	750.5	3.0	[1]	8
	0	0	1.5	240.0	1000.0	800.0	550.0	1.0		
1	150.5	2000+	5.5	239.9	1600.5	1000.5	750.5	3.0	[2]	12
	100.0	0	1.5	200.0	1000.0	800.0	550.0	1.0		
1	99.9	2000+	5.5	239.9	1600.5	1000.5	750.5	3.0	[3]	22
	0	1600	1.5	200.0	1000.0	825.0	550.0	1.0		

2.1.2 Real Data

The volume of data in terms of the number of parameters, number of instances and resolution of parameter needs to be considered in pre-processing. A choice was made about which parameters to input into the LCS, because this is not a closed or bounded problem. Over 200 properties were measured (e.g. power monitor data, guide set-up and actuator settings), with at least another 50 being considered possible to measure with varying degrees of ease (e.g. ambient temperature or strip curve). Pinching of strip is a fault where thin strip is folded over and rolled down flat causing flatness problems. This is measured after the coiling of the strip, resulting in 64 different parameters being considered from domain knowledge. Over 4000 data points were included in the initial data set, but this was reduced to 2000 to balance the base rates of faults. The base rate of the least frequent fault became 10% of the data, which is equivalent to the smallest base rate in the simulated data. Additional data effects, such as contradiction and lack of completeness, caused by removing important parameters could be introduced by incorrectly applying domain knowledge.

3.0 Development Method

The method by which the LCS technique will be developed for the industrial domain is based on the data described above. The simulated data can determine the merits of the LCS on the characteristics of this data (see the 'no free lunch' theorem [9] on applying different search techniques to multiple domains). The real data can then be used to ascertain how well the LCS performs in an actual problem domain.

To overcome a difficult characteristic, such as the need for exception rules, a specific fix could be applied, such as the 'specify' operator [10]. Unfortunately, with so many characteristics in real domains, the number of fixes would overwhelm the technique. Progress in developing the LCS technique has been made by simplifying the LCS structure [11]. Therefore, the philosophy adopted is to develop the core methods, whilst removing existing fixes (such as taxation, see section 4.7).

The starting point was a standard single population 'Michigan' LCS, similar to Goldberg's SCS [12]. The Michigan approach [13] uses single rules as an individual in the population that represents the rule set, whereas the Pittsburgh approach [14] uses the entire rule set as an individual within a population of rule sets. The Michigan approach was selected as it is more flexible, so could autonomously adapt to the required rule structures, which were unknown prior to training in the real domain.

This system used a strength update relying on specificity, life tax, bid tax and profit sharing of rewards. The domain is stimulus response, so the 'bucket brigade' or other such chaining credit transferral mechanisms were not used. The rule discovery mechanism operated panmitically (throughout the population) relying on the standard genetic operators of mutation and crossover. De Jong crowding was used to introduce a generalisation pressure [15]. The crowding factors and strength update parameters were tested with many combinations of values, in an attempt to find the optimum LCS set-up. Much pre-processing allowed the ternary alphabet (1, 0, # [wild card]) to be used for the real domain data.

4.0 Advances in LCS Technique

Initial testing (see section 5.0 and Figure 1) showed the system performance was very poor. The following advances were suggested to improve performance based on these tests: Real Alphabet, Phases of Training, Fitness Measure, Rule Discovery, The Morphing Genetic Operators, Life Limits and Taxation Removal.

The LCS developed is considered to be a balancing act as each LCS component must successfully interact with all others to obtain the overall learning goal. Although each component will be discussed separately, it is the continuum that is important. Changing one feature affects all others. Similarly, using one of the methods discussed in isolation is unlikely to have the desired effect.

4.1 Real Alphabet

The ternary alphabet has potential problems with aliasing and poor rule boundary representation due to the bit encoding representing schemata [16] and often more than one bit string being required for a single real parameter range [17].

Real-valued encoding has the understandable nature of integer and symbolic representations, but adjustment was needed to avoid the boundary and aliasing problems. This work uses segments (i.e., histogram boxes) to partition a continuous range, but importantly does not constrain them to occur at set intervals or set size. Therefore, a rule boundary could occur anywhere in the parameter range, thus avoiding boundary problems. The segment accuracy is termed *resolution*, with any number of resolutions from the start boundary to the end boundary of the active part of the condition. When a stable level of training was reached the resolution was tuned to determine whether or not aliasing had been a problem. An equivalent to the 'wildcard' (do not care) symbol was needed to represent superfluous information. This is achieved by setting the lower and upper values of a condition equal to the lower and upper limits that the environmental value may take.

4.2 Phases of Training

During this project, study of the operation of an industrial LCS has led to the identification of three important phases of training. They are considered to be of major importance to the operation of a LCS, but have not been explicitly utilised in previous work. The three training phases are:

1.　　Searching.
2.　　Combining.
3.　　Stabilising.

The initial population consists of random strings that contain potentially useful pieces of information, mixed equally with incorrect information. The *searching* phase concentrates on identifying the useful pieces of information and removing rules that have no merit. The resulting population contains rules that are partly correct, but require *combining* to form completely correct rules. Incorrect information should be removed from strings and unimportant conditions transformed to wildcards. The LCS now has good genetic material and potentially some complete rules. In order for correct predictions to be made, *stabilisation* of information into knowledge structures is required. Default hierarchies and rule chains need to be created and maintained when appropriate.

Attempting all three stages in one population is very difficult as the LCS set-up is different for each phase (It should be noted that most LCSs attempt all phases in a single population). A population assigned to each phrase is used with a stepping stone system [18] to transfer 'fit' rules in the sequential order. More than one rule (including differing actions) is transferred to attempt to keep formed hierarchies together.

4.3 Fitness Measure

There are no pressures for a hierarchy of rules to form instead of a homomorphic rule system without an additional mechanism, such as 'specificity in bid' or 'niched rule discovery'. However, these mechanisms have required fixes such as 'bid tax' and 'specify' operators. A mechanism was needed that utilised the effects of a general rule instead of a direct measure of generality. The mechanism had to treat each layer of the hierarchy equally if the Rule Discovery (RD) was not to remove important parts of the hierarchy by biased reproduction. It also had to differentiate the levels of the hierarchy so that the auction could select exception rules in preference to default rules. These tasks are a contradiction for a single fitness measure, such as strength. The success of past systems that were based on a single fitness is attributed to the simple nature of hierarchies allowing correct RD in relatively large populations ensuring the survival of the weak niches.

Multiple fitness measures have been studied [19][20], to determine if the auction and the rule discovery could be controlled by two separate parameters. None of the concepts were thought ideal for an industrial LCS so this work proposes to split fitness to separately govern the *preference* for selection in the auction and the likely *fertility* in the RD. The preference assigns values to classifiers that correctly identify their place in hierarchies and chains via a Woodrow-Hoff update [21]. The fertility measure is independent of preference. It is a function of the accuracy of a rule and a 'lust for life' that creates a generalisation pressure without using specificity or a set based RD. 'Lust for life' is a measure of how active and appropriate a classifier has been to the control auctions that have taken place since the last rule discovery period.

4.4 Rule Discovery

The basis of the RD mechanism in the industrial LCS can be either in the Match Set [M], the Action Set [A] or Panmitically (throughout the population). The industrial LCSs must avoid lethal rules, formed by mating two stable classifiers from different niches, that corrupt the population. Therefore, a panmitic-based RD that can select all niches an equal amount during training was chosen. The only adjustment necessary is to restrict mating in stable populations in order to prevent lethals.

4.5 The Morphing Genetic Operators

The novel *Morphing* genetic operator combines the similar information in two classifier strings. The main difference is that crossover randomly explores the domain, whereas morphing exploits known information. The hypothesis is that fit rules must at least partially match the domain. A crude guess is that the parents represent the same niche, but are either both too general, both too specific or one of each. Morphlow tests this by using low subtlety to adjust the specifically of child classifiers, whilst Morphigh uses higher subtlety in an attempt to combine the best of the two parents. For example, in the ternary alphabet, the Morphlow child keeps any identical information between parents and transforms dissimilar information is based onto wildcards. [where '~' is Morph operator]

e.g., $1 \sim 1 = 1$, $0 \sim 0 = 0$, $\# \sim 1 = \#$, $\# \sim 0 = \#$, $1 \sim \# = \#$, $0 \sim \# = \#$ and $1 \sim 0 = \#$.

4.6 Life Limits

The balancing act associated with LCSs is important when considering the life cycle of classifiers. The concept of an evaluation limit (termed 'evalimit') is used to prevent a young classifier controlling the destiny of other classifiers or the system itself. Therefore, a classifier is initially evaluated to find its correct place in the hierarchy of classifiers, whilst other classifiers' properties are kept constant. Following the example of protecting a classifier from deletion before a certain number of evaluations [22], the evalimit is initially set at 20 evaluations, but increased for the stable population.

A child limit (termed 'childlimit') is used to prevent an 'old' classifier from populating its genes to the detriment of other useful genes. The measure of whether a classifier is 'old' is the number of children it has had as this directly corresponds to the likelihood of it spreading its genes. The childlimit is set at 20 offspring per individual and does not vary across the training phases.

4.7 Taxation

The disadvantages of taxation have been listed as the removal of useful classifiers that match infrequent messages especially in changing environments, the removal of chain starting classifiers due to their low strength and the problem of setting the taxation level prior to training in an unfamiliar domain [23]. Therefore, removing taxation in an industrial LCS is advantageous if the benefits could be achieved by other mechanisms.

5.0 Experimental Results

The performance of the basic LCSs on simulated data (hence, the optimum of 1 could be achieved) stays around the 40% level with occasional fluctuations due to the stochastic nature of the process, see Figure 1. Analysis of the rule base shows that the LCS became too greedy as only the largest reward niche was present. Rules representing alternative actions were forced out of the rule base as the rule discovery mechanism deleted them in preference to apparently stronger rules. Adjustments to internal parameters, such as bid tax, could not improve the situation.

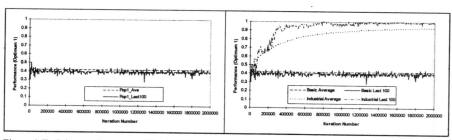

Figure 1 Training performance of Basic LCS in complex simulated domain.

Figure 2 Training performance of Industrial LCS compared with Basic LCS.

Significant improvement in performance is obtained when the advanced mechanisms are used and taxation is removed, see Figure 2. This improvement is a balance of all techniques, if one were removed performance would significantly degrade. Figure 3 indicates that the training with the industrial LCS showed some excellent characteristics as the specificity of the rules increased when *searching* for accurate building blocks, stabilised in *combine* and reduced *stabilising* rule structures. This automatic adjustment meant that rules were becoming more specific and general without additional mechanisms. The accuracy also increased in a desired manner until the optimum of one was reached by most rules; see Figure 4. Only a few rules are required to describe the stable domain; therefore, the majority of the rules were free to search the domain after approximately 750,000 iterations.

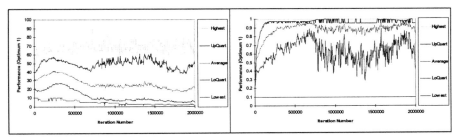

Figure 3 Trend of rule base specificity through training.   Figure 4 Trend of rule base accuracy through training.

Figure 5 shows the desired training effect of three phases in a correctly functioning LCS. In the initial 20,000 iterations the search phases quickly identified good rules, whilst the stable phase contained no good classifiers until they were transferred from other population. By 200,000 iterations all three populations were performing around 70% prediction. After 500,000 iterations the order of stable, combine and search phase had established itself and except for a few periods remained.

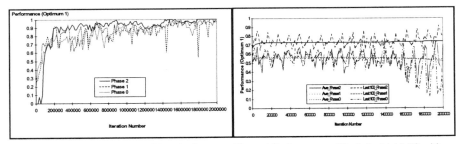

Figure 5 Training performance of functioning LCS showing effect of phases.   Figure 6 Performance of the industrial LCS with complex real data.

Inspecting the rule base showed that correct rules and hierarchies had been formed, shown in Table 3. Instead of a homomorphic hierarchy, where all rules are of equal rank, a default/exception hierarchy was formed. Rules of rank 3 (exception) and of rank 5 (default) interact to allow output 3 (nominally a fault state) to be selected ahead of the normal condition, where appropriate. Often hierarchies of smaller impact, involving additional rules, were formed to ensure crisp rule boundaries, but these gradually disappeared as training improved the precision of the main rules. This was assisted by resolution tuning that reduced the resolution, allowing the genetic operators to make more precise adjustments.

Table 3 Results from simulated data training.

Rank	Parameter - Upper Limit, Lower Limit and Resolution								Output	Specificity
Type	Real	Int.	Real	Real	Hist.	Hist.	Real	Int		
No.	1	2	3	4	5	6	7	8		
1	101.7	1601.4	5.5	236.5	1600.5	1000.5	750.5	3.5	[0]	27
	0	0	1.5	200.0	1000.0	800.0	550.2	0.5		
	10	96	1	5.1	60	20	20	0.8		
4	150.5	2252.8	5.5	250.5	1600.5	980.5	750.5	3.5	[1]	16
	0	0	1.5	240.4	1012.0	800.0	550.0	0.5		
	10	96	1	5.1	60	20	20	0.8		
2	150.5	2120.0	5.5	244.4	1600.5	1000.5	750.5	3.5	[2]	24
	101.5	0	1.5	200.0	1000.1	800.0	550.0	0.5		
	10	96	1	5.1	60	20	20	0.8		
3	101.7	2141.0	5.5	250.5	1600.5	1000.5	750.5	3.5	[3]	20
	0	0	1.5	200.0	1000.1	825.0	550.0	1.3		
	10	96	1	5.1	60	20	20	0.8		
5	100.7	2141.0	5.5	240.5	1600.5	1000.5	750.5	3.5	[0]	22
	0	1600.0	1.5	200.0	1000.1	800	550.0	1.3		
	10	96	1	5.1	60	20	20	0.8		

Although the initial rule base was seeded with all parameter 2 conditions having an upper limit of less than 2000, the LCS generalised from the input supplied to it. Parameters 3, 5, 7 and 8 were all correctly identified as redundant and so could be removed from future engineering analysis of this problem. The search and niche maintenance functions were capable of learning that parameter 6 was critical to output 3 (the most difficult 'fault' to analyse).

The only slight concern was the 'gain reward - make mistake - lose reward - protection - ' cycle which although infrequent was noticeable. However, the chances of real data containing exactly the same accuracy in rules, thus obtaining the same predictive level in a hierarchy and training stopping at the exact moment of equilibrium are thought to be so small as to be insignificant.

The shape of the performance graphs for the industrial LCS with industrial data show that learning of useful information did occur (shown in figure 6 for Ultimate Strength Data). The average performance for the stable population improved from a 'greedy' performance level of 48% to over 75% in the latter stages of training. Performance on all industrial data sets tested was better than a 'random' (pick any output) or 'greedy' (pick the most common output) tactic.

Although the predictive performance was not accurate enough to be useful to plant, it showed that some training had occurred. Decoding the results showed that useful information had indeed been learned. Common knowledge, such as thin gauges are more likely to pinch, had been found. The industrial LCS also discovered uncommon knowledge, in so much as the information was not widely available. A good example was that high crown strip (where the centre of the strip is thicker than the edges) was unlikely to pinch. This was considered accurate as the crown has a centering effect when passed through rolls, similar to a pulley and belt effect. Although experienced operators knew this effect, the LCS could provide a means of discussion and harmonisation of plant knowledge throughout the workforce.

Despite learning useful information, the LCS did not obtain satisfactory performance. Regular peaks and troughs of performance indicate that the LCS was continually searching for better information, which included rejecting previously useful information.

6.0 Future Work

Decoding of the training results is difficult, due to volume, as including all 64 parameters in 100 rules would occupy 22 pages in this font. This task could be made easier by automatically removing conditions that are wild. Similarly, removing and reporting conditions that are the same for all output states would reduce investigation time.

7.0 Conclusions

The advances made for an industrial LCS are:
- Real Numbered Alphabet: Has improved boundary definition and is easier to decode.
- Multiple Populations: Allow each population to concentrate on a specific phase improving effectiveness.
- a the to to the is a is to a the with the than the than in theNovel Fitness Measure: Improved auction and rule discovery selection, by separation of tasks.
- Enhanced Rule Discovery: Increased effectiveness in searching and exploiting sparse domains.
- Life Limits: Allow graceful refinement of hierarchies and prevent stagnation of rule base.
- Removal of Taxation: 'Fixes' no longer needed as advances made overcome many problems within LCS technique.

The industrialised LCS technique can discover useful rules in real data and these rules are often non-obvious. Currently, the industrial LCS can direct areas of study for engineers attempting to solve quality problems. These benefits are due to the combination of co-operation, transparency and discovery in the rule base, but the technique is still hindered by the associated complexity. The technique requires more development to reach optimum performance, especially in the prediction task. This work supports the argument that Learning Classifier Systems are the correct midpoint between symbolic and connectionist approaches to intelligence [24][25]. The industrial LCS developed by this project offers a route through the quagmire caused by the complexity of this machine learning approach.

8.0 References

1. Williams L. R., 1983. *Hot Rolling of Steel.*, Marchel Dekker Inc., New York.

2. Weiss S. M. and Indurkhya N., 1998, *Predictive Data Mining - A Practical Guide*, Morgan Kaufmann Pub. Inc., San Francisco, California.

3. Holland J. H., Ed. 1., 1975 & Ed. 2., 1992, *Adaptation in Natural and Artificial Systems*, University of Michigan Press, Ann Arbor, MI.

4. Bäck T., Fogel D. B. and Michalewicz Z., 1997, *Handbook of Evolutionary Computation*, Oxford University Press, ppA1.3: 1-A1.3: 2.

5. Goldberg D. E., Horn J. and Deb K., 1992, What makes a problem hard for a classifier system? is available from *ENCORE* in file CFS/papers/lcs92-2.ps.gz.

6. Heitkötter J. and Beasley D., eds., 1999, The Hitch-Hikers Guide to Evolutionary Computation: a List of Frequently Asked Questions (FAQ), *USENET*: comp.ai.genetic. Available via anonymous FTP from ftp://rtfm.mit.edu/pub/usenet/news.answers/ai-faq/genetic/ About 110 pages.

7. Smith R. E. and Dike B. A., 1995, Learning Novel Fighter Combat Manoeuvre Rules Via Genetic Algorithms, *Int. Journal of Expert Systems*, JAI Press Inc., Vol. 8 (3), pp 247 - 276.

8. Browne W. N. L., 1999, The Development of an Industrial Learning Classifier System for Application to a Steel Hot Strip Mill, Doctoral Thesis, University of Wales, Cardiff.

9. Wolpert D. H. and Macready W. G., 1996, No Free Lunch Theorems for Search, Technical Report SFI-TR-95_02_010 Santa Fe Institute.

10. Lanzi P., 1997, A Study of the Generalization Capabilities of XCS, Proc. *7th Int. Conf. on Genetic Algorithms*, Morgan Kaufmann, USA, pp 418 - 425.

11. Wilson S. W., 1999, Get Real! XCS with Continuous-Valued Inputs, In: *Festschrift in Honor of John. H. Holland*, Eds., L. Booker, S. Forrest, M. Mitchell and R. Riolo, Centre for the Study of Complex Systems, The University of Michigan, Ann Arbor, MI, Available at http://world.std.com/~sw/pubs.html, May 15-18.

12. Goldberg D. E., 1989, *Genetic Algorithms in Search, Optimization and Machine Learning.*, Addison Wesley USA.

13. Holland J. H. and Reitman J. S., 1978, Cognitive Systems Based on Adaptive Algorithms, in Waterman, D. A. and Hayes-Roth, F. (Eds.). *Pattern-Directed Inference Systems.*, Academic Press, pp 313 - 329.

14. Smith S. F., 1980, A Learning System Based on Genetic Adaptive Algorithms, Ph.D. Dissertation, University of Pittsburgh.

15. De Jong K., 1987, On Using Genetic Algorithms to Search Program Spaces Proc. *2nd Int. Conf. on Genetic Algorithms*, Morgan Kaufmann, USA, pp210 - 216.

16. Goldberg D. E., 1990, The Theory of Virtual Alphabets., *Parallel Problem Solving from Nature 1*, Eds. Schwefel H. P. and Maenner R., Springer-Verlag, Berlin, pp13 - 22.

17. Browne W. N. L., Moore C. J., Holford K. M. and Bullock J. D., 1999, An Industrial Learning Classifier System: The Importance of Pre-Processing Real Data and Choice of Alphabet, *Engineering Applications of Artificial Intelligence*, Ed. R. Vingerhoeds. Submitted for publication, revised manuscript.

18. Whitley L. D., 1993, An Executable Model of a Simple Genetic Algorithm, *Foundations of Genetic Algorithms (FOGA) 2*, Morgan Kaufmann, USA, pp45 - 62.

19. Booker L. B., 1982, Intelligent Behaviour as an Adaptation to the Task Environment., Ph.D. Dissertation, University of Michigan Press, Ann Arbor, MI.

20. Fairley A. and Yates D. F., 1994, Inductive operators and rule repair in a Hybrid Genetic Learning System: Some Initial Results., *Evolutionary Computation, Lecture Notes in Computer Science*, AISB Workshop, Leeds 94, Springer-Verlag, pp166 - 179.

21. Wilson S. W., 1995, Classifier Fitness Based on Accuracy, *Evolutionary Computation*, Vol. 3 (2), pp149- 175.

22. Wilson S. W., 1996, Generalization in XCS, ICML '96, *Workshop on Evolutionary Computing and Machine Learning*, Available at http://netq.rowland.org.

23. Liepins G. E., Hilliard, M. R., Palmer M. and Rangarajan G., 1991, Credit Assignment and Discovery in Classifier Systems., *Int. Journal of Intelligent Systems*, John Wiley & Sons Inc., Vol. 6, pp 55 - 69.

24. Michalewicz Z., 1996, *Genetic Algorithms + Data Structures = Evolution Programs.* 3rd Edition, Springer-Verlag, New York.

25. Wilson S. W., 1998, Structure and Function of the XCS Classifier System, Lecture Presented at the Rowland Institute for Science, Cambridge, Ma, Available at http://world.std.com/~sw/pubs.html.

A Genetic Programming-based Hierarchical Clustering Procedure for the solution of the Cell-Formation Problem

C Dimopoulos and N Mort
Department of Automatic Control & Systems Engineering
The University of Sheffield
Email: cop97cd@sheffield.ac.uk & n.mort@sheffield.ac.uk

Abstract: Cellular manufacturing is the implementation of group technology in the manufacturing process. A key issue during the design of a cellular manufacturing system is the configuration of machine cells and part families within the plant. In this paper we present a hierarchical clustering procedure for the solution of the cell-formation problem which is based on the use of Genetic Programming for the evolution of similarity coefficients between pairs of machines in the plant. The performance of the methodology is illustrated on a number of test problems taken from the literature.

1. Introduction

The traditional flow-line manufacturing layout is not able to cope with the needs of modern manufacturing systems, which require the production of a variety of parts in small or medium-sized batches. The design, scheduling and maintenance of these complex production lines drive the need for the implementation of efficient manufacturing systems.

Group Technology (GT) is a manufacturing philosophy originally introduced by Mitrovanov [1] in the former U.S.S.R. It states that significant benefits can be achieved by grouping similar objects within a corporate structure. At the production line level GT is translated as the grouping of machines into a number of cells, each of them processing a dedicated family of parts. This type of manufacturing layout is known as cellular manufacturing. The intuition behind cellular manufacturing is an attempt to bring the benefits of flow-line production to the batch production processes. Traditionally, batch production was accommodated by grouping machines of similar type in a functional layout. This layout often lead to difficult scheduling and control problems. By processing only a dedicated family of parts, a machine cell approximates a flow-line layout in a smaller scale, thus inheriting some of the benefits of the mass-production effect. Cellular manufacturing has been reported to result in significant benefits for the manufacturing process, like reductions in set-up times, work-in-progress inventories, and improved quality.

The implementation of a cellular manufacturing system is a painstaking procedure that starts with the acquisition of production data and finishes with the configuration and layout of cells. At the heart of this procedure is the problem of creating machine cells and part families, which is usually described as the cell-formation problem.

Genetic Programming [2] belongs to the family of evolutionary computation methods. It uses the concept of Darwinian strife for survival for the evolution of computer programs of variable length. While evolutionary computation methods have been widely used for the solution of manufacturing optimisation problems [3] GP applications have rarely been reported. In this paper we investigate the use of GP for the solution of a simple version of the cell-formation problem.

The rest of this paper is organised as follows: In section 2 the binary cell formation problem is described and some pointers to alternative solution methodologies are given. The hierarchical clustering procedure that forms the basis of our methodology is described in section 3. The operation of Genetic Programming is described briefly in section 4. Section 5 introduces the GP-based algorithm for the solution of the cell-formation problem. An illustrative example of the operation of the algorithm and results on a number of test problems taken from the literature are presented in section 6. Comparisons with alternative solution methodologies are presented in the same section. The conclusions of this paper are summarised in section 7.

2. The cell-formation problem

The complexity of the cell-formation problem depends on the manufacturing data that are included in its formulation. The simplest version of the problem is usually described with the help of the binary machine-component (m/c) matrix [4]. The m/c matrix is comprised of n rows and m columns, where n is the total number of machines in the plant, and m is the total number of parts to be processed. All entries in the matrix are binary values. A positive entry indicates that the part identified by the corresponding column has an operation on the machine on the machine identified by the corresponding row. An example of an m/c matrix $A[3 \times 5]$ is given in figure 2.1.

	p1	p2	p3	p4	p5
m1	0	1	1	0	1
m2	1	0	0	1	0
m3	0	1	1	0	1
m4	1	0	0	1	0

Figure 2.1: An example of an m/c matrix

The value of $A_{1,3}$ is equal to '1', thus part 3 has an operation on machine 1. In contrast, part 2 does not need processing on machine 2, since $A_{2,2} = 0$.

The m/c matrix is constructed by analysing the information available from the route cards of the parts. The cell-formation problem is described as the problem of diagonalising the m/c matrix, i.e. creating a configuration where all positive entries are arranged inside blocks along the main diagonal of the matrix. This diagonalised form allows the easy identification of machine cells and part families, as figure 2.2 illustrates. This form has been achieved by rearranging the rows and the columns of the matrix.

	p4	p1	p5	p3	p2
m2	1	1	0	0	0
m4	1	1	0	0	0
m1	0	0	1	1	1
m3	0	0	1	1	1

Figure 2.2: The diagonalised m/c matrix

From figure 2.2 we can see that there are two clearly identified independent cells, the first one comprising of machines 2 and 4, and the second one comprising of machines 1 and 3. The existence of independent cells is a situation rarely encountered in practice. In most problems some positive entries will remain outside the diagonal blocks after the diagonalisation procedure. This situation, which is illustrated in figure 2.3 using a different m/c matrix, results in parts moving between the manufacturing cells, thus reducing the efficiency of the layout.

	p1	p2	p3	p4
m1	1	1	1	0
m2	0	0	1	1
m3	0	0	1	1

Figure 2.3: m/c matrix with intercell moves

In figure 2.3, $A_{1,3}$ is equal to '1', but the positive entry is not a part of any block diagonal. In cellular manufacturing terminology, machine 1 is called 'bottleneck machine' and part 3 is described as 'exceptional part'. Another undesired situation in a diagonalised matrix is the existence of '0' entries within the block diagonals (voids), since they result in skip moves and low utilisation of machines. The objective of cell-formation algorithms that consider this simple version of the problem is to directly or indirectly minimise the number of exceptional elements and voids in the diagonalised m/c matrix.

The bibliography of the cell-formation problem is vast. Solution methodologies range from simple array-based methods which create cells by simply manipulating the rows and the columns of the m/c matrix, to complex mathematical programming formulations which explicitly consider a wide range of manufacturing data like product demands, processing times, operation sequences, alternative process plans, cell-size constraints, batch sizes, manufacturing costs, etc. Excellent reviews of cell-formation methodologies have been presented by Singh [5], Offodile et al. [6], and Selim et al. [7].

3. McAuley's Single Linkage Cluster Analysis Algorithm

McAuley [8] introduced a hierarchical clustering algorithm for the solution of the cell-formation problem. The algorithm was based on the calculation of a similarity measure for each pair of machines in the plant, which was used for the creation of a pictorial representation of the solutions in the form of a 'dendrogram'. We will describe the operation of the algorithm using the m/c matrix shown in Figure 3.1.

	p1	p2	p3	p4	p5
m1	1	0	1	0	0
m2	0	1	0	1	1
m3	1	0	1	0	0
m4	1	1	0	1	0

Figure 3.1: Example m/c matrix for the illustration of SLCA

The algorithm starts with the calculation of similarity coefficients for each pair of machines. McAuley employed Jaccard's similarity coefficient which, for this particular problem, was defined as follows:

$$S_{ij} = \frac{a_{ij}}{a_{ij} + b_{ij} + c_{ij}}$$

where: S_{ij} : similarity between machines i and j

a_{ij} : number of parts processed by both machines i and j

b_{ij} : number of parts processed by machine i but not by machine j

c_{ij} : number of parts processed by machine j but not by machine i

The value of the similarity coefficient ranges from 0 to 1. The above values are used for the construction of the similarity matrix (figure 3.2).

	m1	m2	m3
m2	0	*	*
m3	1	0	*
m4	0.25	0.5	0.25

Figure 3.2: Similarity matrix for the example problem

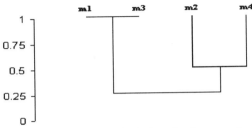

Figure 3.3 : Dendrogram of solutions for the example problem

The calculated similarity matrix is used for the creation of a pictorial representation of solutions known as a 'dendrogram'. Single Linkage Cluster Analysis is employed for the construction of the dendrogram. SLCA assumes that all machines are initially ungrouped. Then, the pair of machines having the highest value in the similarity matrix is recorded and is grouped at this level of similarity. In our example, machines 1 and 3 are grouped at the similarity level of 1. The next highest similarity level is found and the associated machines are merged at this level. In our case, machines 2 and 4 are merged at the similarity level of 0.5. At this point the highest similarity level is 0.25 between machines 1 and 4. Since both machines have already been grouped, their associated groups are merged as well. Thus, at the similarity level of 0.25 all machines have formed a single cell and consequently there is no reason for examining the remaining similarity coefficient values. The constructed dendrogram is illustrated in figure 3.3:

The above dendrogram contains a number of alternative solutions, depending on the value of the threshold level (T) that we wish to choose and the objective of the algorithm. More specifically, there are four alternative solutions:

- Solution 1 (initial)

  cell 1: m_1
  cell 2: m_2
  cell 3: m_3
  cell 4: m_4

- Solution 2 (T=1)

  cell 1: m_1, m_3
  cell 2: m_2
  cell 3: m_4

- Solution 3 (T=0.5)

  cell 1: m_1, m_3
  cell 2: m_2, m_4

- Solution 4 (T=0.25)

  cell 1: m_1, m_2, m_3, m_4

McAuley used as an objective the minimisation of the sum of material handling costs which is calculated by adding the intercellular and intracellular handling costs under a prespecified layout. However, since the output of SLCA is a partition of machines into a number of cells, it can be used in conjunction with any desired objective.

4. Genetic Programming

Genetic Programming is a form of automatic programming, i.e. a method of teaching computers how to program themselves.

As in all evolutionary algorithms, the concept of Darwinian strife for survival drives the search for an optimal solution. Initially, a generation of suitably coded potential solutions of the problem is created randomly. The performance of solutions is evaluated on the problem considered and a fitness value is assigned to each of them. A new generation is then created by probabilistically selecting solutions from the old generation and genetically altering them using standard genetic operators (usually crossover and mutation). The new generation of solutions is again evaluated on the problem and a new fitness value is assigned to each of them. The same procedure is repeated until a prespecified termination criterion has been reached.

The main difference of GP in comparison with other evolutionary algorithms is that it evolves computer programs of variable length, i.e. structures that can be compiled directly or with slight modifications by a computer. GP has attracted considerable research interest during the last decade with a continuously expanding range of applications.

5. Genetic Programming methodology

It has been argued [9] that the performance of similarity coefficients depends heavily on the characteristics of the problem and the grouping measure that is used for the evaluation of solutions. Jaccard's similarity coefficient is less likely to find an optimal solution in the case of ill-structured matrices, since the grouping of machines becomes less straightforward. In addition, SLCA is a 'blind' method, i.e. it produces the same dendrogram irrespective of the optimisation objective. Our methodology employs Genetic Programming for the evolution of similarity coefficients that are used by SLCA for the construction of the dendrogram of solutions. The operation of GP-SLCA in algorithmic form is the following:

Procedure Main
initialise population of randomly created similarity coefficients
run procedure SLCA for each coefficient
loop
        loop
                select individuals for crossover or mutation
                apply genetic operators and form new coefficients
        until a new generation has been formed
        run procedure SLCA for each coefficient

until termination criterion is true

Procedure SLCA
compute similarity matrix
construct dendrogram
loop
  create machine cells for the highest level of similarity coefficient
  assign parts to machine cells
  calculate the fitness value of the cell configuration
  if solution is the best recorded so far, best=current solution
  delete the value of similarity coefficient from the matrix
until a single cell has been formed
assign the best solution as the fitness of the individual

The methodology described in this section belongs to the category of cell-formation methods that group machines into cells and not parts into families. Thus, for each machine-cell configuration created from the dendrogram, the corresponding part families must be formed in order to calculate the value of the objective function. Since no information is available about the sequencing of operations (backtracking, skips, etc.), parts are usually assigned to the cell where the majority of their processing takes part. In case of a tie, the part is assigned to the smallest of the candidate cells. In that way we ensure that the number of voids created by the assignment procedure is minimal. If there is still a tie, the part is assigned randomly to one of the candidate cells. If the allocation of parts to machine cells results in the creation of an empty cell (a cell that processes no parts), then the fitness of this solution is set to '0'. However, there is no limit on the size of machine cells, and consequently no limit on the total number of cells in the plant. If required, the algorithm has the ability to explicitly consider size constraints, by just assigning a zero fitness to each solution that violates the constraints.

The characteristics of the GP algorithm used in the above formulation are described with the help of the Koza tableau (figure 5.1).

Parameters	Values
Objective:	maximisation of grouping efficacy
Terminal set:	a_{ij}, b_{ij}, c_{ij}, (defined earlier), d_{ij} (defined later)
Function set:	+, -, ×, %
Population size:	500
Subtree crossover probability:	.9
Subtree mutation probability:	.1
Selection:	Tournament selection, size 7
Number of generations:	50
Maximum depth for crossover:	17
Initialisation method:	Ramped half and half

Figure 5.1: Koza tableau for the GP-SLCA methodology

Note that with the exception of d_{ij} the same variables were used for the calculation of Jaccard's similarity coefficient. The value of d_{ij} is defined as the number of parts that are processed by neither machine i nor machine j. The grouping efficacy measure, Γ, is calculated using the following formula [10]:

$$\Gamma = 1 - \frac{e_v + e_0}{e + e_v} = \frac{e - e_0}{e + e_v}$$

where:

e: total number of non-zero entries in the m/c matrix

e_0: total number of non-zero entries outside the diagonal blocks (exceptional elements)

e_v: total number of zero entries inside the diagonal blocks (voids)

6. Results

In this paper we have examined 11 problems for the testing of the GP-SLCA methodology. All the problems have been taken from the cellular manufacturing literature and results from alternative cell-formation methods have been reported. The size of the problems ranges from 11×22 to 30×50. Twenty runs of GP-SLCA were conducted for each problem. The problems were chosen so that they represent different levels of difficulty as this is indicated from the reported results. All these problems along with their characteristics and their corresponding references are described in table 6.1. The number in the first column of the table will be used from this point onwards for the identification of these problems. For the reader interested in using the test problems for their own comparisons we should note that problems 1-6 correspond to problems 1-3, 5-7 in the order presented by Chandrasekharan & Rajagopalan.

No.	Reference	Size	e
1	Chandrasekharan & Rajagopalan [11]	24×40	131
2	"	24×40	130
3	"	24×40	131
4	"	24×40	131
5	"	24×40	131
6	"	24×40	130
7	Kumar & Vannelli [12]	30×41	128
8	Seifoddini [13]	11×22	78
9	Stanfel [14]	14×24	61
10	"	30×50	154
11	"	30×50	167

Table 6.1: Test problems[1]

The cumulative results are presented in two parts. First, in table 6.2, the detailed results of the GP-SLCA procedure are illustrated. Then, in table 6.3, the best solution evolved by GP-SLCA is compared with a number of solutions that have been produced by alternative cell-formation methods.

---

[1] All test problems and diagonalised matrices are available in text file format and can be provided by the authors on request.

Pr. No.	max Γ	$\overline{\Gamma}$	σ	e_0	e_v	No. of cells
1	1	1	0	0	0	7
2	0.851	0.851	0	10	11	7
3	0.735	0.735	0	20	20	7
4	0.533	0.531	0.0029	50	21	11
5	0.479	0.476	0.0020	63	11	13
6	0.437	0.435	0.0016	61	28	11
7	0.607	0.607	0.0011	46	7	16
8	0.731	0.731	0	10	15	3
9	0.718	0.718	0.0010	10	10	7
10	0.594	0.583	0.0063	53	16	14
11	0.5	0.488	0.0044	75	17	15

Table 6.2 : Performance of GP-SLCA on the test problems

Pr.No.	GP-SLCA	ZODIAC *(Chandr. & Raj., [15])*	GRAFICS *(Sriniv. & Naren., [16])*	GA-TSP *(Cheng et al., [17])*
1	1	1	1	1
2	0.851	0.851	0.851	0.851
3	0.735	0.730	0.735	0.730
4	0.533	0.204	0.433	0.494
5	0.479	0.182	0.445	0.447
6	0.437	0.176	0.417	0.425
7	0.607	0.337	0.554	0.538
8	0.731	0.731	0.731	-
9	0.718	0.656	0.656	0.674
10	0.594	0.461	0.563	0.566
11	0.5	0.211	0.480	0.459

Table 6.3: Comparison with alternative cell-formation methods

Results from tables 6.1, 6.2 and 6.3 indicate that GP-SLCA is a powerful algorithm for the solution of binary cell-formation problems. More specifically, for the grouping efficacy measure GP-SLCA dominates the performance of ZODIAC, GRAFICS, and the GA-TSP heuristic. It will be interesting to take a closer look at a particular example, so that the operation of the methodology will become clear. The 14×24 m/c component matrix, originally introduced by Stanfel [14], is presented in figure 6.1.

	1	2	3	4	5	6	7	8	9	10	11	12	13	14	15	16	17	18	19	20	21	22	23	24
1						1	1																	
2							1	1																
3						1	1	1										1					1	
4			1	1																				
5			1	1																	1			
6			1																					1
7			1	1																	1			1
8	1	1															1		1	1			1	
9	1	1															1			1			1	
10	1	1					1										1			1				
11									1	1	1	1		1	1	1						1		
12					1				1	1		1		1	1	1						1		
13					1				1			1	1		1									
14									1		1		1		1									

Figure 6.1: 14×24 test problem introduced by Stanfel

20 runs of the GP-SLCA methodology were performed for this test problem with the objective of maximising the grouping efficacy. One of the similarity coefficients that produced the best value of grouping efficacy is the following (in LISP symbolic language):

$$((-(/(b)(/(-(d)(/(d)(c)))(-(-(d)(d))(-(b)(a)))))(-(b)(a))))$$

The evolved program corresponds to the following formula for the coefficient:

$$a - b + \frac{b}{\left(d - \dfrac{d}{c}\right)\Big/(a-b)}$$

It is not expected that Genetic Programming will evolve a similarity coefficient that will be easy to understand how and why it works. However, in this particular case, high values of a and low values of b will result in high values for the similarity coefficient. This does not mean that the evolved coefficient is ideal for every cell-formation problem. We note that the fitness of an individual is measured based on the best solution found from the dendrogram, and it is always possible that an evolved coefficient which does not utilise the variables in a meaningful way from the cellular manufacturing point of view, will produce a fit partition of machines for the objective considered. The above formula is used for the calculation of similarities for each pair of machines in the plant, resulting in the similarity matrix in figure 6.2. The normalised similarity matrix is presented in figure 6.3.

	m1	m2	m3	m4	m5	m6	m7	m8	m9	m10	m11	m12	m13
m2	1												
m3	2	2											
m4	-2.4	-2.4	-7.941176										
m5	-2.315789	-2.315789	-7.34375	3									
m6	-2.4	-2.4	-7.941176	1	0								
m7	-2.296296	-2.296296	-7.222222	2	4	2							
m8	-2.3	-2.3	-4.071428	-2.3	-3.72	-2.3	-5.371428						
m9	-2.294118	-2.294118	-4.066667	-2.294118	-3.703125	-2.294118	-5.333333	4.235294					
m10	1	1	-4.066667	-2.294118	-3.703125	-2.294118	-5.333333	3	4				
m11	-2.326531	-2.326531	-7.597403	-2.326531	-3.791209	-2.326531	-5.523809	-10.114285	-7.597403	-7.597403			
m12	-2.326531	-2.326531	-7.597403	-2.326531	-3.791209	-2.326531	-5.523809	-10.114285	-7.597403	-7.597403	7		
m13	-2.294118	-2.294118	-7.232143	-2.294118	-3.703125	-2.294118	-5.333333	-9.461538	-7.232143	-7.232143	-3.428571	4	
m14	-2.296296	-2.296296	-7.222222	-2.296296	-3.705882	-2.296296	-5.333333	-9.428572	-7.222222	-7.222222	-1	-7.428572	2

Figure 6.2 : Similarity coefficient matrix

	m1	m2	m3	m4	m5	m6	m7	m8	m9	m10	m11	m12	m13
m2	0.649416												
m3	0.707846	0.707846											
m4	0.450751	0.450751	0.126976										
m5	0.455672	0.455672	0.161884	0.766277									
m6	0.450751	0.450751	0.126976	0.649416	0.590985								
m7	0.456811	0.456811	0.168985	0.707846	0.824708	0.707846							
m8	0.456594	0.456594	0.353088	0.456594	0.373623	0.456594	0.277129						
m9	0.456938	0.456938	0.353367	0.456938	0.374609	0.456938	0.279354	0.838456					
m10	0.649416	0.649416	0.353367	0.456938	0.374609	0.456938	0.279354	0.766277	0.824708				
m11	0.455044	0.455044	0.147063	0.455044	0.369462	0.455044	0.268225	0	0.147063	0.147063			
m12	0.455044	0.455044	0.147063	0.455044	0.369462	0.455044	0.268225	0	0.147063	0.147063	1		
m13	0.456938	0.456938	0.168406	0.456938	0.374609	0.456938	0.279354	0.03814	0.168406	0.168406	0.390651	0.824708	
m14	0.456811	0.456811	0.168985	0.456811	0.374448	0.456811	0.279354	0.040067	0.168985	0.168985	0.532554	0.156928	0.707846

Figure 6.3 : Normalised similarity coefficient matrix

If either of these matrices is fed as an input to the SLCA algorithm a dendrogram of potential solutions will be produced. If the dendrogram is cut at the similarity level of 0.707846 (normalised value), the cell configuration will correspond to the diagonalised matrix of figure 6.4, which has a grouping efficacy value of 0.71831.

We have been unable to find a better value of grouping efficacy reported in the literature for this particular problem.

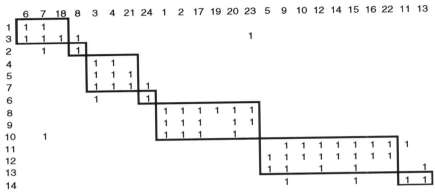

Figure 6.4: Diagonalised matrix

7. Conclusions

In this paper we have investigated the use of Genetic Programming for the solution of binary cell-formation problems. McAuley's Single Linkage Cluster Analysis (SLCA) algorithm has been used as basis for the development of our methodology. SLCA employs Jaccard's similarity coefficient for the creation of a pictorial representation of solutions in the form of a 'dendrogram'. A variety of cell configurations can be created by choosing a particular similarity level in the dendrogram.

Genetic Programming has been employed for the evolution of similarity coefficients for independent cell formation problems. These coefficients were then fed into an SLCA procedure which returned the best solution found in the dendrogram for the desired objective. The methodology was tested on a number of published test problems, performing at least as good as other leading cell-formation algorithms.

The GP-SLCA methodology is quite flexible since it can be used with a variety of grouping objectives without altering its algorithmic structure. On the other hand, as the size of the problem increases the evaluation function becomes computationally expensive since the value of the coefficient is calculated for each pair of machines in the plant and every possible solution from the dendrogram is evaluated.

GP-SLCA was developed for the solution of binary cell-formation problems, thus it has the inherent weakness, as all other similar methods, that it does not consider manufacturing data that are important for the design of production cells, like product demands, processing times, operation sequences etc. However, GP-SLCA can be modified to include some of these data as terminals for the evolution of a similarity coefficient, and current work is being progressed in this direction.

REFERENCES

1. Mitrovanov, S.P., 1966. *The Scientific Principles of Group Technology*, National Lending Library Translation, Boston Spa, Yorkshire, U.K.

2. Koza, J.R., 1992. *Genetic Programming: On the programming of computers by means of natural selection.* MIT Press. Cambridge.

3. Dimopoulos, C., and Zalzala AMS., 1999. Recent developments in evolutionary computation for manufacturing optimisation: problems, solutions and comparisons. *IEEE Transactions in Evolutionary Computation*, in print.

4. Burbidge, J.L., 1971. Production Flow Analysis. *Production Engineer* 50:139-152.

5. Singh, N., 1993. Cellular manufacturing systems: an invited review. *European Journal of Operational Research* 69:284-291.

6. Offodile, O.F., Mehrez, A., and Grznar, J., 1994. Cellular manufacturing: a taxonomic review framework. *Journal of Manufacturing Systems* 13:196-220.

7. Selim, M.H., Askin, R.G., and Vakharia, A.J., 1998. Cell formation in group technology: review, evaluation and directions for future research. *Computers & Industrial Engineering* 34:3-20.

8. McAuley, J., 1972. Machine grouping for efficient production. *Production Engineer* 51:53-57.

9. Sharker, B.R., 1986. The resemblance coeficients in group technology: a survey and comparative study of relational metrics. *Computers & Industrial Engineering* 30:103-116.

10. Kumar, C.S., and Chandrasekharan, M.P., 1990. Grouping efficacy: a quantitative criterion for goodness of block diagonal forms of binary matrices in group technology. *Int.J. of Production Research* 28:603-612.

11. Chandrasekharan, M.P., and Rajagopalan, R., 1989. GROUPABILITY: an analysis of the properties of binary data matrices for group technology. *Int.J. of Production Research* 27:1035-1052.

12. Kumar, K.R., and Vannelli, A., 1987. Strategic subcontrcting for efficient disaggregated manufacturing. *Int.J. of Production Research* 25:1715-1728

13. Seifoddini, H., 1989. Single linkage vs. average linkage clustering in machine cells formation application. *Computers & Industrial Engineering* 16:419-426.

14. Stanfel, L.E., 1985. Machine clustering for economic production", *Engineering Costs & Production Economics* 9:73-81.

15. Chandrasekharan, M.P., and Rajagopalan, R., 1987. ZODIAC - an algorithm for concurrent formation of part families and machine-cells. *Int.J. of Production Research* 25:835-850.

16. Srinivasan, G, and Narendran, T.T., 1991. GRAFICS – a nonhierarchical clustering algorithm for group technology. *Int.J. of Production Research* 29:463-478.

17. Cheng, C.H., Gupta, Y.P., Lee, W.H., and Wong, K.F., 1998. A TSP-based heuristic for forming machine groups and part families. *Int.J. of Production Research* 36:1325-1337.

Chapter 5

Multi-objective Satisfaction

Multi-Objective Evolutionary Optimization: Past, Present, and Future

Kalyanmoy Deb
Department of Mechanical Engineering,
Indian Institute of Technology Kanpur,
PIN 208 016, India, E-mail: deb@iitk.ac.in

Abstract

Many real-world optimization problems truly involve multiple objectives and, therefore, are better posed as multi-objective optimization problems. Interestingly, most multi-objective optimization problems give rise to a set of optimal solutions, called Pareto-optimal solutions, instead of a single optimal solution. The optimal solutions are different from each other, primarily varying in the relative importance of each objective. In the absence of any priori information of different objectives, it becomes important to find as many such Pareto-optimal solutions as possible. Since classical optimization methods are not efficient in finding multiple optimal solutions and evolutionary algorithms are ideal for finding multiple optimal solutions, it is natural that there has been a considerable interest among evolutionary algorithmists to concentrate on solving these problems. In this paper, we present a brief overview of the past research activities, discuss current salient methodologies, and highlight some immediate future research in this area.

1 Introduction

For the past decade or so, multi-objective evolutionary optimization (MOEO) has gained increasing popularity. There are primarily two main reasons. Firstly, the classical methods of handling multiple objectives have been mainly to scalarize objectives to a single aggregate objective and use a single-objective optimizer. The obtained optimal solutions become subjective to the user and, moreover, these algorithms can only find a single optimum in one simulation. Secondly, evolutionary algorithms (EAs) are capable of finding multiple optima in one simulation run and are, therefore, ideal candidates for finding multiple Pareto-optimal solutions.

In this paper, we review the past and present evolutionary algorithms that have been developed for multi-objective optimization and later discuss a few important yet immediate research topics that should be pursued for better understanding and evaluation of multi-objective evolutionary algorithms.

2 Principles of Multi-Objective Optimization

In a single-objective optimization, the primary goal is to find the global optimal solution. In multi-criterion optimization, the goal is more than just finding the global optimal solution. In order to understand what other aspects are of interest in multi-objective optimization, let us refer to Figure 2.1, where a typical two-objective scenario is illustrated. The figure considers two objectives—cost and accident rate—both

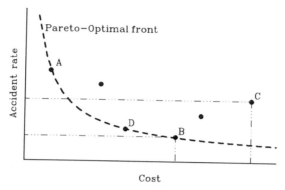

Figure 2.1: The concept of Pareto-optimal solutions is illustrated.

of which are to be minimized. The point A represents a solution which incurs a near-minimal cost, but is highly accident-prone. On the other hand, the point B represents a solution which is costly, but is near least accident-prone. If both objectives (cost and accident rate) are important goals of design, one cannot really say whether solution A is better than solution B, or vice versa. One solution is better than other in one objective, but is worse in the other. In fact, there exist many such solutions (like solution D), about which one cannot conclude an absolute hierarchy of solutions without any further information. All these solutions (in the front marked by the dashed line) are known as *Pareto-optimal* or *non-inferior* solutions. Looking at the figure, we also observe that there exists solutions which are not Pareto-optimal solutions, like the point C. This is because there exists at least one solution such as D in the search space, which is better than solution C in *both* objectives. That is why solutions like C are known as *dominated* or *inferior* solutions. This concept of dominated and Pareto-optimal solutions in a search space also carries over in problems with more than two objectives.

It is now clear that the concept of optimality in multi-criterion optimization deals with a number (or a set) of solutions, instead of just one solution. For a problem having more than one objective function (say, f_j, $j = 1, \ldots, M$ and $M > 1$), a solution $x^{(1)}$ (of P dimension) is said to dominate another solution $x^{(2)}$, if both the following conditions are true [22]:

1. The solution $x^{(1)}$ is no worse (say the operator $\prec$ denotes worse and $\succ$ denotes better) than $x^{(2)}$ in all objectives, or $f_j(x^{(1)}) \not\prec f_j(x^{(2)})$ for all $j = 1, 2, \ldots, M$ objectives.

2. The solution $x^{(1)}$ is strictly better than $x^{(2)}$ in at least one objective, or $f_{\bar{j}}(x^{(1)}) \succ f_{\bar{j}}(x^{(2)})$ for at least one $\bar{j} \in \{1, 2, \ldots, M\}$.

If any of the above condition is violated, the solution $x^{(1)}$ does not dominate the solution $x^{(2)}$. A more generic definition of non-dominance of solutions is given in [14]. Recently, the concept of local and global Pareto-optimal solutions has been better understood [2, 13, 24].

Local Pareto-optimal Set: If for every member x in a set $\underline{P}$, there exist no solution y satisfying $\|y - x\|_\infty \leq \varepsilon$, where ε is a small positive number (in principle, y is obtained by perturbing x in a small neighborhood), which dominates any member in the set $\underline{P}$, then the solutions belonging to the set $\underline{P}$ constitute a local Pareto-optimal set.

Global Pareto-optimal Set: If there exists no solution in the search space which dominates any member in the set $\bar{P}$, then the solutions belonging to the set $\bar{P}$ constitute a global Pareto-optimal set.

From the above discussions, it is clear that there are primarily two goals that a multi-criterion optimization algorithm must try to achieve:

1. Guide the search towards the global Pareto-optimal region, and

2. Maintain population diversity in the Pareto-optimal front.

The first task is desired for any optimization algorithm. The second task is unique to multi-objective and multi-modal optimization. Since no one solution in the Pareto-optimal set can be said to be better than the other, what an algorithm can do best is to find as many different Pareto-optimal solutions as possible.

2.1 Classical Approaches

The classical ways of tackling multi-objective optimization problems have primarily ignored the second goal as mentioned above. Most methods convert multiple objectives into one objective using different heuristics [13, 20, 22]—weighted sum approach, ε-perturbation method, Tchybeshev method, min-max method, goal programming method. Since multiple objectives are converted into one objective, the resulting solution to the single-objective optimization problem is usually subjective to the parameter settings (for example, weights chosen for each objective in the weighted sum approach). Since most classical methods use point-by-point approach, it is expected that one unique solution (hopefully a Pareto-optimal solution) will be found in each application of the optimization algorithm. Thus, in order to find multiple Pareto-optimal solutions, the chosen optimization algorithm must have to be used a number of times. Furthermore, the ability of a classical optimization method to find a different Pareto-optimal solution in each simulation is found to be sensitive to the convexity and continuity of the Pareto-optimal region.

2.2 Evolutionary Approaches

Since evolutionary approaches use a population-based approach, it is intuitive that a number of Pareto-optimal solutions can, in principle, be developed in one single simulation run. Reasonably enough, there exists a number of past studies and a growing number of current approaches in this direction, which is providing the evolutionary computing field a new dimension.

3 Multi-Criterion Evolutionary Optimization: Past

As early as in 1967, Rosenberg suggested, but did not simulate, a genetic search method for finding the chemistry of a population of single-celled organisms with multiple properties or objectives [16].

However, the first implementation of a real multi-objective evolutionary algorithm (vector-evaluated GA or VEGA) was suggested by David Schaffer in the year 1984 [19]. Schaffer modified the simple tripartite genetic algorithm GAs with selection, crossover, and mutation) by performing independent selection cycles according to each objective. The selection method is repeated for each individual objective to fill up a portion of the mating pool. Then the entire population is thoroughly shuffled to apply crossover and mutation operators. This is performed to achieve the mating of individuals of different subpopulation groups. The algorithm worked efficiently for some generations but in some cases suffered from its bias towards some individuals or regions (mostly individual objective champions). This does not fulfill the second goal of MOEO.

Ironically, no significant study was performed for almost a decade after the pioneering work of Schaffer, until a revolutionary 10-line sketch of a new non-dominated sorting procedure suggested by David Goldberg [7]. Since an EA needs a fitness function for reproduction, the trick was to find a single metric from a number of objective functions. Goldberg's suggestion was to use the concept of domination to assign more copies to non-dominated individuals in a population. Since diversity is another concern, he also suggested the use of a *niching* strategy among solutions of a non-dominated class. Getting this clue, at least three independent groups of researchers developed different versions of multi-objective evolutionary algorithms. Basically, these algorithms differ in the way a fitness is assigned to each individual. We discuss them one by one in the following.

Fonesca and Fleming [5], in their multi-objective GA (MOGA), first classify the whole population according to different non-dominated classes. Thereafter, individuals of the first (best) class are all assigned a rank '1'. Other individuals are ranked by calculating how many solutions (say k) dominate a particular solution. That solution is assigned a rank one more than k. Therefore, at the end of this ranking procedure, there may exist many solutions having the same rank. The selection procedure then uses these ranks to select or delete blocks of points to form the mating pool. As discussed elsewhere [8], this type of blocked fitness assignment is likely to produce a large selection pressure which might cause premature convergence. However, MOGA also uses a niche-formation method to distribute the population over the Pareto-optimal region. But instead of performing niching on the parameter values, they have used niching on objective function values. This causes the MOGA difficulty to find multiple solutions in problems where different Pareto-optimal points correspond to the same objective function value [2]. However, the ranking of individuals according to their non-dominance in the population is an important aspect of the work.

Horn, Nafploitis, and Goldberg [10] used *Pareto domination tournaments* instead of non-dominated sorting and ranking selection method in their niched-Pareto GA (NPGA). In this method, a *comparison set* comprising of a specific number (t_{dom}) of individuals is picked at random from the population at the beginning of each selection process. Two random individuals are picked from the population for selecting a winner according to the following procedure. Both individuals are compared with the members of the comparison set for domination with respect to objective functions. If one of them is non-dominated and the other is dominated, then the non-dominated point is selected. On the other hand, if both are either non-dominated or dominated, a *niche count* is found for each individual in the entire population. The niche count is calculated by simply counting the number of points in the population within a certain distance (σ_{share}) from an individual. The individual with least niche count is selected. Since this non-dominance is computed by comparing an individual with a randomly chosen population set of size t_{dom}, the success of this algorithm highly depends on the parameter t_{dom}. If a proper size is not chosen, true non-dominated (Pareto-optimal) points may not be found. Nevertheless, the concept of niche formation among the non-dominated points using the tournament selection is an important aspect of the work.

Srinivas and Deb [21] developed a non-dominated sorting GA (NSGA), which is similar to the MOGA. NSGA differs from MOGA in two ways: fitness assignment and the way niching is performed. After the population is classified for the best class of non-domination, they are assigned a dummy fitness value equal to N (population size). A sharing strategy is then used on parameter values (instead of objective function values) to find the a niche count for each individual of the best class. Niche count measures a qualitative number of individuals in the vicinity of a solution. For each individual, a shared fitness is found by dividing the assigned fitness N by the niche count. The smallest shared fitness value $\mathcal{F}_1^{min}$ is noted for further use. Thereafter, The second class of non-dominated solutions are found and a dummy fitness value equal to $\mathcal{F}_1^{min} - \varepsilon_1$ (where ε_1 is a small positive number) is assigned to all individuals. Niche counts of individuals within this class are found and shared fitness values are calculated. This process is continued till all solutions are assigned a fitness value. This fitness assignment procedure ensures two aspects: (i) a dominated solution is assigned a smaller shared fitness value than any solution which dominates it and (ii) In each non-dominated class, diversity of ensured. On a number of test problems and real-world optimization problems, NSGA has found wide-spread Pareto-optimal or near Pareto-optimal solutions. One difficulty of NSGA is to choose the niching parameter, which signifies the maximum distance between two neighboring Pareto-optimal solutions. Although most studies used a fixed value of the niching parameter, there exists a study where an adaptive sizing strategy has been suggested [6].

To demonstrate the working of the NSGA, we consider a welded beam design problem with two apparently conflicting objectives: (i) minimize fabrication cost and (ii) minimize end deflection of the beam. A beam needs to be welded on another beam and must carry a certain load F (Figure 3.1). There are four design parameters:

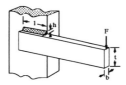

Figure 3.1: The welded beam design problem. Minimizations of cost and end deflection are two objectives.

thickness of the beam, b, width of the beam t, length of weld ℓ, and weld thickness h. The overhang portion of the beam has a length of 14 inch and $F = 6,000$ lb force is applied at the end of the beam. It is intuitive that an optimal design for cost will make all four design variables to take small values. When the beam dimensions are small, it is likely that the deflection at the end of the beam is going to be large. Thus, the design solutions for minimum cost and minimum end deflection are conflicting to each other. The resulting problem is as follows:

$$
\begin{aligned}
\text{Minimize} \quad & f_1(\vec{x}) = 1.10471 h^2 \ell + 0.04811 t b (14.0 + \ell), \\
\text{Minimize} \quad & f_2(\vec{x}) = \delta(\vec{x}), \\
\text{Subject to} \quad & g_1(\vec{x}) \equiv 13,600 - \tau(\vec{x}) \geq 0, \\
& g_2(\vec{x}) \equiv 30,000 - \sigma(\vec{x}) \geq 0, \\
& g_3(\vec{x}) \equiv b - h \geq 0, \\
& g_4(\vec{x}) \equiv P_c(\vec{x}) - 6,000 \geq 0.
\end{aligned}
\tag{1}
$$

The deflection term $\delta(\vec{x})$ is given as follows: $\delta(\vec{x}) = 2.1952/(bt^3)$. The expression for other terms can be found in the literature [15]. The variables are initialized in the following range: $(0.125 \leq h, b \leq 5.0)$ and $(0.1 \leq \ell, t \leq 10.0)$. Constraints are handled using the bracket-operator penalty function [1]. Penalty parameters of 100 and 0.1 are used for the first and second objective functions, respectively. We use real-parameter GAs with simulated binary crossover (SBX) operator [3] to solve this problem. Unlike in the binary-coded GAs, variables are used directly and a crossover operator that creates two real-valued children solutions from two parent solutions is used. We use a σ_{share} of 0.281, calculated from guidelines [4]. Figure 3.2 shows that the population after 500 generations (marked with diamonds) has truly come near the Pareto-optimal front. The dots in the figure are randomly chosen feasible solutions.

4 Multi-Criterion Evolutionary Optimization: Present

The field of multi-objective optimization is relatively new for classifying the studies according to past and present. However, in this section, we outline algorithms which

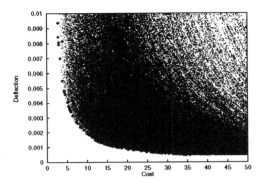

Figure 3.2: Population at generation 500 shows that a wide range of Pareto-optimal solutions are found for the welded beam design problem.

are suggested in the past couple years or so.

Zitzler and Thiele [26] suggested an elitist multi-criterion EA with the concept of non-domination in their strength Pareto EA (SPEA). They suggested maintaining an external population at every generation storing all non-dominated solutions discovered so far beginning from the initial population. This external population participates in genetic operations. At each generation, a combined population with the external and the current population is first constructed. All non-dominated solutions in the combined population are assigned a fitness based on the number of solutions they dominate. To maintain diversity and in the context of minimizing the fitness function, they assigned more fitness to a non-dominated solution having more dominated solutions in the combined population. On the other hand, more fitness is also assigned to solutions dominated by more solutions in the combined population. Care is taken to assign no non-dominated solution a fitness worse than that of the best dominated solution. This assignment of fitness makes sure that the search is directed towards the non-dominated solutions and simultaneously diversity among dominated and non-dominated solution are maintained. On knapsack problems, they have reported better results than any other method used in that study. However, such comparisons of algorithms is not appropriate, simply because SPEA approach uses a inherent *elitism* mechanism of using the best non-dominated solutions discovered up to the current generation, whereas other algorithms do not use any such mechanism. Nevertheless, an interesting aspect of that study is that it shows the importance of introducing elitism in evolutionary multi-criterion optimization.

Knowles and Corne [11] suggested a simple possible MOEO using evolution strategy (ES). In their Pareto-archived ES (PAES) with one parent and one child, the child is compared with respect to the parent. If the child dominates the parent, the child is accepted as the next parent and the iteration continues. On the other hand, if the parent dominates the child, the child is discarded and a new mutated solution (a new child) is found. However, if the child and the parent do not dominate each other, the choice of child or a parent considers the second task of keeping diversity among obtained solutions. To maintain diversity, an archive of non-dominated solutions found so far is maintained. The child is compared with the archive to check if it dominates any member of the archive. If yes, the child is accepted as the new parent and the dominated solution is eliminated from the archive. If the child does not dominate any member of the archive, both parent and child are checked for their *nearness* with the solutions of the archive. If the child resides in a least crowded region in the parameter space among the members of the archive, it is accepted as a parent and a copy of added to the archive. It is interesting to note that both features of (i) emphasizing non-dominated solutions, and (ii) maintaining diversity among non-dominated solutions are present in this simple algorithm. Later, they suggested a multi-parent PAES with similar principles as above.

Van Veldhuizen and Lamont [23] suggested a multi-objective messy GA (MOMGA), which works in two phases: (i) primordial phase finds important *building blocks* associated with the problem, and (ii) juxtapositional phase combines these building blocks to form optimal or near-optimal solutions [9]. In the primordial phase, partial strings are initialized. Unlike in the original messy GA, here, multiple templates, each corresponding to an individual objective function, are needed. MOMGA begins with random templates and then finds best templates for each objective from the best solutions obtained at the end of each *era*, thereby finding the competitive template for each objective for an era, a matter which is important for proper evaluation of partial solutions in an era. His selection operator is exactly the same as in [10] and the recombination operator is exactly identical as that of the original messy GAs [9]. Since the dominance measure is used for selection and diversity preserving mechanism of NPGA is used, the resulting algorithm has both the desired properties of a multi-objective optimizer. The study also suggests a parallel version of MOMGA to overcome the computational burden of messy GAs.

Rudolph [17] suggested, but did not simulate, a simple elitist multi-objective EA based on systematic comparison of individuals from parent and offspring populations. The non-dominated solutions of the offspring population are compared with that of parent solutions to form an overall non-dominated set of solutions, which becomes the parent population of the next iteration. If the cardinality of this set is not the same as the desired population size, other individuals from the offspring population are included. With this strategy, he has been able to prove the convergence of this algorithm to the Pareto-optimal set. Although this is an important achievement in its own right, the algorithm lacks motivation for the second task of maintaining diversity of Pareto-optimal solutions. An explicit diversity preserving mechanism must be added to make it more usable in practice.

Laumanns, Rudolph, and Schwefel [12] suggested a predator-prey approach which is closer to the nature. In a toroidal graph, preys with the solution vectors are kept in the vertices of the graph and predators move along the edges in search of preys. At least M predators chase a prey. A predator is associated with an independent objective function and catches a prey with worst objective function value. When the prey is caught, it is deleted from the vertex and is replaced by a mutated solution. Although no explicit non-domination concept is used, the above concept is shown to force the prey solutions to converge to the Pareto-optimal solutions in a couple of test problems using a mutation-based evolution strategy. In order to obtain a better distribution of Pareto-optimal solutions, 100 predators per objective function is deployed in the neighborhood of each prey. The idea is nice and different from the rest of the multi-objective evolutionary algorithms, but it remains to be investigated how this concept scales to more complex problems.

5 Multi-Criterion Evolutionary Optimization: Future

In this section, we discuss a number of issues that must be attempted to answer in order to understand the working principles of multi-objective evolutionary optimization better and to make them more useful in practice. A list of them are outlined and discussed in the following:

1. Comparison of existing multi-objective GA implementations

2. To understand the dynamics of GA populations with generations

3. Scalability of multi-objective GAs with number of objectives

4. Introduction of elitism in multi-objective EAs

5. Convergence to Pareto-optimal front

6. Definition of metrics for comparing two populations

7. Application to real-world problems

As mentioned earlier, there exists a number of different MOEO implementations primarily varying in the way non-dominated solutions are emphasized and in the way the diversity in solutions are maintained. Although some studies have compared different implementations [26, 25], there is a need of more studies comparing computational complexities and accuracies of finding solutions close to the Pareto-optimal front.

An interesting study would be to investigate how an initial random population of solutions move from one generation to the next. In an initial random population, it is expected to have solutions belonging to many non-domination levels. One hypothesis about the working of a multi-objective GA would be that most population members soon collapse to a single non-dominated front and each generation thereafter proceeds by improving this large non-dominated front. On the other hand, GAs may also thought to work by maintaining a number of non-domination levels at each generation. Both these modes of working should provide enough diversity for the GAs to find new and improved solutions and are thus likely candidates.

Most research in the area use only two objectives, although extensions of these test problems for more than two objectives can also be done. It is intuitive that as the number of objectives increase, the Pareto-optimal region is represented by multi-dimensional surfaces. With more objectives, multi-objective GAs must have to maintain more diverse solutions in the non-dominated front in each iteration. Whether GAs are able to find and maintain diverse solutions, as demanded by the search space of the problem with many objectives would be a matter of interesting study.

Elitism ensures that the best solutions in each generation will not be lost. In multi-objective optimization, all non-dominated solutions of the first level are the best solutions in the population. There is no way to distinguish one solution from the other in the non-dominated set. The decision of which solutions to be carried over is an important one. There exists a few studies [11, 17, 25], but more studies are needed in this direction.

It would be interesting to introduce special features (such as elitism, mutation, or other diversity-preserving operators), which may help us to prove convergence of a GA population to the global Pareto-optimal front [18]. An additional burden here is to prove that a diversity of solutions can be maintained among the Pareto-optimal solutions.

There are two goals in a multi-objective optimization—convergence to the true Pareto-optimal front and maintenance of diversity among Pareto-optimal solutions. A multi-objective GA may have found a population which has many Pareto-optimal solutions, but with less diversity among them. How would such a population be compared with respect to another which has fewer number of Pareto-optimal solutions but with a wide diversity?

As with other growing fields, there is always a need for more applications of multi-objective EAs to real-world problems. Along with the development of new and improved methods, applications must proceed side by side to demonstrate the efficiency of the proposed methods.

6   Conclusions

The popularity of multi-objective evolutionary algorithms over the past decade has been mainly due to the ability of an evolutionary algorithm to find multiple optimal solutions in one single simulation run. This makes the task unbiased and allows a number of trade-off solutions to be found. In this paper, we have categorized the growing list of evolutionary algorithms into past and present approaches. The sheer number of different algorithms exist till to date supports for its popularity and importance in practice. This also demands studying a growing number of immediate tasks related to multi-objective optimization, a list of which has also been highlighted in the paper.

References

[1] Deb, K. (1995) *Optimization for engineering design: Algorithms and examples.* Prentice-Hall, New Delhi.

[2] Deb, K. (1999) Multi-objective genetic algorithms: Problem difficulties and construction of test Functions. *Evolutionary Computation, 7*(3), 205–230.

[3] Deb, K. and Agrawal, R. B. (1995) Simulated binary crossover for continuous search space. *Complex Systems, 9* 115–148.

[4] Deb, K. and Goldberg, D. E. (1989) An investigation of niche and species formation in genetic function optimization. In Schaffer, D., editor, *Proceedings of the Third International Conference on Genetic Algorithms*, pages 42–50, Morgan Kauffman, San Mateo, California.

[5] Fonseca, C. M. and Fleming, P. J. (1993) Genetic algorithms for multi-objective optimization: Formulation, discussion and generalization. In Forrest, S., editor, *Proceedings of the Fifth International Conference on Genetic Algorithms*, pages 416–423, Morgan Kauffman, San Mateo, California.

[6] Fonseca, C. M. and Fleming, P. J. (1998) Multiobjective optimization and multiple constraint handling with evolutionary algorithms–Part II: Application example. *IEEE Transactions on Systems, Man, and Cybernetics: Part A: Systems and Humans*. 38–47.

[7] Goldberg, D. E. (1989) *Genetic algorithms for search, optimization, and machine learning*. Addison-Wesley, Reading, Massachusetts.

[8] Goldberg, D. E. and Deb, K. (1991) A comparison of selection schemes used in genetic algorithms, *Foundations of Genetic Algorithms*, 69–93.

[9] Goldberg, D. E., Korb, B., and Deb, K. (1989) Messy genetic algorithms: Motivation, analysis, and first results, *Complex Systems*, 3:93–530.

[10] Horn, J. and Nafploitis, N., and Goldberg, D. E. (1994) A niched Pareto genetic algorithm for multi-objective optimization. In Michalewicz, Z., editor, *Proceedings of the First IEEE Conference on Evolutionary Computation*, pages 82–87, IEEE Service Center, Piscataway, New Jersey.

[11] Knowles, J. and Corne, D. (1999) The Pareto archived evolution strategy: A new baseline algorithm for multiobjective optimisation. *Proceedings of the 1999 Congress on Evolutionary Computation*, Piscatway: New Jersey: IEEE Service Center, 98–105.

[12] Laumanns, M., Rudolph, G., and Schwefel, H.-P. (1998) A spatial predator-prey approach to multi-objective optimization: A preliminary study. In Eiben, A. E., Bäck, T., Schoenauer, M., and Schwefel, H.-P., editors, *Parallel Problem Solving from Nature, V*, pages 241–249, Springer, Berlin, Germany.

[13] Miettinen, K. (1999) *Nonlinear multiobjective optimization*, Boston: Kluwer.

[14] Parmee, I., Cvetkovic, D., Watson, A. H., and Bonham, C. R. (in press) Multi-objective satisfaction within an interactive evolutionary design environment. *Evolutionary Computation, 8*.

[15] Reklaitis, G. V., Ravindran, A. and Ragsdell, K. M. (1983) *Engineering optimization methods and applications*. New York: Wiley.

[16] Rosenberg, R. S. (1967) *Simulation of genetic populations with biochemical properties*. PhD dissertation. University of Michigan.

[17] Rudolph, G. (1999) Evolutionary search under partially ordered sets. Technical Report No. CI-67/99, Dortmund: Department of Computer Science/LS11, University of Dortmund, Germany.

[18] Rudolph, G. and Agapie, A. (2000) Convergence properties of some multi-objective evolutionary algorithms. Technical Report No. CI-81/00, Dortmund: Department of Computer Science/LS11, University of Dortmund, Germany.

[19] Schaffer, J. D. (1984) Some experiments in machine learning using vector evaluated genetic algorithms. Doctoral Dissertation, Vanderbilt University, Nashville, Tennessee.

[20] Sen, P. and Yang, J.-B. (1998) *Multiple criteria decision support in engineering design.* London: Springer.

[21] Srinivas, N. and Deb, K. (1995) Multi-Objective function optimization using non-dominated sorting genetic algorithms, *Evolutionary Computation*, 2(3):221–248.

[22] Steuer, R. E. (1986) *Multiple criteria optimization: Theory, computation, and application.* John Wiley, New York.

[23] Van Veldhuizen, D. and Lamont, G. B. (1998) Multiobjective evolutionary algorithm research: A history and analysis. Technical Report Number TR-98-03. Wright-Patterson AFB, Department of Electrical and Computer Engineering, Air Force Institute of Technology, Ohio.

[24] Zitzler, E. (1999). Evolutionary algorithms for multiobjective optimization: Methods and applications. Doctoral thesis ETH NO. 13398, Zurich: Swiss Federal Institute of Technology (ETH), Aachen, Germany: Shaker Verlag.

[25] Zitzler, E., Deb, K., and Thiele, L. (in press) Comparison of multiobjective evolutionary algorithms: Empirical results. *Evolutionary Computation, 8.*

[26] Zitzler, E. and Thiele, L. (1998) Multiobjective optimization using evolutionary algorithms—A comparative case study. In Eiben, A. E., Bäck, T., Schoenauer, M., and Schwefel, H.-P., editors, *Parallel Problem Solving from Nature, V,* pages 292–301, Springer, Berlin, Germany.

City Planning with a Multiobjective Genetic Algorithm and a Pareto Set Scanner

Richard J. Balling
John T. Taber
Kirsten Day
Scott Wilson
Department of Civil and Environmental Engineering
Brigham Young University
email: balling@byu.edu

Abstract. A genetic algorithm was used to search for optimal future land use and transportation plans for a pair of high-growth cities. Hundreds of thousands of plans were considered. Constraints were imposed to insure adequate housing for future residents. Objectives included the minimization of traffic congestion, the minimization of costs, and the minimization of change from the status quo. The genetic algorithm provided planners and decision-makers with a rich set of optimal plans known as the Pareto set. An interactive computer tool was developed to assist decision-makers in scanning the information in the Pareto set as they "shop" for a plan.

1. Introduction

The Wasatch Front Metropolitan Region (WFMR) is located along the western base of the Wasatch mountain range in the state of Utah, USA. The region is about 130 kilometers long and 24 kilometers wide and includes all of the major cities in the state of Utah including Salt Lake City. The WFMR has experienced unprecedented growth in the past decade. Polls show that growth management is the political issue of highest concern in the state right now. Three years ago the Governor convened a three-day Growth Summit which was televised on prime time on all of the major television stations. He focused the discussion on the issues of transportation, open space, and water. The urgency of developing a plan for growth has been heightened by the anticipated Winter Olympics in 2002.

The authors have been involved in a U.S. National Science Foundation project to develop future land-use and transportation plans for the WFMR. In meeting with city and state officials, it became clear that this is a problem with multiple competing objectives including:

1) minimize traffic congestion,
2) provide adequate and affordable housing,
3) preserve open spaces, 4) minimize cost,
5) minimize crime,
6) maximize opportunity for economic development,

7) minimize air pollution,
8) minimize change,
9) provide adequate water and utilities.

It also became clear that it is impossible to quantify the relative importance of these objectives. There are many politicians involved in this politically-charged problem, and each is trying to satisfy the preferences of his or her constituency. The public preference may be irrational and difficult to assess. The most important objective today may not be the most important objective tomorrow. Thus, obtaining weights or acceptable values for the above objectives is an exercise in futility. How then, can one rationally search for an optimum plan?

2. The Approach

Our approach is to produce a set of plans rather than a single optimum plan. The set should consist of competitive but widely different plans. We will then put this set before the decision-makers who will go through the value-driven process of selecting a plan. The set we intend to place before the decision-makers is a Pareto set of non-dominated plans. A plan is a member of this set if no other single plan has been found which is better in every objective. Thus, no matter how the objectives are weighted, the optimum plan should be a member of the Pareto set. In this way we perform the computational legwork necessary to keep the decision-makers focused on competitive plans and divert their attention away from wasteful dominated plans.

This approach utilizes two different computational tools. First, a multiobjective genetic algorithm is used to generate the Pareto set. We have devised a fitness function which is a quantitative measure of Pareto-optimality. Second, an interactive, graphical tool has been developed to assist decision-makers in scanning through the plans in the Pareto set enabling the vast amount of data therein to be quickly assimilated.

3. The Application Problem

Earlier the approach was applied to the city of Provo in the WFMR [1]. The city of Provo is adjacent to its sister city of Orem. We decided to apply the approach to both cities simultaneously in order to obtain better plans. The population of Provo/Orem is currently about 183,000. By the year 2020, the population is projected to increase to 352,000.

The land in the cities of Provo and Orem was divided into 199 zones. A design variable was associated with each zone. The values of these 199 design variables are integers ranging from 1 to 17 corresponding to the land uses listed in Table 1.

The major streets in both cities were divided into 60 transportation corridors. All other streets were assumed to remain fixed at their status quo capacities. A design variable was associated with each corridor. The values of these 60 design variables are integers ranging from 1 to 10 corresponding to the street types listed in Table 2.

	Land Use	Description	housing density (homes/km^2)
1	FARM	farm land	21
2	VLDRm	very low density residential -- medium income	371
3	VLDRh	very low density residential -- high income	371
4	LDRl	low density residential -- low income	716
5	LDRm	low density residential -- medium income	716
6	LDRh	low density residential -- high income	716
7	MDRl	medium density residential -- low income	2150
8	MDRm	medium density residential -- medium income	2150
9	MDRh	medium density residential -- high income	2150
10	HDRl	high density residential -- low income	4990
11	HDRm	high density residential -- medium income	4990
12	HDRh	high density residential -- high income	4990
13	CBD	central business district	
14	SC	shopping center	
15	GC	general commercial	
16	LI	light industrial	
17	HI	heavy industrial	

Table 1: Possible Land Uses

Thus, there are 199+60 = 259 design variables for both cities. The total number of possible future plans is $17^{199}10^{60}=10^{305}$. Some plans were deemed unacceptable. For example, it was decided that the corridors could only be allowed to upgrade in capacity and not downgrade. In addition, the following housing constraints were imposed:

low income housing capacity $\geq$ 87,000 persons
medium income housing capacity $\geq$ 134,000 persons
high income housing capacity $\geq$ 114,000 persons
total housing capacity $\geq$ 352,000 persons

	corridor	description	ave. speed (km/hr)	capacity (veh/hr)
1	C0	unused		
2	C2	2-lane collector	53	920
3	C3	3-lane collector	56	1380
4	A2	2-lane arterial	61	2400
5	A3	3-lane arterial	64	2600
6	C4	4-lane collector	60	3160
7	C5	5-lane collector	64	5200
8	A5	5-lane arterial	72	6100
9	C7	7-lane collector	64	7800
10	A6	6-lane arterial	72	9150

Table 2: Possible Street Types

The housing capacity of a future plan is easily evaluated by summing the areas of the appropriate residential zones multiplied by their respective housing densities, and multiplying the sum by an average 3.68 persons per home. A future plan which does not satisfy the all of the housing constraints is labelled an "infeasible" plan. The status quo zoning plan for Provo/Orem is infeasible since its total housing capacity is only 289,644 persons.

Three competing objectives were identified for this problem. The first objective was the minimization of traffic congestion. A traffic analysis model was developed. This model prescribed daily trips between zones and assigned them to the transportation corridors and existing streets. Trips were rerouted by the model when congestion develops. The sum of the travel time for all trips in a 24-hour period was minimized as the objective. Evaluation of this objective for a single future plan on a 200MHz Pentium processor required about 105 seconds.

The second objective was the minimization of cost. The cost of corridor upgrades was determined. This included both construction cost and right-of-way purchase. Sales tax and property tax revenues are also affected by changing land use. Thus, it is desirable to both minimize cost and maximize revenue. The second objective was taken as the minimization of annualized cost minus revenue. This was divided by the total housing capacity to get a per capita value. Since it is likely that the considered revenues exceed the considered costs, this objective function usually had a negative value. This objective could be evaluated in less than a second for a particular future plan on a 200MHz pentium processor.

The third objective was the minimization of change from the status quo. This objective was included because future plans which are radically different from the status quo may be politically unacceptable to the public. The change objective was calculated as the sum of change over the zones and corridors. The change for a single zone was taken as the status quo land value of the zone multiplied by a degree-of-change factor between zero and one. A matrix of degree-of-change factors between status quo land use and future plan land use was pre-specified. A value of one indicated the most extreme case of land use change, while a value of zero indicated no change. The change for a single corridor was taken as the status quo land value of a strip of land on both sides of the corridor multiplied by a street degree-of-change factor. The change objective was measured in dollars. This objective could be evaluated in less than a second for a particular future plan on a 200MHz pentium processor.

4. Multiobjective Genetic Algorithm

Research on the development of multiobjective genetic algorithms has received considerable attention in recent years. Survey papers are available [2-3]. In this work, we used a genetic algorithm to search for a Pareto set of optimum plans. The genetic algorithm begins with a randomly chosen starting generation of feasible plans, and proceeds from generation to generation for a specified number of generations.

Suppose we temporarily ignore the change objective, and consider only the travel time and cost objectives. We could plot the values of travel time and cost for each member of the starting generation as shown hypothetically in Figure 1. Note that no single plan has both the lowest travel time and the lowest cost. This is because these two objectives are competing. The 7 plans on the Pareto front in Figure 1 represent the Pareto set for the starting generation. All other plans

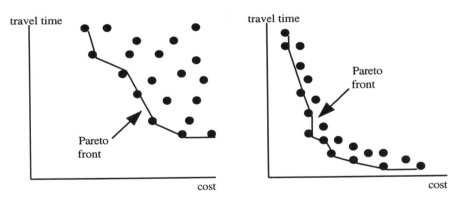

Figure 1: Starting Generation   Figure 2: Final Generation

are dominated because for each such plan there exists a Pareto plan which has both lower travel time and cost. As the genetic algorithm proceeds from generation to generation, we would hope that the Pareto front advances both leftward and downward to a final generation which is hypothetically plotted in Figure 2. Hopefully, the Pareto set for the final generation is close to the Pareto set for the universe of possible plans.

In order for the genetic algorithm to improve Pareto optimality from generation to generation, we must devise a fitness function which measures Pareto optimality in a generation. Goldberg [4] proposed a discrete fitness function which simply ranks plans on the Pareto front with a rank of one and dominated plans with a rank of two. The fitness is then taken as the reciprocal of the rank.

Instead, we devised a continuous-valued fitness function. Suppose there are N objective functions to be minimized. Let f_{ki} be the value of the kth objective evaluated for the ith plan in the generation. The ith plan will be dominated by the jth plan if:

$$f_{ki} > f_{kj} \quad \text{for } k=1,N \tag{1}$$

This is equivalent to:

$$\min_{k}(f_{ki} - f_{kj}) > 0 \tag{2}$$

Thus, the ith plan is dominated if:

$$\max_{j \neq i}\left(\min_{k}(f_{ki} - f_{kj})\right) > 0 \tag{3}$$

Let f_{kmax} and f_{kmin} be the maximum and minimum values of the kth objective over all plans in the generation. The fitness of the ith plan is:

$$\text{fitness}_i = \left[1 - \max_{j \neq i}\left(\min_{k}\left(\frac{f_{ki} - f_{kj}}{f_{kmax} - f_{kmin}}\right)\right)\right]^p \tag{4}$$

This fitness function will be less than one for dominated plans, and greater than one for Pareto plans. The positive exponent p amplifies the difference between dominated and Pareto plans. In our study, a value of p=15 was used.

Our implementation of the genetic algorithm used a generation size of 100 feasible plans. Each plan in the starting generation was created by randomly changing the land uses in up to 30% of the zones from their status quo land uses, and randomly upgrading the street types in up to 50% of the corridors from their status quo street types. Future plans which did not satisfy the housing constraints were immediately aborted. The random changes in land use were biased towards residential land use in order to more quickly find feasible plans. Nevertheless, 393,309 future plans were aborted during the process of finding 100 feasible plans for the starting generation.

The process for creating a new generation from a previous generation will now be explained. First, ten plans in the previous generation were automatically cloned to the new generation. These plans included the best plans for each objective minimized individually as well as the most fit plans. The remaining 90 plans in the new generation were generated by repeatedly executing the following four step process:

1) selection
2) crossover
3) mutation
4) abortion

The selection step involved randomly selecting a father plan and a mother plan from the previous generation. This random selection was biased according to the fitness function. Plans can be represented by chromosomes with 259 genes corresponding to the 199 zone variables and 60 corridor variables. In the crossover step, a random integer between 1 and 199 was generated, and a random integer between 1 and 60 was generated. These two integers defined the crossover genes in the zone and corridor portions of the chromosome, respectively. The chromosomes corresponding to the father and mother plans were cut immediately after the crossover genes, and the segments following the cut were swapped in each portion of the chromosome. Thus, two new chromosomes were created corresponding to two new children plans. In the mutation step, each gene in the two new children chromosomes was randomly changed with a low probability. The probability of mutation for zone genes was 0.5%, and the probability of mutation for corridor genes was 5%. In the abortion step, the housing constraints were evaluated. Any child plan which violated any of the housing constraints was immediately aborted. On average, 19.5 plans were aborted in order to obtain the remaining 90 feasible plans for each generation. These 90 plans were then analyzed to calculate the values of the three objective functions and the fitness function.

The genetic algorithm was executed for 100 generations. Thus, 10,000 feasible plans were analyzed for travel time, cost, and change. This required about 12 days. In the process, 395,235 infeasible plans were aborted. Significant improvements were achieved in all three objectives as shown in Table 3.

objective function	minimum value starting generation	minimum value final generation	percent reduction
travel time	275,955 veh-hr/day	204,808 veh-hr/day	26%
cost minus revenue	-$25,917/capita/yr	-$45,342/capita/yr	75%
change	$6,127,320,000	$4,125,340,000	33%

Table 3: Results from Genetic Algorithm

We lumped all 10,000 feasible plans from the 100 generations into a single "global generation" and found the Pareto set for this global generation. The global Pareto set contained 209 distinct plans. Most of these plans were discovered in the final generations. This global Pareto set of 209 plans represents the best plans found by our multiobjective genetic algorithm.

4. Pareto Set Scanner

The amount of data contained in the global Pareto set of 209 plans is overwhelming for decision-makers. We developed an interactive tool which allows decision-makers to graphically explore the global Pareto set. We call this tool a Pareto set scanner.

This tool displays a map of the city for a particular Pareto plan on the computer screen as shown in Figure 3 (black and white version). Land uses for each zone are indicated by color, and street types for each corridor are indicated by line width. Numerical values of the three objectives for the particular Pareto plan being displayed are also shown on the screen. In addition, there is a slider bar for each of the three objectives. The user can use the mouse to move the slider bar for each objective to any position between zero and one. The positions of these slider bars indicate the relative importances of the objectives. A relative importance of one for a particular objective is maximum, and a relative importance of zero is minimum. These relative importances can be translated into objective values. A relative importance of one for a particular objective corresponds to the minimum value of that objective among all plans in the global Pareto set. A relative importance of zero for a particular objective corresponds to the maximum value of that objective among all plans in the global Pareto set. Thus, the positions of the slider bars define a coordinate in objective space. The Pareto plan whose map and numerical data are displayed on the screen is the plan whose coordinate in objective space is closest in the Euclidean sense to the coordinate defined by the slider bar positions.

All of this occurs in real time. Thus, as the user moves the slider bars re-adjusting the relative importances of the objectives, the map and numerical data displayed on the screen change from Pareto plan to Pareto plan. It is possible to execute this in real time because the values of the design variables and objectives for each plan in the Pareto set were stored in a database. In fact, bitmap images of the map of each Pareto plan were stored in the database. Thus, in real time, the computer recalls the appropriate bitmap image. We have even placed this pareto set scanner on the web so that the citizens of Provo and Orem may scan through future plans.

This tool allows decision-makers to observe trends as they interactively shift the relative importances of the objectives. They can place maximum importance on one particular objective and no importance on the other objectives and observe the resulting plan on the screen. They may observe that the optimum plan of some parts of the city may be insensitive to the relative importance placed on some objectives, and very sensitive to the relative importance placed on other objectives. If two objectives are not competing, then the map displayed on the screen would not change as the relative importance is shifted between these two objectives. This tool allows decision-makers to rapidly scan through plans in the global Pareto set. As they do so, they will see what is achievable, and they will sharpen their preferences. Ultimately, this can help them make a decision and select a final plan.

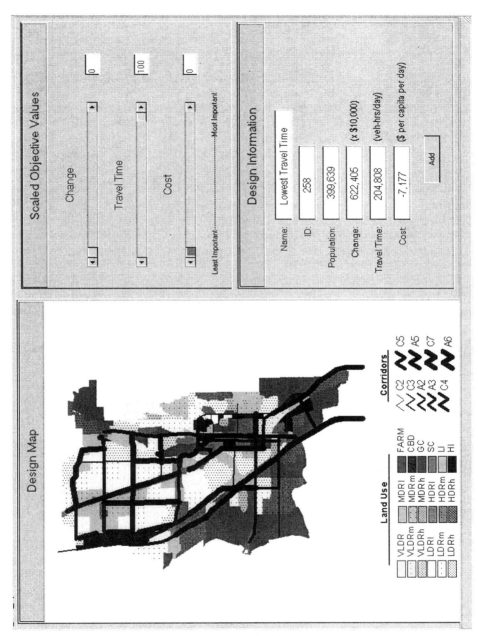

Figure 3: Pareto Set Scanner

5. Conclusions

The future land use and transportation plan for two adjacent high growth cities was optimized with a genetic algorithm. The algorithm produced a set of 209 Pareto plans. An interactive Pareto set scanner was developed to assist decision-makers in examining the Pareto set and selecting a final plan.

Our approach can be extended to other engineering design and planning problems. We call our approach "design by shopping". Compare this to the traditional optimization approach which requires that decision-makers quantify the relative importance of objectives before executing an optimization algorithm which produces a single optimum design. The design by shopping approach does not require decision-makers to grapple with relative importances until after the optimization algorithm produces a rich set of good designs (the Pareto set). Thus, the traditional optimization approach is classified as an "a priori articulation of preference" approach while the design by shopping approach is classified as an "a posteriori articulation of preference" approach [5].

The design by shopping approach as two distinct advantages over the traditional optimization approach. Both are psychological in nature. The first advantage is that most people can make better decisions regarding preferences after they are able to examine good designs. The shopping process is essential to articulation of preference and good decision-making. The second advantage is that with design by shopping, the decision-makers are in control of the final step of the process. With the traditional optimization approach, decisions are made up front and then control is turned over to the optimization algorithm to produce the final design. Most decision-makers are uncomfortable with this order of events, and the likelihood is high that they will not be satisfied with the design produced by the optimization algorithm. On the other hand, since decision-makers make the final selection in the design by shopping approach, the likelihood of satisfaction is much higher.

6. Acknowledgements

This work was funded by the National Science Foundation under Grant No. CMS-9526018 for which the authors are grateful.

7. References

1. Balling, R.J., Taber, J.T., Brown, M.R., and Day, K., 1999. Multiobjective urban planning using a genetic algorithm, *ASCE Journal of Urban Planning and Development*, 125:86-99

2. Coello Coello, C.A., 1999. A comprehensive survey of evolutionary-based multiobjective optimization techniques, *Knowledge and Information Systems*, 1:269-308

3.   Fonseca, C.M. and Fleming, P.J., 1995. An overview of evolutionary algorithms in multiobjective optimization, *Evolutionary Computation*, 3:1-16

4.   Goldberg, D.E., 1989. *Genetic Algorithms in Search, Optimization, and Machine Learning*, Addison-Wesley, Reading, MA, USA

5.   Hwang, C.-L. and Masud, A.S.M, 1979. *Multiple Objective Decision Making -- Methods and Applications: A State-of-the-Art Survey; Lecture Notes in Economics and Mathematical Systems No. 164*, Springer-Verlag, Berlin

Designer's Preferences and Multi–objective Preliminary Design Processes

Dragan Cvetković and Ian Parmee

Plymouth Engineering Design Centre, University of Plymouth, UK

email: {dcvetkovic,iparmee}@plymouth.ac.uk

Abstract

In this paper we present a method based on preference relations for transforming non–crisp (qualitative) relationships between objectives in multi–objective optimisation into quantitative attributes (i.e. numbers). This is integrated with two multi–objective genetic algorithms (GAs): weighted sums GA and a method for combining the Pareto method with weights. Examples of the preference relations application together with traditional genetic algorithms are also presented. Some complexity issues are discussed and analysed.

1 Introduction

When dealing with industrial design problems, it rapidly becomes apparent that there are significant differences between so called 'textbook optimisation problems' and 'real–world applications'. In both cases, in multi–objective optimisation we have a multi–objective function to optimise under a set of constraints:

$$\max_{x}(f_1(x),\ldots,f_k(x)) \quad \text{s.t.} \quad g_1(x,p) \leq 0,\ldots,g_l(x,p) \leq 0 \tag{1}$$

This problem is well known and a number of approaches, both non–genetic [1] and genetic algorithm [2] approaches exist. An additional problem is that not all objectives are equally important which necessitates the use of weights or preferences. We have applied appropriate genetic algorithms for multi–objective optimisation and described a number of the design problems in [3].

According to the psychological study in [4, p. 254], people generally have problems in reporting accurately on the weights they assign to the various stimulus factors in making evaluations. The designer cannot always completely objectively define the preferences regarding the objectives which have to be optimised. A common situation is a subjective statement "objective A is much more important than objective B" but without any quantitative representation. In this paper we address this problem and integrate our developed preference methods to different preliminary design techniques. Some of the work presented here has been presented in [5].

2 Fuzzy preferences

One general problem of the (very popular) weighted sums based multi–objective optimisation approach is how to specify weights. Figure 2.1 illustrates the problem that we face when optimising 2 objectives. The example is based on objectives y_4 and y_9 of the BAe airframe problem briefly described in section 4.3 and in more detail in [3]. The rightmost point has been obtained by setting the weight w_4 of y_4 to 0 and the weight w_9 of y_9 to 1, whereas the leftmost point has been obtained for $(w_4, w_9) = (1, 0)$. The points in–between are obtained by varying w_4 and w_9 under the constraint $w_4 + w_9 = 1$. Note that in the case of convex Pareto fronts, we can obtain every point of Pareto front just by varying those weights. In this figure there are only two objectives but more usually there are 20 or more objectives to consider at the same time. We will present an approach here that can simplify the weights assignment problem.

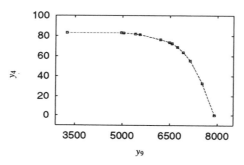

Figure 2.1: Influence of weights on BAe function

In our approach, for every two objectives we ask the designer to specify one of the following characterisations:

- Much less important ($\ll$)
- Less important ($\prec$)
- Equally important ($\approx$)
- Don't care (#)
- More important ($\succ$)
- Much more important ($\gg$)

We can easily extend the number of degrees of importance (such as slightly more important, vastly more important etc.).

2.1 Treatment of *don't care* relation

In the following text, *don't care* is treated exactly as *equally important*. There is a psychological justification for this: if we do not care in respect of two objectives, then we also do not care which one provides better results. Alternatively, we can use the following: "Given a partial preorder structure, it is always possible to replace the incomparabilities by preferences or indifferences in order to obtain a complete preorder structure" [6, p. 113].

Roy [7] distinguishes two types of hesitations:

weak preference: can not decide between $a \succ b$ and $a \approx b$, being sure that $b \not\succ a$;

incomparability: can not decide between $a \succ b$ and $b \succ a$.

He states that any hesitation may come from:

- the existence in the decision–maker's mind of zones of uncertainty, half–held belief, or conflicts and contradictions;
- the vaguely defined quality of the decision–maker;
- the fact that the engineers who built the model ignore, in part, how the decision–maker compares a and b;
- the imprecision, uncertainty, inaccurate determination of the maps $g(a)$ and $g(b)$ by means of which a and b are compared.

2.2 Method used and properties of relations

In general, k objectives require in the worst case $k(k-1)/2$ questions, but usually less is enough to construct the fuzzy preference relation R. The question of complexity will be discussed in section 3.2 below.

Once we know the fuzzy preference relation $R \subseteq A^2$, we can compute the complete ordering on A using the relation graph $G = (A, R)$ and the concept of 'leaving score' [8, p. 151] and the corresponding order:

$$S_L(a, R) \stackrel{\text{def}}{=} \sum_{c \in A \setminus \{a\}} R(a, c) \tag{2}$$

$$a \geq_L b \text{ if and only if } S_L(a, R) \geq S_L(b, R) \tag{3}$$

We define the following relations:

relation	intended meaning
$\approx$	is equally important
$\prec$	is less important
$\ll$	is much less important
$\neg$	is not important
$!$	is important

The required properties are:

- Relation $\approx$ is an *equivalence relation* (reflexive, symmetric and transitive);
- Relations $\prec$ and $\ll$ are *strict orders* (irreflexive and transitive);
- Relation $\approx$ is *congruent* with $\ll$ and $\prec$;
- Relation $\ll$ is sub–relation of $\prec$ i.e. $x \ll y \Rightarrow x \prec y$;
- Miscellaneous properties:

$$!x \vee \neg x, \quad !y \wedge \neg x \Rightarrow x \ll y, \quad \neg x \wedge \neg y \Rightarrow x \approx y, \quad x \prec y \wedge y \ll z \Rightarrow x \ll z$$

Predicates $\succ$ (is more important) and $\gg$ (is much more important) are then defined in the obvious way.

2.2.1 *Some philosophical aspects, contradictions and group preferences*

Transitivity, yes or no? Whether preferences should be transitive is a very much discussed question in the Logic of Preferences field [9]. Some of the authors argue that human preferences are not necessarily transitive and give very good examples of non–transitivity of preference relations [10].

However, being pragmatic, we can argue that transitivity reduces the number of questions (e.g. if A is more important than B, and B more important than C, it is automatically inferred that A is more important than C). Transitivity is also used to avoid cyclic preferences $A \prec B$, $B \prec C$, $C \prec A$ or indifferences $A \approx B$, $B \approx C$ and $A \prec C$.

Further, intransitivity of preferences could yield to contradictions. Contradiction (in classical mathematical logic) is something we try to avoid because by having contradictions, every possible conclusion is derivable. There are some branches of logic that try to bypass this situation i.e. *logic of relevance* [11] and *Paraconsistent logics* [12].

Group preferences In most real–world cases, the designer is not just *one* person but a group of people usually with different and sometimes contradictory opinions relating to what is important and what is not. This causes an already hard decision–making process to be even harder.

We can use group preferences to aggregate single designer's preferences, but there are some problems as stated by Arrow's Impossibility Theorem [13].

Arrow's problem is roughly as follows: Given the ranking of a set of alternatives by each individual in a decision making group, what should the grouping ranking for these alternatives be? He postulated some very reasonable assumptions concerning the aggregation of individuals' rankings, and then he investigated their composite implications. These assumptions are as follows [14, p. 523]: Complete Domain, Positive Association of Social and Individual Ordering, Independence of Irrelevant Alternatives, Individual's Sovereignty, and Nondictatorship. Arrow proved that there is no rule for combining the individual's rankings that is consistent with these seemingly innocuous assumptions.

One interpretation of the Theorem is that, in general, there is no procedure for combining individual rankings into a group ranking that does not explicitly address the question of interpersonal comparison of preferences.

3 Algorithm and Complexity

3.1 Description of the algorithm

- Let the set of objectives be $O = \{o_1, \ldots, o_k\}$. Construct the equivalence classes $\{C_i \mid 1 \leq i \leq m\}$ of the equivalence relation $\approx$ and choose one element x_i from each class C_i giving set $X = \{x_1, \ldots, x_m\}$ where $m \leq k$.
- Use the following valuation v:

    - If $a \ll b$ then $v(a) = \alpha$ and $v(b) = \beta$

 – If $a \prec b$ then $v(a) = \gamma$ and $v(b) = \delta$
 – If $a \approx b$ then $v(a) = v(b) = \varepsilon$.[1]

Note: Taking into account the intended meaning of the relations, we can assume that $\alpha < \gamma < \varepsilon = 1/2 < \delta < \beta$ and $\alpha + \beta = \gamma + \delta = 1$. We are not limited to constants α, β, γ, δ and ε. The above valuation is the simplest possible case but we can as well use the more complex one $\alpha = h(u, t)$ etc, where u is some parameter, t is time (as we can change our preferences as the time goes on), and h is some real-valued function. Their order and property $\alpha + \beta = \gamma + \delta = 1$ is what matters.

- Initialise two matrices R and R_a of size $m \times m$ to the identity matrix $\mathbf{E}_m$. They will be used in the following way:

$$x_i \ll x_j \Leftrightarrow R(i, j) = \alpha, R(j, i) = \beta \Leftrightarrow R_a(i, j) = 0, R_a(j, i) = 2$$
$$x_i \prec x_j \Leftrightarrow R(i, j) = \gamma, R(j, i) = \delta \Leftrightarrow R_a(i, j) = 0, R_a(j, i) = 1 \quad (4)$$
$$x_i \approx x_j \Leftrightarrow R(i, j) = \varepsilon, R(j, i) = \varepsilon \Leftrightarrow R_a(i, j) = 1, R_a(j, i) = 1$$

Note: This valuation already provides an indication of how to generalise preferences to have s stages instead of only 5 (from "much less important" to "much more important"): if x_i is (say) s' times more important the x_j, we will simply assign $R_a(i, j) = s'$ and $R_a(j, i) = 0$ etc.

- Perform the following procedure:

 1. For each pair of objectives x_i and x_j s.t. $R_a(i, j) + R_a(j, i) = 0$ and $i \neq j$:

  (a) ask whether $x_i \ll x_j$, $x_i \prec x_j$, $x_j \ll x_i$ or $x_j \prec x_i$ and using (4) set $R_a(i, j)$ and $R_a(j, i)$ accordingly;

  (b) Using Warshall's algorithm [15], compute transitive closure of R_a;

 2. Using (4), calculate matrix R from R_a;

- For each $x_i \in X$ compute weight as a normalised leaving score:

$$w(x_i) = \frac{S_L(x_i, R)}{\sum_{x_j \in X} S_L(x_j, R)}.$$

and for each $y \in C_i$ set $w(y) = w(x_i)$.

Example 1 Let $O = \{o_1, \ldots, o_6\}$, and $o_1 \approx o_2$ and $o_3 \approx o_4$. We have

$$C_1 = \{o_1, o_2\}, \quad C_2 = \{o_3, o_4\}, \quad C_3 = \{o_5\}, \quad C_4 = \{o_6\}$$

and $X = \{x_1, x_2, x_3, x_4\}$ where $x_i \in C_i$ for $1 \leq i \leq 4$. Let $R = R_a = \mathbf{E}_4$ — identity 4×4 matrix.

 Suppose that the first question gives the answer $x_2 \ll x_1$, the second question gives the answer $x_3 \prec x_1$ and the third one $x_1 \prec x_4$. The fourth question gives the answer $x_2 \ll x_3$ and since for each pair (i, j) we have $R_a(i, j) + R_a(j, i) \neq 0$, we have enough information (without computing transitive closure we would have to ask 6 questions

---

[1] Since we work with class represents, this is only possible if $a = b$.

and additionally handle non–consistent answers), and we can construct the matrix R. Suppose that $\alpha = 0.05$, $\beta = 0.95$, $\gamma = 0.35$, $\delta = 0.65$ and $\varepsilon = 0.5$. Then

$$R = \begin{bmatrix} \varepsilon & \beta & \delta & \gamma \\ \alpha & \varepsilon & \alpha & \alpha \\ \gamma & \beta & \varepsilon & \gamma \\ \delta & \beta & \delta & \varepsilon \end{bmatrix} = \begin{bmatrix} 0.50 & 0.95 & 0.65 & 0.35 \\ 0.05 & 0.50 & 0.05 & 0.05 \\ 0.35 & 0.95 & 0.50 & 0.35 \\ 0.65 & 0.95 & 0.65 & 0.50 \end{bmatrix}.$$

Further,

$$S_L(x_1, R) = 1.95, \ S_L(x_2, R) = 0.15, \ S_L(x_3, R) = 1.65 \text{ and } S_L(x_4, R) = 2.25$$

and the order of importance is $x_2 \ll x_3 \prec x_1 \prec x_4$. Weights w_i are finally calculated and normalised so that they sum to 1:

$$w_1 = w_2 = 0.2407, w_3 = w_4 = 0.0185, w_5 = 0.2037, w_6 = 0.2778.$$

3.2 Scalability and complexity issue

One important question is "how does the preference method scale with increased number of objectives?". It is easy to see that in the worst–case scenario for k objectives we need $n_q^*(k) = k \cdot (k-1)/2$ questions, but usually transitivity and the other preference properties reduce this number. In order to find the average number of questions needed we have performed the following test: for a given number of objectives k random answers to each of the questions is given, and the number of questions needed to construct the complete preference matrix is counted. The tests have been performed for $4 \leq k \leq 100$, each repeated 100 times, and average value $\bar{n}_q$ and its standard deviation $\sigma_q(k)$ have been calculated. The results are presented in Table 3.1. For a comparison, the maximal theoretical number of questions n_q^* is presented in column 4. As we can see whereas $n_q^* \sim O(k^2)$, we have the following approximate formula

$$\bar{n}_q \approx \left(0.85 + \frac{1}{40} \cdot \sqrt{k}\right) \cdot k \cdot \ln k \tag{5}$$

i.e.

$$\bar{n}_q \sim O(\sqrt{k} \cdot k \cdot \ln k + k \cdot \ln k) \sim O(\sqrt{k} \cdot k \cdot \ln k) \tag{6}$$

The value of (5) is presented as a last column in Table 3.1.

Note: The worst–case scenario (i.e. the designer has to answer all n_q^* questions) happens when the designer answers all the questions with the answers in the 'same direction' (e.g. $y_1 \succ y_2, y_1 \gg y_3, \ldots, y_1 \succ y_k, y_2 \gg y_3, \ldots, y_2 \succ y_k, \ldots, y_{k-1} \succ y_k$) so that the transitivity can not reduce the number of questions. In some cases asking the questions in different order might help e.g. for order $x_1 \succ x_2 \succ x_3$ instead of asking $x_1 ? x_2, x_1 ? x_3, x_2 ? x_3$ which all give answer $\succ$ we can just ask $x_1 ? x_2, x_2 ? x_3$ and since both answers are $\succ$ we automatically infer $x_1 \succ x_3$. In this case, using 'depth first' traversing instead of 'breadth first' traversing does help. Of course, we do not know in advance the designer's opinions. Dynamically changing the order of questions as they are being answered is a matter of further research.

k	$\bar{n}_q$	$\sigma_q(k)$	$n_q^*(k)$	(5)
4	4.36	0.6745	6	4.99
5	6.37	0.4852	10	7.29
6	8.89	0.5842	15	9.80
7	11.78	0.7047	21	12.48
8	14.83	1.5378	28	15.32
9	17.61	1.0040	36	18.29
10	20.84	2.0137	45	21.39
11	23.80	1.2144	55	24.61
12	26.48	1.1145	66	27.93
15	37.30	1.7552	105	38.46

k	$\bar{n}_q$	$\sigma_q(k)$	$n_q^*(k)$	(5)
18	50.04	2.6282	153	49.74
21	62.25	2.8793	210	61.67
25	79.51	2.6723	300	78.46
30	101.48	2.3376	435	100.70
35	125.68	3.0281	595	124.18
40	150.77	3.3900	780	148.75
50	202.42	3.2820	1225	200.84
60	257.78	3.2492	1770	256.38
75	345.49	4.7958	2775	345.35
100	501.31	5.6527	4950	506.57

Table 3.1: The number of questions $\bar{n}_q$ needed for k objectives with random answers. Average over 100 runs.

4 Applications of preferences

The concept of preferences and of the relative importance of the objectives can be integrated with Genetic Algorithms in at least two different situations:

1. weighted sum based optimisation, and

2. Pareto optimisation.

In both cases the preference method is used to calculate the weights as required.

Our method integrated with weighted sums has a significant advantage over the traditional weighted sum based optimisation methods since the user doesn't have to express the weights quantitatively but qualitatively (within a few categories) which is much easier. More details are given in [5].

However, we are not only limited on objectives: we can also use preferences to assign different priorities and/or importance to constraints, scenarios, agents in agent–based design etc.

4.1 Pareto optimisation based methods

Comparing Pareto principle based multi-objective optimisation with lexicographic order based optimisation, we see two extremes concerning the objective importance: in the case of Pareto optimisation, all objectives are considered simultaneously (and equally important) whereas in the case of lexicographical order the first objective is the most important one and only if we get the same results for the first objective, do we then consider the second objective etc. In this section we try to develop an optimisation method based on the Pareto principle where we can specify the relative importance of objectives.

As in the case of weighted sums based methods, relative importance of objectives in this weighted Pareto method could be specified using weights (quantitatively) or they could be combined with the above developed preference method that would translate qualitative specification into quantitative. Without this combination, our modified Pareto method suffers from the same problem as the weighted sum method: how to specify weights in the case of 15–20 or more objectives.

4.2 Definition of weighted Pareto method

Lin [16] distinguishes between orders on k–dimensional vectors:

$$x \geq y \quad \text{if and only if} \quad (\forall i \leq k)(x_i \geq y_i) \tag{7}$$

$$x \gtrless y \quad \text{if and only if} \quad (x \geq y) \wedge (\exists j \leq k)(x_j > y_j) \tag{8}$$

He notes that the orders (7) and (8) are definable in terms of each other and therefore ordering a set in $\mathbf{R}^k$ by $\gtrless$ is equivalent to ordering a set by $\geq$ [16, p. 46]. With this remark in mind, we define *non–dominance* (usually defined using (8)) in the following way:

Definition 1 We say that (in object space) the vector $x = (x_1, \ldots, x_k)$ is *non–dominated* by vector $y = (y_1, \ldots, y_k)$, denoted $x \succeq y$, if $x_i \geq y_i$ for all $1 \leq i \leq k$. In other words,

$$x \succeq y \iff \frac{1}{k} \sum_{i=1}^{k} I_{\geq}(x_i, y_i) \geq 1, \quad \text{where } I_{\geq}(x, y) = \left\{ \begin{array}{ll} 1, & x \geq y \\ 0, & x < y \end{array} \right. \tag{9}$$

We can generalise (9) (assuming $\sum_{i=1}^{k} w_i = 1$):

$$x \succeq_w y \quad \text{if and only if} \quad \sum_{i=1}^{k} w_i \cdot I_{\geq}(x_i, y_i) \geq 1, \tag{10}$$

or we can even put some threshold $\tau \leq 1$:

$$x \succeq_w^{\tau} y \quad \text{if and only if} \quad \sum_{i=1}^{k} w_i \cdot I_{\geq}(x_i, y_i) \geq \tau. \tag{11}$$

Note: The relation $\succeq_w^{\tau}$ is transitive as a product of transitive (component–wise) orders and has all the usual features of an order relation. Also, we assume that the weights do not change during the optimisation process.

Definition 2 We call relation $\succeq_w$ defined by (10) *w–non-dominance* and the relation $\succeq_w^{\tau}$ defined by (11) *(w, τ)–non-dominance*. The *Pareto front* is defined as a maximal set of non–dominated elements (according to a given order $\succeq$) and this definition is naturally extended to *w*–Pareto front and to *(w, τ)*–Pareto front for a given vector of weights *w* and threshold τ i.e. according to the order $\succeq_w$ and $\succeq_w^{\tau}$ given by (10) and (11) respectively. We assume that at least one of the inequalities is strict.

Note: The standard dominance relation is a special case of (11) for $w = (\frac{1}{k}, \frac{1}{k}, \ldots, \frac{1}{k})$ and $\tau = 1$.

Vector *w* could be either specified directly by the designer or it can be calculated from preferences which would help the designer to work in more qualitative terms without the burden to reason if the weight should be set of 0.1 or to 0.09.

The Pareto front method combined with genetic algorithms is a very powerful optimisation method since it maintains the diversity of population. However, it can be computationally expensive.

4.3 BAe Function and GA/Pareto Optimisation

This section describes applications of the weighted Pareto method described in section 4.2. The British Aerospace (BAe) design problem is presented in [3]. Briefly, we have 9 input variables $x_1, \ldots, x_9$ and 13 objectives $y_1, \ldots, y_{13}$. The interaction between objectives y_4 and y_9 is specially interesting as they strongly conflict.

However, the BAe design problem is not only an optimisation problem, in fact optimisation is a rather small part of it. The problems of conceptual design relate to the fuzzy nature of initial design concepts and the many different variants that engineer wishes to try. Computers should be able to help him explore these variants whilst also suggesting some others as well [3, 17]. Therefore, interaction with the design team is very important. Our goal is to assist the designer during conceptual design process more in the sense of MCDA (multiple criteria decision aid), than MCDM (multiple criteria decision making) [18]. See [19] for more details.

The core of the genetic algorithm we have used is based on the Breeder Genetic Algorithm [20]. It utilises genetic operators suitable for real–valued chromosomes (arithmetic crossover, exponential mutation etc.), and is adapted to use techniques for multi–objective optimisation [3].

Figure 4.1(a)&(b) show two w–Pareto fronts of y_3 versus y_4 for different preferences i.e. for different weights, whereas Figure 4.1(c) show the shape of the complete Pareto front. Figure 4.1(d)&(e) show two w–Pareto fronts of y_4 versus y_9 for different preferences i.e. for different weights, whereas Figure 4.1(f) show the shape of the complete Pareto front. Varying the Pareto threshold and the weights of each objective, we can obtain different Pareto fronts. Knowing the behaviour of the Pareto front with respect to those parameters, would enable us to vary the parameters during the genetic algorithm run to identify those parts of the Pareto front that are of special interest (considering the density of points in given regions etc.). There are some other approaches that use biased sharing in order to explore parts of the Pareto front [21].

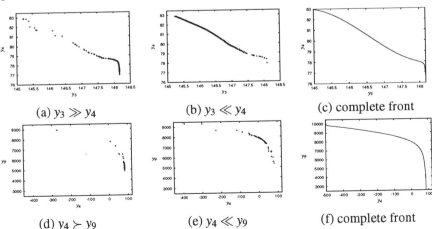

$$\text{(a) } y_3 \gg y_4 \qquad \text{(b) } y_3 \ll y_4 \qquad \text{(c) complete front}$$

$$\text{(d) } y_4 \succ y_9 \qquad \text{(e) } y_4 \ll y_9 \qquad \text{(f) complete front}$$

Figure 4.1: (a)–(c) w–Pareto front of (y_3, y_4) of the BAe function and complete Pareto front. (d)–(f) w–Pareto front of (y_4, y_9) of the BAe function and complete Pareto front.

5 Numerical Examples

5.1 Example with 8 objectives

Let us consider the case where we are interested in the results of more then 2 objectives. Suppose that we are optimising the following 8 objectives of the BAe function: $y_3, y_4, y_5, y_6, y_7, y_8, y_9, y_{13}$ and that we specify the following answers concerning preferences (those are the exact questions the program asks — there are 6 questions only for 5 distinctive classes of objectives):

$$y_3 \approx y_4,\ y_5 \approx y_6,\ y_7 \approx y_8,$$
$$y_3 \succ y_5,\ y_3 \gg y_7,\ y_3 \prec y_9,\ y_3 \succ y_{13},\ y_5 \succ y_7,\ y_5 \prec y_{13}$$

This give us the following preference order:

$$y_9 \succ y_3 \approx y_4 \succ y_{13} \succ y_5 \approx y_6 \gg y_7 \approx y_8$$

and using the same valuation as in Example 1, we have the following weights:

$$w_3 = w_4 = 0.1722,\ w_5 = w_6 = 0.1126,\ w_7 = w_8 = 0.053,\ w_9 = 0.1921,\ w_{13} = 0.1325$$

Performing weighted Pareto GA optimisation (with Pareto set size limited to 400) we end up with an 8–dimensional surface with some 3D slices presented in Figure 5.1. For the comparison, the right hand side contains the same 3D slice without preferences (i.e. using standard Pareto method). Please note that these are just 3–dimensional slices of an 8–dimensional surface. We are still facing the problem of n–dimensional surface presentation for $n > 4$. Although we are actually looking at a very small part of the 8D Pareto front, we can nevertheless notice the shift towards the smaller values of y_7 in Figure 5.1(a) as the objective y_7 is the least important of all. Similarly, the values of y_5 are less in Figure 5.1(a) than in Figure 5.1(b). This could be easily explained by noticing that the weight factor for y_9 (0.1921) is almost twice the weight of y_5 (0.1126) and almost four times the weight of y_7 (0.053).

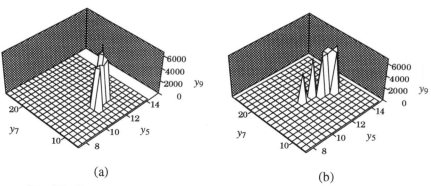

(a)   (b)

Figure 5.1: 3D slices of 8D Pareto front: (a) (y_5, y_7, y_9) with preferences, (b) (y_5, y_7, y_9), without preferences.

5.2 Example with 13 objectives

Another large experiment involves *all* 13 objectives $y_1, y_2, \ldots, y_{12}, y_{13}$ of the BAe function. Suppose that the designer has the following preferences (as asked by the program — 24 questions for 13 objectives (instead of theoretically maximal 78), or 21 question for 10 non–equivalent objectives instead of 45 — agreeing with results from section 3.2):

$y_3 \approx y_4, \; y_5 \approx y_6, \; y_7 \approx y_8,$

$y_1 \succ y_2, \; y_1 \prec y_3, \; y_1 \succ y_5, \; y_1 \succ y_7, \; y_1 \ll y_9, \; y_1 \succ y_{10}, \; y_1 \gg y_{11}, \; y_1 \succ y_{12}, \; y_1 \prec y_{13},$

$y_2 \prec y_5, \; y_2 \succ y_7, \; y_2 \prec y_{10}, \; y_2 \prec y_{11}, \; y_2 \succ y_{12}, \; y_3 \ll y_9, \; y_3 \prec y_{13},$

$y_5 \succ y_{10}, \; y_5 \succ y_{11}, \; y_7 \succ y_{12}, \; y_9 \succ y_{13}, \; y_{10} \gg y_{11}$

This gives the following order:

$$y_9 \succ y_{13} \succ y_3 \approx y_4 \succ y_1 \succ y_5 \approx y_6 \succ y_{10} \gg y_{11} \succ y_2 \succ y_7 \approx y_8 \succ y_{12}$$

and the weights vector: $(0.1003, 0.0334, 0.1054, 0.1054, 0.0951, 0.0951, 0.0283, 0.0283, 0.1414, 0.09, 0.0386, 0.0231, 0.1157)$.

The run was performed using a GA with a population size of 500 for 500 generations and limiting the size of Pareto front to 1000. Objectives $\{y_1, y_2, y_{10}, y_{11}, y_{12}\}$ are being minimised, all others are maximised. Each objective is normalised to $[0, 1]$ range. Limited space prohibits the inclusion of graphs similar to those of Figure 5.1, but as before, results ilustrate shift towards the larger values of y_9 and y_5 in the case of preferences (since $y_9 \succ y_5 \gg y_{12}$).

We must stress that in our method the increase in number of objectives decreases the difference in importance of particular objectives (i.e. the weights–difference of A and B in preference $A \gg B$ is much higher if we are working with two objectives instead of 10) so asymptotically weighted Pareto method performs the same as the 'classical' Pareto method i.e. with no preferences at all.

6 Conclusion

In this paper one method for transforming qualitative characterisation of objective relative importance into quantitative characterisation is presented. One algorithm is given that implements the transformation. Integration with traditional and GA based multi–objective optimisation method is discussed and a novel Pareto optimisation method combination with weights via preferences developed. Some real–world applications of preferences in the new Pareto based method are presented. Complexity of the procedure is analysed. In the future work we will try to further develop the preference model, to reduce the number of questions, and to integrate it more tightly into the real world applications.

References

[1] Ching-Lai Hwang and Abu Syed Md. Masud. *Multiple Objective Decision Making – Methods and Applications*. Springer Verlag, Berlin, 1979.

[2] David A. Van Veldhuizen and Gary B. Lamont. Multiobjective evolutionary algorithm research: A history and analysis. Technical Report TR-98-03, Air Force Institute of Technology, Wright–Paterson AFB, October 1998.

[3] Dragan Cvetković, Ian C. Parmee, and Eric Webb. Multi–objective optimisation and preliminary airframe design. In Ian C. Parmee, editor, *Adaptive Computing in Design and Manufacture*, pages 255–267. Springer–Verlag, 1998.

[4] Richard E. Nisbett and Timothy DeCamp Wilson. Telling more then we can know: Verbal reports on mental processes. *Psychological Review*, 84(3):231–259, May 1977.

[5] Dragan Cvetković and Ian C. Parmee. Use of preferences for ga–based multi–objective optimisation. In W. Banzhaf and J. Daida et al., editors, *GECCO–99: Proceedings of the Genetic and Evolutionary Computation Conference*, pages 1504–1510, Orlando, Florida, USA, July 1999. Morgan Kaufmann.

[6] Philippe Vincke. Basic concepts of preference modelling. In Bana e Costa [18], pages 101–118.

[7] Bernard Roy. The outranking approach and the foundation of ELECTRE method. In Bana e Costa [18], pages 156–183.

[8] János Fodor and Marc Roubens. *Fuzzy Preference Modelling and Multicriteria Decision Support*. Kluwer Academic Publishers, Dordrecht, The Netherlands, 1994.

[9] Georg Henrik von Wright. The logic of preference. An essay. Edinburgh, 1963.

[10] A. Tversky. Intransitivity of preferences. *Psychological Review*, 76:31–48, 1969.

[11] A. R. Anderson and N. D. Belnap, Jr. *Etailment. The Logic of Relevance and Necessity*, volume I. Princeton University Press, 1975.

[12] G. Priest, R. Routley, and J. Norman, editors. *Paraconsistent Logic. Essays on the Inconsistent*. Philosophia Verlag, München, Germany, 1989.

[13] K. J. Arrow. *Social Choice and Individual Values*. John Wiley & Sons, 1951.

[14] Ralph L. Keeney, Howard Raifa, and Richard F. Meyer. *Decisions with Multiple Objectives: Preferences and Value Tradeoffs*. John Wiley & Sons, 1976.

[15] Stephen Warshall. A theorem on Boolean matrices. *Journal of the ACM*, 9(1):11–12, 1962.

[16] J. G. Lin. Maximal vectors and multi–objective optimization. *Journal of Optimization Theory and Application*, 18(1):41–64, 1976.

[17] Ian C. Parmee. Exploring the design potential of evolutionary search, exploration and optimisation. In P. J. Bentley, editor, *Evolutionary Design by Computers*, pages 119–143. Morgan Kaufmann, San Francisco, CA, 1999.

[18] Carlos A. Bana e Costa, editor. *Readings in Multiple Criteria Decision Aid*. Springer–Verlag, Berlin, 1990.

[19] Ian C. Parmee, Dragan Cvetković, Andrew H. Watson, and Christopher R. Bonham. Multi–objective satisfaction within an interactive evolutionary design environment. *Evolutionary Computation*, 8(2), 2000.

[20] Heinz Mühlenbein and Dirk Schlierkamp-Voosen. Predictive models for the breeder genetic algorithm I: Continuous parameter optimization. *Evolutionary Computations*, 1(1):25–49, 1993.

[21] Kalyanmoy Deb. Multi–objective evolutionary algorithms: Introducing bias among Pareto–optimal solutions. KanGAL report 99002, Indian Institute of Technology, Kanpur, India, 1999.

Chapter 6

Algorithm Comparison and Development

Improving the Robustness of COGA: The Dynamic Adaptive Filter
C.R. Bonham, I.C. Parmee

Efficient Evolutionary Algorithms for Searching Robust Solutions
J. Branke

A Comparison of Semi-deterministic and Stochastic Search Techniques
A.M. Connor, K. Shea

A Multi-population Approach to Dynamic Optimization Problems
J. Branke, T. Kaussler, C. Smidt, H. Schmeck

Short Term Memory in Genetic Programming
K. Bearpark, A.J. Kean

Improving the Robustness of COGA: The Dynamic Adaptive Filter

Christopher R. Bonham & Ian C. Parmee
Plymouth Engineering Design Centre, University of Plymouth
cbonham@soc.plym.ac.uk   iparmee@plymouth.ac.uk

Abstract: Considerable research has focussed upon understanding and improving the performance of the diverse search engine within COGA. However, the performance of COGA is also dependent upon the extraction of high performance solutions by the adaptive filter. This paper introduces a new filtering approach that significantly increases the robustness of the adaptive filter and adds to the generic appeal of the COGA approach.

1. Introduction

Cluster Oriented Genetic Algorithms (COGAs) have been developed to assist the designer during the conceptual design stage by supporting the initial exploration of complex design spaces [1, 2]. This is achieved by a rapid decomposition of the search space into succinct regions of high performance (HP). Extracted qualitative and quantitative information, coupled with designer expertise and more specific "in-house" knowledge (such as manufacturing criteria, material costs etc.) allow the design team to assess the relative utility of each decomposed region. Successive COGA runs may then investigate specific areas of the search space for novel design directions. Throughout this iterative process the designer continuously gains general knowledge relating to the overall design domain, e.g. design variable sensitivity, variable / objective interaction, constraint violation, degree of multi-objective satisfaction etc., and more specific information relating to the regions of high performance (location, shape, average fitness etc).

Variable mutation COGA (vmCOGA) search is split into a number of sequential stages. After each search stage the fittest solutions within the population are extracted via an adaptive filter and stored in the Final Clustering Set (FCS). The Adaptive Filter (AF) consists of two components: Explicit filtering involves the normalisation of individual population fitnesses in terms of mean and standard deviation of overall population fitness. If the scaled fitness of a chromosome is greater than a pre-defined filtering threshold (Rf), the solution is copied to the FCS. Implicit filtering occurs during every search stage (excluding the first); here a solution also enters the FCS if its fitness exceeds that associated with the previous filtering threshold. Consequently, the filter adapts to the fitness landscape of the current population and allows a high degree of explorative interaction between the designer and the evolutionary process through the variation of Rf settings [3,4].

The generation, mutation and filtering vectors (figure 1) define the extraction generation of each search stage, mutation probability per search stage and filtering thresholds used during search.

$$\mathbf{g}^T = \begin{bmatrix} end \ of \ search \ stage \ 1 \ (gen_1) \\ end \ of \ search \ stage \ 2 \ (gen_2) \\ end \ of \ search \ stage \ 3 \ (gen_3) \\ \cdot \\ \cdot \\ \cdot \\ end \ of \ search \ stage \ n \ (gen_n) \end{bmatrix} \qquad \mathbf{m}^T = \begin{bmatrix} p(mutation) \ in \ search \ stage \ 1 \ (mp_1) \\ p(mutation) \ in \ search \ stage \ 2 \ (mp_2) \\ p(mutation) \ in \ search \ stage \ 3 \ (mp_3) \\ \cdot \\ \cdot \\ \cdot \\ p(mutation) \ in \ search \ stage \ n \ (mp_n) \end{bmatrix}$$

$$\mathbf{m}^T = \begin{bmatrix} p(mutation) \ in \ search \ stage \ 1 \ (mp_1) \\ p(mutation) \ in \ search \ stage \ 2 \ (mp_2) \\ p(mutation) \ in \ search \ stage \ 3 \ (mp_3) \\ \cdot \\ \cdot \\ p(mutation) \ in \ search \ stage \ n \ (mp_n) \end{bmatrix}$$

Figure 1: VmCOGA search and filtering vectors

The AF models the population fitness by approximating it to the standardised normal distribution. AF performance is therefore dependent upon the assumption that population fitness at any generation is normally distributed and remains so throughout the search. However, in certain search scenarios this assumption may be inappropriate or false. An example being a heavily constrained search space, where the fitness distribution may be distorted by the action of the constraint equation(s). Such inaccuracies may seriously degrade the decision support potential of COGA since inconsistent information extraction during search may affect the interpretation of presented results and may reduce knowledge gain during the investigative stages of conceptual design.

The Dynamic Adaptive Filter (DAF) increases both the accuracy and robustness of COGA by reducing the effects of model drift and model mismatch. This is achieved by hybridising both explicit and implicit filtering into a single filtering operation which occurs at the end of each generation (after the first search stage). When the population is filtered using the DAF, a library of probability density functions is used to model the current fitness distribution. The theoretical distribution that produces the closest match to the actual fitness distribution, according to a "closeness of fit" metric is used to model the population fitness. A probability mapping then converts the value of Rf to an equivalent fitness threshold (f_{Rf}) corresponding to the chosen distribution. A population member is then copied to the FCS if the fitness exceeds the f_{Rf} threshold.

2. Model drift and mismatch

In certain cases implicit filtering is prone to a degree of inaccuracy since the actual population mean and standard deviation change during the generations between the calculation of the fitness associated with the previous filtering threshold. This phenomenon is referred to as model drift and may easily be eliminated by explicitly filtering the population at the end of each generation (after the first search stage).

Model mismatch occurs when the theoretical population distribution no longer adequately models the true population. The problem of model mismatch may further augment the inaccuracies caused by model drift. There are two scenarios where the actual population distribution may depart from that described by the normal distribution, kurtosis and skewness. Kurtosis is a measure of the peaked nature of a distribution and is measured relative to the Normal distribution (figure 2). Kurtosis may be either leptokurtic (where there are a greater proportion of solutions at the extremes of the distribution) or platykurtic (where there are a greater number of solutions located about the mean of the distribution). If the degree of kurtosis becomes too great the normal distribution may no longer adequately model the true distribution and model mismatch occurs. The degree of symmetry of a distribution is measured by the skewness of a distribution (figure 3). If the distribution has a greater proportion of solutions located at the head of the distribution it is positively skewed. If there is a higher proportion of solutions at the tail, the distribution is negatively skewed. If Skewness becomes too great, model mismatch may result.

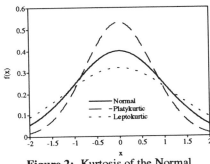

Figure 2: Kurtosis of the Normal
distribution

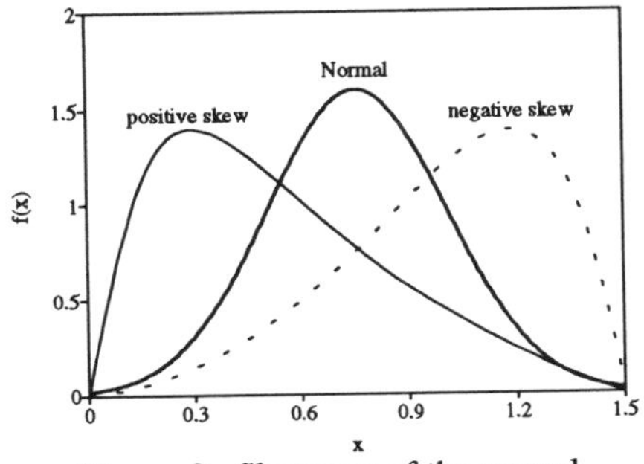

Figure 3: Skewness of the normal
distribution

3. The Library of Probability Density Functions (PDFs)

PDFs are used to probabilistically model any continuous variable (such as fitness) from any sufficient number of observations (sample) of that variable (individual chromosome fitness) [5]. The DAF reduces model mismatch by selecting the most accurate population fitness distribution model from a library of five theoretical Probability Density Functions (PDFs). The following section justifies the inclusion of each PDF by giving examples where the true population fitness distribution may closely relate to the distribution given by the library PDF.

Uniform distribution: During the initial generations of search, the evolving population may contain diverse solutions that have not converged upon the fitter regions of the search space. As a consequence of random initialisation, the population is also likely to be evenly dispersed throughout the search space and may result in a fitness distribution that is uniform in nature.

Exponential distribution: In many real world design domains, models are constrained to avoid operation within undesirable or potentially dangerous regions of the search space. This may result in an exponentially distributed population fitness distribution where large numbers of solutions lie at the lower quartiles of the fitness range, with relatively lower proportions lying at the higher quartiles.

Normal distribution: As search begins to converge upon high performance regions, the number of solutions populating infeasible regions decreases and the fitness distribution may assume the "bell shape" of the normal distribution.

Lognormal distribution: As search continues to converge upon high performance regions, it is possible that the fitness distribution will not assume the perfectly symmetrical appearance of the normal distribution but may be positively skewed producing a greater proportion of solutions that lie at the lower quartiles of the fitness range. As the name suggests, the lognormal distribution is an adaption of the normal distribution, allowing for such skewing.

Weibull distribution: The assumption that the fitness distribution will develop into either a symmetrical or positively skewed bell shape, does not account for the possibility of a negatively skewed distribution where there are large numbers of solutions that lie at higher quartiles of the fitness range. This is more likely to occur at latter stages of search where the population converges upon optimal areas of the search space. The Weibull distribution may provide a more accurate model in such cases since it allows for positive and a degree of negative skewing.

4.   Closeness of fit metrics

The closeness of fit or d-metric metric is used to determine which of the five theoretical distributions gives the closest match to the actual population fitness. If the theoretical distribution perfectly models the actual distribution the d-metric approaches zero, larger values indicate less accurate modelling. Two closeness of fit metrics are presented, Chi-squared (χ^2) and Kolmogorov-Smirnov.

The χ^2 test [7], groups the sample distribution into k discrete groups. The probability of falling into each class (p_i) is then compared with the actual number of points in each class to give an overall d-metric. The main drawback of the χ^2 test is the need to select both the number and boundary values of the discrete groups. As its name suggests, the equiprobable approach [8] offers a solution by automatically selecting the class boundaries to ensure equal p_i values across all classes. Yarnold [9] argues that the equiprobable test is valid only if, $k \geq 3$ and $np_i \geq 5$. The equiprobable χ^2 test used within the DAF has 10 discrete classes. Assuming a population size of 100, np_i equals 10, the smallest population size permitted is 50.

Unlike the χ^2 metric, the Kolmogorov-Smirnov (KS) test does not group the sample distribution into discrete classes [7]. Furthermore, the KS test is valid for any sample size n, an important factor since the KS test may be used when the $np_i < 5$. However, the application of the KS test has been shown to be more limited than the χ^2 test. In certain cases the test becomes conservative and may give larger d-statistic values [10]. This is less critical in the case of the DAF since the conservatism will be evident (to the same degree) in all cases and will therefore have little effect when a relative comparison is made. Nevertheless, within the DAF the KS test is only used only when $np_i < 5$.

5.   Application of PDFs to multi-dimensional test functions

To justify the replacement of the AF by the DAF, a series of tests are executed which indicate search scenarios where the ability of the library PDFs to model the fitness distribution supersede that of the AF. In each example, the d-metric clearly illustrates which PDF gives the closest modelling of the fitness distribution at any particular generation (smaller d-metric values indicate more accurate modelling). Four test functions are investigated, each test explores different model and search environments.

In each test case the d-metrics corresponding to each PDF are averaged over 100 independent vmCOGA runs. The generation and mutation vectors are {50, 100, 150, 200, 250} and {0.08, 0.06, 0.04, 0.02, 0.01} respectively. Note that during this investigation, the value of the filtering vector is arbitrary since filtering in no way affects the evolving population and will therefore have no effect upon the fitness distribution at any generation.

5.1.   DeJong F5

Test function five of the DeJong five function test bed [11], also referred to as Shekel's foxholes, is a 2d test function of the form,

$$f(x_1, x_2) = 0.002 + \sum_{j=1}^{25} \frac{1}{j + \sum_{i=1}^{2}(x_i - a_{ij})^6} \qquad -65.536 \le x_i \le 65.536$$

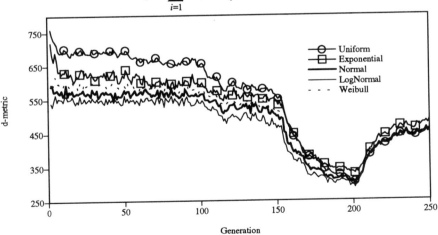

Figure 4: DeJong f5

The function consists of multiple, discrete peaks of equal magnitude, and is therefore highly discontinuous.

Referring to figure 4, it is quite apparent that the Lognormal distribution outperforms[1] the normal distribution and every other PDF at every generation upto the 200th generation where p(mutation) drops from 0.02 to 0.01. After this point the normal, lognormal, uniform and Weibull distribution all produce near identical d-metrics. An interesting characteristic of this plot is the large drop in all d-metrics in the generations between 150 and 200 corresponding to a mutation probability of 0.02.

5.2.   MiniCAPS (maximising SEP2)

The miniCAPS model is an abridged version of the CAPS model (Computer Aided Project Studies) developed by British Aerospace to support preliminary airframe design [12].   The COGA technique has been successfully integrated with miniCAPS via a Graphical User Interface [13]. This provides an efficient medium between the designer and design model and allows for the rapid exploration and subsequent decomposition of the design domain described by miniCAPS.

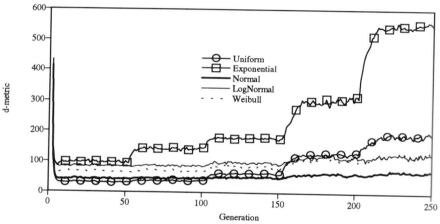

Figure 5: MiniCAPS (maximising SEP1)

MiniCAPS contains a number of subroutines, each of which calculates properties relating to criteria such as performance, wing geometry, propulsion, fuel capacity, structural integrity and so on. It consists of eight inputs and eleven outputs and may be configured for multi-objective optimisation or to maximise / minimise any one of the eleven output variables. In this test case, miniCAPS is used to maximise Subsonic Specific Excess Power (SEP1).

Immediately noticeable is the stepped nature of the uniform and exponential curves in figure 5. In contrast to the other test functions, the Uniform distribution produces the most accurate model for all the generations during the first

---

[1] Performance in this context is taken as the closeness of fit (give by the d-metric) between the actual population distribution and the theoretical population distribution obtained from the relevant PDF.

and second search stages. As expected, as search starts to converge to upon high performance regions, the performance of both the uniform and exponential distributions decrease and are finally superseded by the "bell" type PDFs, with the normal distribution dominating after the 100th generation.

5.3.   Keane'2 20d Bump function

Keane's Bump function is a high dimensional non-linear function consisting of multiple optima,

$$f(x_i) = \frac{\left| \sum_{i=1}^{n} \cos^4(x_i) - 2\prod_{i=1}^{n} \cos^2(x_i) \right|}{\sqrt{\sum_{i=1}^{n} ix_i^2}} \qquad 0 < x_i < 10 \qquad i = 1,...,n$$

Subject to,

$$\prod_{i=1}^{n} x_i > 0.75 \qquad and \qquad \sum_{i=1}^{n} x_i < \frac{15n}{2}$$

It contains one linear and one non-linear constraint. Added complexity in inherent since the global optimum is defined by the presence of a constraint boundary [14].

The performance of the exponential and uniform PDFs are comparatively poor (located at d-metric values of 200 and 100 respectively), hence to increase the resolution of figure 6 both curves are omitted.

Throughout the 250 generations, the normal distribution is outperformed by both the lognormal and Weibull distributions, suggesting that the 20d bump function tends to produce higher numbers of solutions that lie at the lower quartiles of the fitness range (positive skew). As the mutation probability decreases the performance of the Weibull distribution no longer competes with the lognormal distribution. This result is surprising since it has been shown that the Weibull distribution can model both positively and negatively skewed populations, and should therefore compete with the lognormal distribution in this instance.

Arguably, the most salient point to be drawn from this investigation is the large variation in the optimal PDF, since cases where the uniform, lognormal, normal and Weibull distributions individually produce the lowest d-metric are all evident. The normal distribution has been shown to perform adequately well over the whole range of functions and therefore justifies its inclusion as the single PDF used in the AF. However in many cases the normal distribution is outperformed by one or more of the remaining PDFs. The uniform and exponential distributions perform poorly over the majority of functions, however it is a dangerous to assume that this will be the case for all future test functions (generations 0 to 100 of figure 5 suggest that this is certainly the case for the uniform PDF), hence both distributions remain in the DAF.

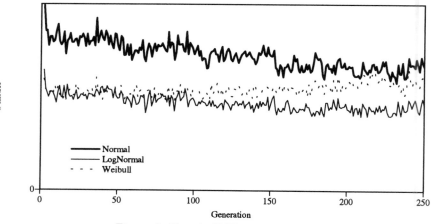

Figure 6: Keane's 20d Bump function
(uniform and exponential curves omitted)

6. Filtering threshold probability mapping

The filtering threshold value (Rf) is based upon application to the standardised normal distribution. To maintain consistency across all library PDFs, a mapping must be made to determine the equivalent Rf value if one of the library PDFs are selected for the filtering process.

Consider a standardised normal fitness distribution (figure 7). For a given filtering threshold Rf, the probability of a random solution with standardised fitness f_i passing the filtering process (P(Pass$_{Rf}$)) and entering into the FCS is given by,

$$P(Pass_{Rf}) = 1 - \int_{-\infty}^{Rf} \frac{1}{\sigma\sqrt{2\pi}} e^{-\frac{1}{2}\left(\frac{Rf-\mu}{\sigma}\right)^2}$$

The equation may be reduced since $\mu = 0$ and $\sigma = 1$.

$$P(Pass_{Rf}) = 1 - \int_{-\infty}^{Rf} \frac{1}{\sqrt{2\pi}} e^{-\frac{Rf^2}{2}}$$

The equivalent fitness (f_{Rf}) for any PDF (figure 8) that gives a probability of passing filtering of P(Pass$_{Rf}$) may be obtained by rearranging,

$$1 - P(f_i \leq f_{Rf}) = P(Pass_{Rf})$$

Where P($f_i \leq f_{Rf}$) for each PDF is given in table 1.

Distribution	P(f$_i$ ≤ f$_{Rf}$)	Parameters
Uniform	$\begin{cases} 0 & f_{Rf} \leq a \\ \dfrac{f_{Rf}-a}{b-a} & a \leq f_{Rf} \leq b \\ 1 & f_{Rf} > b \end{cases}$	a = minimum fitness b = maximum fitness
Exponential	$1 - e^{\frac{-f_{Rf}}{\mu}}$	$\mu = \dfrac{\sum f_i}{popsize}$
Normal	$\int_{-\infty}^{f_{Rf}} \dfrac{1}{\sigma\sqrt{2\pi}} e^{-\frac{1}{2}\left(\frac{f_{Rf}-\mu}{\sigma}\right)^2}$	$\mu = \dfrac{\sum f_i}{popsize}$ $\sigma = \sqrt{\dfrac{\sum(f_i - \mu)}{popsize-1}}$
Lognormal	$\int_{-\infty}^{x} \dfrac{1}{x\sigma\sqrt{2\pi}} e^{\frac{-(\ln(x)-\mu)^2}{2\sigma^2}}$	$\mu = \dfrac{\sum \ln(f_i)}{popsize}$ $\sigma = \sqrt{\dfrac{\sum(\ln(f_i)-\mu)}{popsize-1}}$
Weibull	$1 - e^{\frac{-f_{Rf}^{\alpha}}{\theta}}$	θ and α estimated from the actual population fitness distribution

Table 1: Calculating P(f$_i$ ≤ f$_{Rf}$) for each library PDF

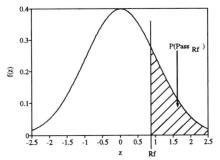

Figure 7: Calculating P(Pass$_{Rf}$) from a Standardised Normal Distribution

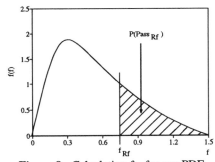

Figure 8: Calculating f$_{Rf}$ for any PDF

Consequently, a solution is copied to the FCS if its true fitness exceeds the value of f$_{Rf}$. This mapping ensures that approximately similar (allowing for stochastic error) proportions of solutions are copied to the FCS for any given value of Rf, independent of the PDF selected.

7. Assessing the performance of the DAF

To assess the performance of the DAF, vmCOGA algorithms using both the AF and the DAF filter were applied to the test function presented in [14]. This six dimensional function describes a search space containing 18 fully defined regions of high performance. As with all investigations the results were averaged over 100 independent trials. To maintain consistency, the mutation, generation and filtering vectors were initially set to values used in [15],

$$\mathbf{m} = \{0.08, 0.06, 0.04, 0.02, 0.01\}$$
$$\mathbf{g} = \{25, 50, 75, 100, 125\}$$
$$\mathbf{Rf} = \{1.75, 1.75, 1.75, 1.75, 1.75\}$$

The effects of eliminating model drift can be seen by comparing the results of the control filter and filter A. The control filter is the AF introduced in section 1, and uses both explicit and implicit filtering. Filter A is a standard AF with implicit filtering replaced by explicit filtering at every generation (excluding the first search stage) effectively isolating the effects of model drift. Reduced set cover (number of solutions in the FCS), increased Global Decomposition Efficiency (GDE) (percentage of FCS in any of the defined high performance regions) and increased average FCS fitness all indicate a higher levels of convergence present in the FCS when model drift is eliminated. When looking at Regional Decomposition (RD) (figure 9), it can be seen that smaller set cover and higher GDE combine to produce fewer results in all regions when compared with the AF control, however this decrease in RD must be offset by a dramatic decrease in both set cover and GDE sensitivity (standard deviation calculated over 100 trials).

Filter	Filter	Set Cover	GDE	$\overline{f}$	f_{best}
Control	AF (Explicit and Implicit Filtering) $\mathbf{Rf} = \{1.75, 1.75, 1.75, 1.75, 1.75\}$	285.72 (28.00%)	42.11% (30.16%)	1.344	1.471
A	AF (Explicit Filtering every generation) $\mathbf{Rf} = \{1.75, 1.75, 1.75, 1.75, 1.75\}$	178.06 (12.54%)	51.35% (16.84%)	1.354	1.471
B	DAF $\mathbf{Rf} = \{1.75, 1.75, 1.75, 1.75, 1.75\}$	208.64 (10.82%)	47.75% (18.57%)	1.349	1.472
C	DAF (Match GDE) $\mathbf{Rf} = \{1.70, 1.70, 1.70, 1.70, 1.70\}$	248.48 (8.66%)	41.99% (16.93%)	1.341	1.471
D	DAF (Match Set Cover) $\mathbf{Rf} = \{1.65, 1.65, 1.65, 1.65, 1.65\}$	285.98 (8.39%)	38.44% (20.27%)	1.337	1.469

Table 2: Statistical analysis of FCS (standard deviation of measured value is given in brackets)

Model drift is eliminated and model mismatch reduced by using the DAF (filter B). The degree of convergence within the FCS reduces when compared with filter A, indicating slightly less discriminatory filtering, subsequently increasing the overall number of regional hits. Furthermore, the set cover and GDE sensitivities have decreased and increased respectively; both sensitivities are again considerably smaller than those associated with the control filter.

It may be concluded that reducing model drift and model mismatch have the side effect of producing slightly more discriminatory filtering (increased GDE and reduced set cover). It is therefore suggested that when using the DAF, the filtering vector (**Rf**) may be reduced slightly to produce comparable decomposition results whilst still maintaining the large reduction in both set cover and GDE sensitivities. Empirical studies indicate that there is no single reduction factor that yields an exact match for the set cover and GDE results of the control filter, however if these two measures are considered independently, comparable results are achieved.

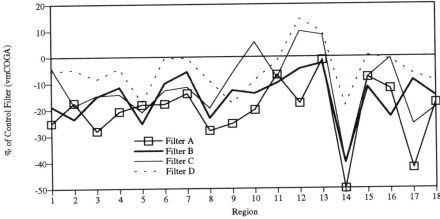

Figure 9: Regional Decomposition results for the 18 high performance regions (given as a percentage increase or decrease of the control filter)

Filter examples C and D are obtained by relaxing **Rf** to a point where the GDE (filter C) and then the set cover (filter D) approximately match the corresponding values for the control filter. However the attainment of these values results in a much lower value for the remaining measure. If filter D is taken as an example, approximately equal set cover is achieved but at the expense of a reduction in the corresponding GDE. The characteristic is also noted with filter C. Although the Regional Decomposition of both filters is still lower than those associated with the control filter, they offer notable improvement over both filter A and B, and produce similar performance in terms of both set cover and GDE sensitivity.

8. Conclusions

The role COGA plays in supporting the conceptual design process has been well documented [3,4,16], however the utility of such a technique lies heavily upon the ability to generate consistent results across successive COGA runs and during application to higher and more complex design domains. If such robustness is not evident, inconsistent information extracted during search may degrade the design team understanding of the design domain, and reduce knowledge gain during the investigative stages of conceptual design. It is therefore vital that the robustness of COGA is maximised to achieve acceptable generic applicability.

This paper has addressed the role of the adaptive filter relating to COGA robustness. The sensitivity of both set cover and GDE have been greatly reduced by removing implicit filtering and explicitly filtering every generation after the first search stage. This increased accuracy has the side effect of producing characteristically more discriminatory filtering which reduces set cover and increases GDE. This may be alleviated by reducing the effect of model mismatch, and is achieved by replacing the Adaptive Filter with the Dynamic Adaptive Filter. In the real-world environment the decrease in set cover and increase in GDE may be further addressed by relaxing the value of the filtering vector with little negative effect upon the robustness of the DAF.

References

[1] Parmee, I. C. (1996). "The Maintenance of Search Diversity for Effective Design Space Decomposition using Cluster Oriented Genetic Algorithms (COGAs) and Multi-Agent Strategies (GAANT)". Proceedings of ACDM 96, University of Plymouth, UK, PP 128-138.

[2] Parmee, I. C. (1996). "Cluster Oriented Genetic Algorithms (COGAs) for the identification of High Performance Regions of Design Spaces". Proceedings of EvCA 96, Moscow, pp 66-75.

[3] Bonham, C. R. & Parmee, I. C. (1998). "Using Variable Mutation Cluster Oriented Genetic Algorithms (vmCOGAs) to Support Conceptual Engineering Design", Proceedings of the 6th European Congress on Intelligent Techniques & Soft Computing, Aachen, pp 402-406.

[4] Bonham, C. R. & Parmee, I. C. (1999). "Cluster Oriented Genetic Algorithms (vmCOGAs) and Conceptual Engineering Design", Proceedings of the 12th International Conference on Engineering Design 99, pp, 1019-1022.

[5] Rothschild, V. & Logothetis, N. (1986). "Probability Distributions". John Wiley & Sons.

[6] Scheaffer, L. S. & McClave, J. T. (1995). "Probability and Statistics for Engineers". Duxbury Press.

[7] Kanji, G. K. (1993). "100 Statistical Tests". SAGE Publications.

[8] Law, A. M. & Kelton, W. D. (1990). "Simulation Modelling and Analysis". McGraw Hill.

[9] Yarnold, J. K. (1970). "The Minimum Expectation in χ^2 Goodness-of-Fit Tests and the Accuracy of Approximation for the Null Distribution". Journal of the American Statistical Association, 66, pp 864-886.

[10] Conover, W. J. (1980). "Practical Non-Parametric Statistics 2nd Edition". John Wiley & Sons.

[11] DeJong, K. A. (1975). "An Analysis of the Behaviour of a Class of Genetic Adaptive Systems". PhD Dissertation, Department of Computer and Communication Sciences, University of Michigan, Ann Arbor, MI.

[12] Webb, E. (1995). "MINICAPS – A Simplified Version of CAPS for use as a Research Tool". Unclassified Report, Bae-WOA-RP-GEN-11313, British Aerospace, Warton.

[13] Bonham, C. R. & Parmee, I. C. (1999). "The Graphical User Interface", ACDMnet & EvoDES Joint Workshop, 14th- 15th April, University of Bath

[14] Keane, A. J. (1994). "Experiences with Optimisers in Structural Design". Proceedings of the Conference on Adaptive Computing in Engineering design and Control, pp 14-27.

[15] Bonham, C. R. & Parmee, I. C. (1999). "An Investigation of Exploration and Exploitation Within Cluster Oriented Genetic Algorithms (COGAs)", Proceedings of the Genetic and Evolutionary Computation Conference, July 13-17, Orlando Florida, USA, pp 1491-1497.

[16] Parmee, I. C. & Bonham, C. R. (1989). "Supporting Innovative and Creative Design using Interactive Designer / Evolutionary Computing Strategies". Computation Models of Creative Design, University of Sydney, Australia.

Efficient Evolutionary Algorithms for Searching Robust Solutions

Jürgen Branke

Institute AIFB
University of Karlsruhe
D-76128 Karlsruhe, Germany
Email: branke@aifb.uni-karlsruhe

Abstract. For real world problems it is often not sufficient to find solutions of high quality, but the solutions should also be robust. By robust we mean that certain deviations from the solution should be tolerated without a total loss of quality.
One way to reach that goal is to evaluate each individual under a number of different scenarios. But although this method is effective, it requires significant computational power. In this paper, we continue some previous work aimed at minimizing the search effort while still providing the desired robustness. In particular, we examine the effect of varying the number of samples per individual over the course of the run, and look closely at the effect of certain EA parameters. Besides, a refined set of test functions is presented that allows to examine interesting aspects more specifically.

Keywords: evolutionary algorithm, robust solution

1 Introduction

For real world problems, it is often important to create robust solutions to a problem, i.e. solutions that are insensitive to changes in the environment or disturbance of the decision variables.

For example, a product design should usually allow certain tolerances in production and yield good results over all settings within these tolerances.

From a more abstract perspective, this means that not only the solution should be good, but also that the (phenotypic) neighborhood of the solution should have a high average quality.

Looking at the fitness landscape, a solution on a high plateau should be preferred over a solution on a thin peak: if the environment changes slightly (e.g. shifts in a random direction) or if it can not be guaranteed that the exact parameters of the solution are actually implemented (but a solution close to the original solution), the solution on the plateau will yield much better expected quality than the solution on the peak. This *effective* fitness function, $f_{eff}(x)$, depends on the distribution of the disturbances to the input and could be calculated as $\int_{-\infty}^{\infty} p(\delta) \cdot f(x + \delta) \, d\delta$, with $p(\delta)$ being the probability density function for the disturbance δ (which is assumed to be known or easy to estimate). Of course, for problems of relevant complexity, the calculation of $f_{eff}(x)$ will not be possible, thus other ways have to be found. This paper assumes that $f_{eff}(x)$ is estimated by averaging over a number of evaluations with random disturbances δ, i.e. by taking a number of *samples* in the neighborhood of individual x.

Note that the problem of creating robust solutions as defined here has some similarities to optimizing noisy functions, which have already been examined in combination with Evolutionary Algorithms (EAs) (e.g. [1–4]). However there are two main differences:

- with noisy functions, some noise is added to the output (quality) of the solution. In the settings regarded here, noise is added to the decision variables (or phenotype) of the solution. I.e. if $f(x)$ is the fitness function and δ is some (e.g. normally distributed) noise, then a noisy fitness function would mean $f'(x) = f(x) + \delta$, while in our case $f'(x) = f(x+\delta)$.
- noisy functions can not be evaluated without noise, the EA has to find good solutions despite the noise. In the settings considered in this paper, it is assumed that only the decision variables of the final solution are disturbed, usual function evaluations during the EA run can be performed without disturbance. This is justified because evaluation is usually done on the basis of a theoretical, computerized model, while the final solution is then actually implemented and has to face all the uncertainties present in reality.

One method to search robust solutions, suggested by Parmee et al. [5–7], involves searching for several high performance regions in the fitness landscape, and then examining each of these regions more closely e.g. in terms of their sensitivity to parameter changes. After that, further search can be restricted to those regions that are most promising in terms of robustness. Since only some small regions are assessed extensively, and then the search is restricted to areas that promise sufficiently robust solutions, the computational overhead is comparatively small. However, the approach does not optimize the desired criteria directly and may thus not yield optimal results. Furthermore, as Wiesmann et al. show in [8], the optimal solution in terms of robust quality does not necessarily lie in a high performance region. An example for such a (perhaps special) case can be seen in Figure 1.

A much simpler idea is to evaluate an individual under several scenarios (disturbances of the input values) and to use the average performance as estimator for $f_{eff}(x)$. This has already proven successful in a number of different applications like producing fault tolerant neural networks [9], the search for electronic circuits insensitive to temperature changes [10,11], flight control under changing condition [12], multilayer optical coatings insensitive to manufacturing tolerances [13,14,8], or job shop scheduling [15–17].

But although this method seems to be quite effective, it requires substantial computational power, since every individual has to be evaluated several times. Tsutsui and Gosh [18] were the first to examine the idea of disturbing the phenotype more thoroughly and showed, using the schema theorem, that given an infinitely large population size, an EA with single disturbed evaluations is actually performing as if it would

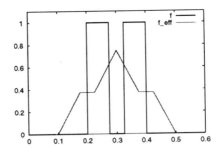

Fig. 1: Function for which the maximum of f_{eff} is at a minimum of f, disturbance is assumed equally distributed in the interval $[-0.1, 0.1]$

work on the effective fitness function. Nevertheless, as has been shown in [19], increasing the number of samples per individual also increases the quality of solutions found, with decreasing return.

In our previous paper on robustness [19], we have explored a number of ways to increase solution quality more efficiently than by just increasing the number of samples. Among those were the ideas to reevaluate the better individuals more often than the bad individuals (since those are the ones which the EA will use to generate the future population), and to use a memory to store individuals, taking into account their quality when evaluating individuals nearby (e.g. when an individual has many low quality neighbors, its fitness is degraded).

Here we examine additional ways to achieve efficiency, for example by assigning less computational power in terms of reevaluations to early generations, and more to later generations. Besides, we examine the effect of population size and test the island model for robust optimization.

The paper is organized as follows:
Section 2 describes in more detail the test functions and standard settings used in the following tests. In Sections 3 to 6 we propose a number of modifications to the standard EA, aimed at delivering robust solutions efficiently, and test them empirically. The paper concludes with a summary and some ideas for future work.

2   Test Functions

The test functions have been chosen carefully to reflect certain problem characteristics and to test certain aspects of the approaches. On the other hand, they have to be easy to analyze and computationally inexpensive.

For computational efficiency, we here restrict ourselves to a disturbance equally distributed over the interval $[-0.2, 0.2]$. Note, however, that the results should be independent of the kind of disturbance used. In previous experiments [19] for example, we have considered normally distributed disturbances.

The mathematical formulation of the two test functions is given below, while Figures 2 and 3 provide visualizations for one dimension.

$$f_1(x) = \sum_{i=1}^{10} \hat{f}_1(x_i) \quad \text{and} \quad f_2(x) = \sum_{i=1}^{10} \hat{f}_2(x_i) \quad \text{with} \tag{1}$$

$$\hat{f}_1(x_i) = \begin{cases} -(x_i+1)^2 + 1.4 - 0.8|\sin(6.283 \cdot x_i)| &:& -2 \le x_i < 0 \\ 0.6 \cdot 2^{-8|x_i-1|} + 0.958887 - 0.8|\sin(6.283 \cdot x_i)| &:& 0 \le x_i < 2 \\ 0 &:& otherwise \end{cases} \tag{2}$$

$$\hat{f}_2(x_i) = \begin{cases} x_i + 0.8 &:& -0.8 \le x_i < 0.2 \\ 0 &:& otherwise \end{cases} \tag{3}$$

$$\tag{4}$$

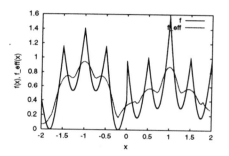

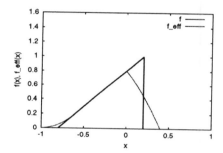

Fig. 2: Test Function f_1 and effective fitness $f_{1,eff}$ when disturbed.

Fig. 3: Test Function f_2 and effective fitness $f_{2,eff}$ when disturbed.

Function f_1 has been derived from f_1 as proposed in [19]. It offers two main alternatives in each dimension: a high sharp peak at $x = 1.0$ and smaller not so sharp peak at $x = -1.0$. The sharp peak would be preferable in a deterministic setting, but the other peak actually has a higher effective fitness. In order to avoid an initialization bias, both hill's basin of attraction is of the same size and has the same average quality (i.e. the x-interval and the area below the curve is the same on both sides). This basic function is overlayed with a strong low frequency oscillating function to make optimization more challenging.

Function f_2 is designed to show the influence of asymmetry on the optimization performance. The function has only one peak that drops sharply on one side (cf. Figure 3). The maximum for $f_{2,eff}(x)$ is at $x = 0.0$. This function allows to rate the approaches according to their risk awareness: the further they approach the peak, the greater their acceptance of the "risk" to drop down the edge in case of disturbances.

The standard settings for the EA applied in the experiments were real-valued, direct encoding of the decision variables, generational reproduction, ranking selection, two-point-crossover with probability 0.8, mutation rate of 0.1, and population size of 50 individuals. Each reported result is the average of 50 runs with different random seeds.

3 How to select the final solution?

There are two fundamental potential errors due to the stochastic nature of evaluation under disturbance:

1. the population may converge to the wrong peak, and
2. the wrong individual is picked from the final population.

In this section, we will try to assess the relative importance of these two errors. While for deterministic problems, it is clear that the best solution found during the run should be implemented, this is not so clear in our settings, since all fitness values computed are merely stochastic estimations for f_{eff}. Therefore, the best individual in terms of fitness does not necessarily mean it is really the best individual according to f_{eff}, but it might just have been lucky during evaluation. Here we compare three possible ways to pick the solution from the final population:

- selecting the best according to fitness (same fitness used for selection throughout the run)
- selecting the best individual according to an undisturbed evaluation (note that as opposed to noisy fitness functions, we may evaluate an individual accurately according to f). This approach has also been used in [18].
- reevaluating each individual in the final population several times and selecting the best according to the average. This, of course, incurs additional computational costs, but will often be the only way to actually find the individual with maximum f_{eff}-value. And, since it is only applied to the last generation, the additional computational cost is limited.

Figure 4 compares the effective quality of an individual selected from the final population, depending on the number of samples (evaluations in the neighborhood of a single individual) during the run and the number of samples used in the final generation. Test function used was f_1 and the EA was run for 7,500,000 evaluations (the algorithm had fully converged by then). As can be seen, even with an equal number of total function evaluations (note that a larger number of samples during the run means fewer generations), and an equal number of samples in the final population, using a larger number of samples throughout the run is advantageous, with decreasing return. Using more evaluations throughout the run reduces the probability of converging to the wrong peak (error 1). Also clearly visible is the benefit of multiple evaluations in the final population, which reduces error 2. In fact, using 50 evaluations in the final population is about as good as using 3 samples throughout the run, but computationally much cheaper.

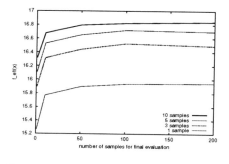

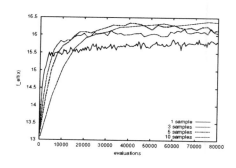

Fig. 4: Effective fitness of the selected individual after 7,500,000 evaluations, depending on the number of evaluations per individual during the run and during the final generation.

Fig. 5: Convergence curves on Function f_1 for different number of evaluations per individual, early generations.

For Function f_1, the evaluation of an individual without disturbance in the final population seems advantageous when compared to just using the averaged fitness values, especially for small sample numbers (cf. Figure 6). However that is basically due to the peak's symmetry. For the asymmetric function f_2, using undisturbed fitness is misleading and yields largely inferior results except for only one sample per individual (Figure 7). The same may be true when the population has not yet converged to only one peak, or even worse, when the fitness landscape is symmetric, but the optimum actually lies in a trough (as in Figure 1).

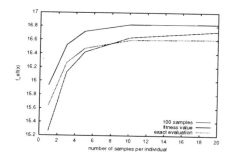

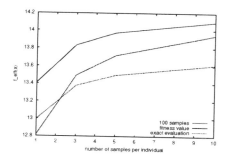

Fig. 6: Effective fitness of the selected individual after 7,500,000 evaluations, depending on number of evaluations per individual during the run and selection strategy in the final generation. Results for Function f_1

Fig. 7: Effective fitness of the selected individual after 7,500,000 evaluations, depending on number of evaluations per individual during the run and selection strategy in the final generation. Results for Function f_2

Obviously, unless there is some explicit knowledge about the shape and symmetry of peaks, it seems recommendable to spend some extra computation time to pick the best individual from the final population.

This is done in all experiments reported in the subsequent sections: the best individual from the final population is selected on the basis of 100 evaluations per individual.

Note that the above results assume that there is actually enough time for convergence. If that is not the case, using fewer evaluations per individual throughout the run may be advantageous, since fewer evaluations allow to run more generations in the same time. Figure 5 shows the convergence plots for different sample numbers per individuals for the first 10000 evaluations. At every point in time, the best individual from the population has been selected by using 100 evaluations. Clearly, the less samples per individual, the faster the convergence, but eventually more samples lead to better solutions.

4 Influence of the Population Size

As has been proved by Tsutsui and Gosh [18] for an infinitely large population size, an EA with single disturbed evaluations is actually performing as if it would work on the effective fitness function. The actual importance of the population size can be seen in Figure 8: Using a population size of 100 but only single evaluations yields better results than using a population size of 25, but 10 evaluations per individual. As can be seen, the difference between different sample numbers is reduced with increasing population size, which confirms the result by Tsutsui and Gosh, according to which the number of evaluations per individual should not make any difference when using an infinite population size.

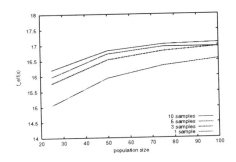

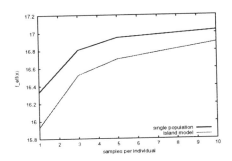

Fig. 8: Effective fitness of the best individual after 7,500,000 evaluations, depending on number of evaluations per individual during the run and population size. Results for Function f_1

Fig. 9: Performance of a single population of size 75 compared to an island model with three populations of size 25 each, tested on Function f_1.

5 The Island Model

In an island EA, the population is distributed onto several independent subpopulations which evolve independently, only exchanging individuals from time to time. As it turned out, this island model is advantageous in deterministic settings, not only because it can be parallelized efficiently, but also because it may yield better results due to a smaller risk of early convergence [20–22]. That was the reason why in [19], the island model was also used as a default.

Here we compare a single population EA with population size of 75, to an island EA using 3 populations with 25 individuals each. Surprisingly, the superiority of the island model could not be confirmed in our settings. As can be seen in Figure 9, the single population model significantly outperforms the island model no matter whether only one or 10 evaluations were used per individual. This may be due to the negative impact of small population sizes as reported in the previous section. Another reason might be the smoothness of the effective fitness function (due to averaging), for which the island model might not be particularly helpful.

Note however, that the difference between island model and single population model is decreasing with increasing sample number. It remains to be seen whether the island model might be advantageous for very large population sizes and large sample numbers.

6 Changing the Number of Samples over the Course of the Run

In Section 3 we have seen that it is especially important to use a large number of samples in the last generation, in order to be able to pick the optimal individual. This observation naturally extends to the question whether it is a useful strategy to use a different number of samples in different generations. Here we present a preliminary investigation into that aspect by comparing a number of different heuristic strategies.

Intuitively, there seem to be two possibly critical stages of an EA run: in the beginning, when the population quickly converges towards a region of the search space that looks promising, and in the end, when it finally has to climb to the top of a peak. Corresponding to these assumptions, we tested the following six strategies, all using an average of 3 evaluations per individual.

1. a linear increase of the number of samples over the course of the run.
2. an exponential increase of the number of samples per individual over the course of the run.
3. an emphasis on early as well as late generations.
4. a linear decrease in the number of samples over the course of the run. This has basically been included as a counter example, expected to decrease performance.

5. a strong emphasis on early as well as late generations.
6. a strong emphasis on late generations.

The different strategies are displayed visually in Figure 10.

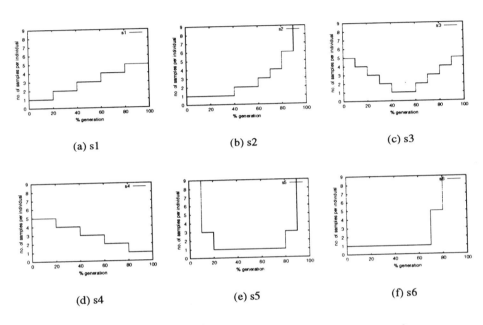

(a) s1   (b) s2   (c) s3

(d) s4   (e) s5   (f) s6

Fig. 10: Different strategies on how to distribute evaluations on generations.

These strategies have been tested on f_1, for varying numbers of generations (note that for these strategies, an equal number of generations means an equal number of evaluations as well). As Figure 11 shows, all strategies (except the counter strategy 4) clearly outperform the simple strategy of using an equal number of samples throughout the run, for all numbers of generations tested. The differences among the strategies are not as clear, but strategies 2 and 6 seem to be the top performers, followed by strategy 5, and then 1 and 3. This indicates that using extra samples for evaluating the individuals is particularly useful during the late generations.

One reason for the smaller importance in early generations may be that at that time, the population consists of a wide variety of different individuals, with a large variance of fitness values, and that at that time, small variations in fitness values due to disturbed evaluations have less effect than in later generations, when the population has converged.

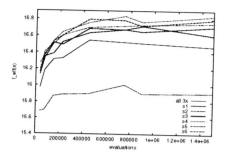

Fig. 11: Performance of different strategies to distribute evaluations over the generations, measured on Function f_1.

7 Conclusion and Future Work

In this paper, we have extended our previous work from [19] and examined a number of additional strategies to make the search for robust solutions more efficient and to ·better understand the influence of different EA parameters.

We have shown that the population size is a very critical parameter, and should never be too small. The island model, a successful EA variant in deterministic settings, does seem to be detrimental when using disturbed evaluations to search for robust solutions.

During the run, a large number of evaluations per individual slows down convergence (since evaluation becomes more costly), but eventually leads to better solutions. Picking the right individual from the final population is difficult, and it is worth to evaluate each individual a large number of times at that stage. But not only during the last generation do additional evaluations pay off, it seems that in general more evaluations should be used in later generations than during early generations.

There remain ample opportunities for future work. First of all, it should be tested how the reported results transfer to real world problems like e.g. scheduling. It would be desirable to have explicit rules on how to distribute the evaluations to different generations. Then, combinations of the ideas presented here and in our previous paper [19] should be examined.

References

1. A. N. Aizawa and B. W. Wah. Scheduling of genetic algorithms in a noisy environment. *Evolutionary Computation*, pages 97–122, 1994.
2. J. Michael Fitzpatrick and John J. Greffenstette. Genetic algorithms in noisy environments. *Machine Learning*, 3:101–120, 1988.
3. U. Hammel and T. Bäck. Evolution strategies on noisy functions, how to improve convergence properties. In Y. Davidor, H. P. Schwefel, and R. Männer, editors, *Parallel Problem Solving from Nature*, number 866 in LNCS. Springer, 1994.

4. P. Stagge. Averaging efficiently in the presence of noise. In A. E. Eiben, T. Bäck, M. Schoenauer, and H.-P. Schwefel, editors, *Parallel Problem Solving from Nature V*, volume 1498 of *LNCS*, pages 188–197. Springer, 1998.

5. I. C. Parmee. The maintenance of search diversity for effective design space decomposition using cluster-oriented genetic algorithms (COGAs) and multi-agent strategies (GAANT). In *Proceedings of ACEDC'96*, 1996.

6. I. C. Parmee. The maintenance of search diversity for effective design space decomposition using cluster-oriented genetic algorithms (cogas) and multi-agent strategies (gaant). In *Proceedings of ACEDC'96*, 1996.

7. I. C. Parmee, M. Johnson, and S. Burt. Techniques to aid global search in engineering design. In *Proceedings of International Conference on Industrial and Engineering Applications of AI and Expert Systems*, 94.

8. D. Wiesmann, U. Hammel, and T. Bäck. Robust design of multilayer optical coatings by means of evolutionary algorithms. *IEEE Transactions on Evolutionary Computation*, 2(4):162–167, 1998.

9. A.V. Sebald and D.B. Fogel. Design of fault tolerant neural networks for pattern classification. In D.B. Fogel and W. Atmar, editors, *1st Annual Conference on Evolutionary Programming*, pages 90–99, San Diego, 1992. Evolutionary Programming Society.

10. A. Thompson. Evolutionary techniques for fault tolerance. In *Proc. UKACC Intl. Conf. on Control*, pages 693–698. IEE Conference Publications, 1996.

11. Adrian Thompson. On the automatic design of robust elektronics through artificial evolution. In A. Peres-Urike M. Sipper, D. Mange, editor, *Proceedings of the 2nd International Conference on Evolvable Systems*, pages 13 – 24. Springer - Verlag 1998, 1998.

12. Philip W. Blythe. Evolving robust strategies for autonomous flight : A challenge to optimal control theory. In Ian Parmee, editor, *Adaptive Computing in Design and Manufacture*, pages 269 – 283. Springer - Verlag London, 1998.

13. H. Greiner. Robust filter design by stochastic optimization. In F. Abeles, editor, *Optical Interference Coatings, Proc. SPIE*, pages 150–161, 1994.

14. H. Greiner. Robust optical coating design with evolutionary strategies. *Applied Optics*, 35(28):5477–5483, 1996.

15. C. R. Reeves. A genetic algorithm approach to stochastic flowshop sequencing. In *IEE Colloquium on Genetic Algorithms for Control and Systems Engineering*, number 1992/106 in Digest, pages 13/1–13/4. IEE, London, 1992.

16. M. Tjornfelt-Jensen and T. K. Hansen. Robust solutions to job shop problems. In *Congress on Evolutionary Computation*, volume 2, pages 1138–1144. IEEE, 1999.

17. C. Ventouris. Gestaltung robuster Maschinenbelegungspläne unter Verwendung evolutionärer Algorithmen. Master's thesis, Institute AIFB, University of Karlsruhe, 1998.

18. S. Tsutsui and A. Ghosh. Genetic algorithms with a robust solution searching scheme. *IEEE Transactions on Evolutionary Computation*, 1(3):201–208, 1997.

19. J. Branke. Creating robust solutions by means of an evolutionary algorithm. In A. E. Eiben, T. Bäck, M. Schoenauer, and H.-P. Schwefel, editors, *Parallel Problem Solving from Nature V*, volume 1498 of *LNCS*, pages 119–128. Springer, 1998.

20. U. Kohlmorgen, H. Schmeck, and K. Haase. Experiences with fine-grained parallel genetic algorithms. *Annals of Operations Research*, (90):203–219, 1999.

21. H. Schmeck, U. Kohlmorgen, and J. Branke. Parallel implementations of evolutionary algorithms. In *Solutions to Parallel and Distributed Computing Problems*. Wiley, to appear.

22. D. Whitley, S. Rana, and R. B. Heckendorn. The island model genetic algorithm: On separability, population size and convergence. *Journal of Computing and Information Technology*, (1):33–47, 1999.

A Comparison of Semi-deterministic and Stochastic Search Techniques

A M Connor and K Shea

Engineering Design Centre, University of Cambridge

email: amc50@eng.cam.ac.uk, ks273@eng.cam.ac.uk

Abstract. This paper presents an investigation of two search techniques, tabu search (TS) and simulated annealing (SA), to assess their relative merits when applied to engineering design optimisation. Design optimisation problems are generally characterised as having multi-modal search spaces and discontinuities making global optimisation techniques beneficial. Both techniques claim to be capable of locating globally optimum solutions on a range of problems but this capability is derived from different underlying philosophies. While tabu search uses a semi-deterministic approach to escape local optima, simulated annealing uses a complete stochastic approach. The performance of each technique is investigated using a structural optimisation problem. These performances are then compared to each other as well as a steepest descent (SD) method.

1. Introduction

The purpose of this paper is to investigate the relative merits of two optimisation techniques applied to an engineering design problem. Tabu search [1] is an aggressive metaheuristic that guides a local search out of local optima while simulated annealing [2] uses a probabilistic approach to obtain the same end. Comparing heuristic techniques is difficult since performance is highly dependent on the specific formulation of the general method. In this comparison, the implementation of each technique is sufficiently mature such that a direct comparison is unlikely to lead to results that favour one particular method due to bias in development effort. The tabu search algorithm used in this study was originally developed for application to the optimisation of fluid power circuits [3]. The simulated annealing algorithm is the underlying search technique used in an approach to size, shape and topology optimisation of structures [4].

The problem chosen to assess the relative merits of each algorithm is the shape and size optimisation of a ten bar truss. This is a simple extension of the standard problem, which only considers size optimisation, and a step towards applying tabu search to structural topology optimisation.

2. Optimisation Techniques

A variety of techniques can be used to tackle the optimisation problems in engineering design. Traditional methods, such as steepest descent and conjugate gradient methods have recently fallen into disfavour due the advent of claimed global optimisation methods such as genetic algorithms, tabu search and simulated annealing. A wealth of studies exist in the literature which compare different methods [5,6,7] although no general conclusions can be made as sample problems come from domains and large variations in algorithm implementation and representations exist.

2.1. Tabu Search

The tabu search concept is a heuristic procedure designed to guide other search methods to avoid local optimality. Tabu search has been shown to be effective for a wide variety of classical optimisation problems, such as graph colouring and travelling salesman problems, as well as practical problems such as scheduling and electronic circuit design. The method uses constraint conditions, for example aspiration levels and tabu restrictions, and a number of flexible attribute based memories with different time cycles. The flexible memories allow search information to be exploited more thoroughly than rigid memory or memory-less systems. Flexible memories can be used to either intensify or diversify the search to force the method to find optimum solutions.

The underlying search method in the current implementation is a variable step size steepest descent algorithm. This has been chosen as it allows a direct search to be carried out without requiring gradient information. Two memory lists are used to control the search algorithm. The short term memory contains representations of recently visited solutions that are classed as tabu. When the search algorithm locates an optimum it is forced to make a move and it is not allowed to return to a solution that is contained in the list. As the search progresses the list is updated by removing the oldest member of the list before adding a new member so that the list remains a fixed size. It is the short term memory and the notion of tabu restriction that provide the capability to escape local optima.

The intermediate term memory is similar to the short term memory, but it contains a list of previously visited best solutions. This list of solutions is used to provide a means of focussing on good regions of the solution space. Intensification accelerates the search by examining trends in good solutions and proposing new solutions based on extending these trends. In addition, intensification can locate new good solutions by investigating the centroid of several disparate good solutions. The final search control mechanism is diversification. This is often implemented by using an additional long term memory cycle but in this implementation is based on a simple refreshment involving a scattering of new random solutions.

2.2. Simulated Annealing

Simulated annealing algorithms are based on the analogous process of annealing in solids to optimise complex functions or systems. In the physical world, annealing is accomplished by heating a solid to an elevated temperature and then allowing it to cool slowly enough so that the thermal equilibrium is maintained. Atoms in the material then assume a globally minimum energy state. Simulated annealing algorithms have been successfully applied to a variety of problems. The algorithm starts with an initial design and generates a new design by changing one or more of the design variables. The objective function is then evaluated for the new design. While a better design is always accepted there is a possibility that a worse design may be accepted based on a probability function [8].

The change in energy is expressed as the change in objective function value, whereas the temperature is a control parameter that sets the probability of selecting an inferior design. In the general method, the temperature is held constant for a prescribed number of iterations to allow the system to gain "thermal equilibrium" and is then decreased in accordance with a cooling curve. As the temperature decreases, so does the probability that an inferior design is accepted. This forces the algorithm to converge to an optimal, or near optimal, solution.

Simulated annealing algorithms are reasonably robust if the parameters controlling the cooling curve are assigned values that reflect the complexity of the problem. In this implementation the modified Lam-Delosme schedule is used [9] along with rules selected based on dynamic quality metrics [10] and dynamic constraint weights [11].

3. Optimisation of a Ten Bar Truss

The problem considered in this paper is the optimisation of a ten bar truss. In the standard approach to this problem, the spatial layout of the truss is constrained as shown in Figure 3.1 by fixing the position of the nodes.

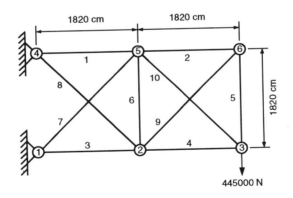

Figure 3.1: Ten bar truss

The truss is assumed to consist of an idealised set of pin jointed bars connected together at the nodes. The design optimisation problem is to find the cross sectional areas of each member such that the mass is minimised. The problem has been expanded in this paper to allow the spatial layout of the truss to be adjusted as well as the cross sectional area of the members. The built in nodes are fixed in position, as is the loaded node. This variation of the problem therefore introduces six new parameters that determine the position of nodes 2, 5 and 6 in Cartesian space. It is important to realise that the introduction of new parameters does not create a topology optimisation problem. Even if member areas are allowed to drop to zero, the removal of members is only a reduction in the connectivity of a given topology. A true topology optimisation can only be achieved if nodes are added and removed during the optimisation process.

Essentially, all of the design parameters are continuous. However, the nodal positions have a minimum allowable change of 1 cm while the cross sectional areas have a minimum allowable change of 0.01 cm^2. In reality, these cross sectional areas would be limited to discrete values corresponding to available stock material. In this example, the material for the truss is aluminium with Young's modulus of 6.88x10^6 N/cm^2 and material density is 2.7^{-3} kg/cm^3. Each member is modelled as a solid circular cross section.

3.1. Constraints and Convergence

Despite the simplicity of the ten bar truss example, it is still a reasonably constrained problem due to the difficulty in finding high quality, i.e. low mass, solutions that do not violate either the buckling or stress constraints. The constraints on the problem are that each member should not be stressed in excess of 17,200 N/cm^2, not buckle under Euler buckling criteria and have a length of at least 15 cm.

The tabu search and simulated annealing implementations used in this paper take different approaches to dealing with constraint violations. In order to reflect the aggressive nature of the search, the tabu search method uses a simple rejection of infeasible solutions. In comparison, the simulated annealing approach used in this study uses a dynamically weighted penalty function that decreases the allowable violation as the search progresses [11]. This approach allows the method to track through infeasible regions in order to locate new feasible solutions whereas the tabu search implementation relies on the intensification of trends to carry the search through infeasible regions.

Both methods have the capability to test convergence of the search and induce a premature termination. However, the implementations are significantly and have not been used. The implication is that the simulated annealing approach will carry out all of the evaluations specified by the cooling strategy. For the tabu search a potentially large number of the evaluations used by the will only be producing very small decreases in mass.

4. Results

An initial investigation of optimising only member size resulted in solutions in the range of 2900 to 3300 kg. These results have been used to set a target mass threshold of 2900 kg to determine the degree of success of the methods on the expanded problem.

Results are presented for ten optimisation runs of each method. In the first instance, statistics are presented that summarise the results for all three methods. Both the tabu search method and the underlying steepest descent method are initialised by generating a random scatter of solutions with the search being started from the best feasible solution. The simulated annealing method is started from the same point but has an initial random walk that takes it to a different region of the solution space before the probability of accepting a worse solution is reduced below 1. Therefore, the multiple runs of the problem indicate how the methods perform when started from different regions of the solution space. The results summary is shown in Table 4.1.

	SD	TS	SA
Best mass (kg)	2299	1598	1491
Best mass evals	4206	12004	34000
Worst mass (kg)	7208	2948	2307
Worst mass evals	3252	8896	34000
Average mass (kg)	4162	2401	1967
Std. Dev. of mass (kg)	1796	495	323
Lowest evals	2711	8441	34000
Highest evals	5676	36806	34000
Average evals	4386	13455	34000
No. runs below 2900kg threshold	3/10	8/10	10/10

Table 4.1: Statistical comparison of methods

The summary of the results shows the best and worst masses achieved by each method, along with the number of evaluations required in each case. The simulated annealing algorithm always carries out the same number of evaluations where as for the other methods the number of evaluations is dependent on how the search progresses. In addition, the summary shows the average mass of the ten runs and the standard deviation away from that value. This is essentially an indication of the consistency of the performance. Finally, the lowest, highest and average number of evaluations is shown.

Due to the dynamic penalty function used by the simulated annealing approach, a number of solutions found exhibited minor stress violations including the best solution found. Manually adjusting these solutions to remove the violations led to a slight increase in mass. The following results first compare the best initially feasible solution found by the simulated annealing approach to those found by the tabu search and steepest descent methods. The adjusted solution that has the lowest mass is then described.

Table 4.2 shows the numeric values for each of the design parameters where the x,y coordinates of the node positions are expressed relative to the lower fixed

292

node. The solution shown for the simulated annealing approach is the best solution found with no residual constraint violations. The shading in the table indicates minimum area members. Such minimum area members can be removed from the structure to produce a reduced topology solution if the change does not produce a violation of constraints in other members or create a mechanism.

	SD	TS	SA
x_2,y_2 (cm)	488,89	445,-61	568,-151
x_5,y_5 (cm)	840,581	807,408	-13,920
x_6,y_6 (cm)	1436,-44	1197,-112	1252,-176
A_1 (cm^2)	59.71	60.39	65.29
A_2 (cm^2)	28.17	16.6	22.92
A_3 (cm^2)	204.14	183.17	242.63
A_4 (cm^2)	0.01	0.01	0.01
A_5 (cm^2)	132.31	239.9	193.82
A_6 (cm^2)	125.66	3.04	13.78
A_7 (cm^2)	0.01	0.01	0.19
A_8 (cm^2)	145.94	1.42	0.01
A_9 (cm^2)	374.43	310.26	259.65
A_{10} (cm^2)	42.52	47.9	35.66
Mass (kg)	2299	1598	1526

Table 4.2: Comparison of best solutions

The best truss found by the steepest descent algorithm is shown in Figure 4.1.

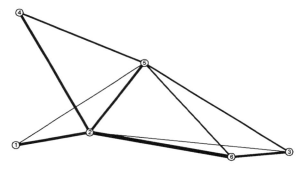

Figure 4.1: SD solution

Removing the minimum area members of the solution shown in Figure 4.1 produces a minor buckling violation in the member between nodes 2 and 3. Increasing the area of this member from 145.94 cm^2 to 146 cm^2 removes the violation and makes the reduced topology a feasible design as shown in Figure 4.2.

The fact that a tiny increase in area removes the buckling violation suggests that removing the behavioural constraints or introducing a tolerance for allowable violation may be of benefit. Accepting solutions with minor violations may allow the search to track through infeasible regions to find new feasible solutions.

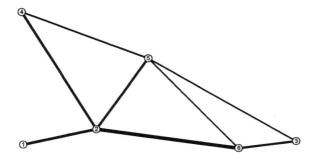

Figure 4.2: Adjusted SD solution

The solution in Figure 4.3 resulted from the tabu search method.

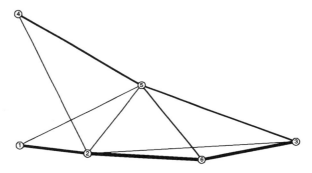

Figure 4.3: TS solution

Removing the minimum area members produces the solution shown in Figure 4.4. The reduced topology has no constraint violations and needs no adjustment.

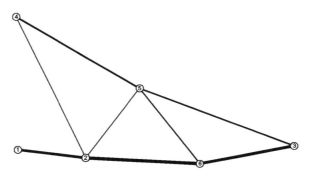

Figure 4.4: Adjusted TS solution

This solution has the same reduced topology as that found using the steepest descent method but is considerably lighter. Due to the different joint positions the solution has smoother load transfer and the variance of member lengths is noticeably smaller than in the solution found by the steepest descent method. Whilst a low variance of member lengths is a desirable feature it has not been explicitly included in the objective function formulation and so this may be a random occurrence.

The solution shown in Figure 4.5 is the best feasible solution found by the simulated annealing algorithm. The position of node 5 has moved just behind the support node since no constraint on spatial design boundary was formulated. This may cause problems in practice depending on how the truss is supported and the physical nature of fixed support. The support node, node 4, and the member joining nodes 4 & 5 is obscured in the Figure 4.6 but the length of the member is longer than the minimum length. An expanded view of this region (not to scale) is shown in Figure 4.6.

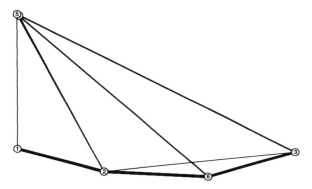

Figure 4.5: Feasible SA solution

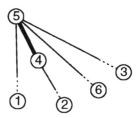

Figure 4.6: Expanded region of solution

From this expanded view it can be seen that the positioning of node 5 above and behind the fixed support node has resulted in a compression member between nodes 4 & 5. This member transfers the forces applied to node 5 from the other members to the support node. Removing the minimum area members does not produce violations. Although it is not shown in the adjusted solution in Figure 4.7, it is also possible to remove the member between nodes 1 and 5. This member has a small area, .19 cm^2, and its removal produces no adverse impact on the behaviour of the structure.

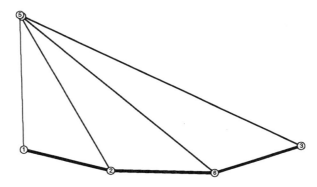

Figure 4.7: Adjusted feasible SA solution

The simulated annealing approach did locate a number of solutions that were slightly infeasible due to the soft constraint formulation. The best solution found after small violations were removed has the parameter values given in Table 4.3.

	SA
x_2,y_2 (cm)	612,-92
x_5,y_5 (cm)	93,850
x_6,y_6 (cm)	1259,-96
A_1 (cm^2)	61.36
A_2 (cm^2)	11.64
A_3 (cm^2)	252.61
A_4 (cm^2)	0.01
A_5 (cm^2)	202.02
A_6 (cm^2)	8.05
A_7 (cm^2)	0.01
A_8 (cm^2)	.74
A_9 (cm^2)	251.89
A_{10} (cm^2)	43.47
Mass (kg)	1491

Table 4.3: Initially infeasible SA solution (adjusted to become feasible)

The minimum area members, four and seven, were removed while the area of member ten was altered from 28.41 cm^2 to 43.47 cm^2. The area of member six was slightly altered from 8.03 cm^2 to 8.05 cm^2 to obtain this solution. This solution is shown in Figure 4.8.

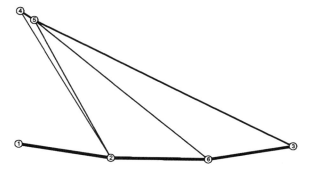

Figure 4.8: Adjusted SA solution that initially had violations

5. Discussion

A goal of this study was to assess the merit of each method in light of the computational expense required and the quality of the solution obtained. Using steepest descent as the basis, the best solution found by tabu search is 31% lighter with 2.85 times the number of evaluations. Simulated annealing resulted in a solution that is 35% lighter but required eight times the number of evaluations. While for the ten bar truss problem the evaluation is not time intensive, for larger problems and situations where the evaluation is more computationally expensive, reducing the number of evaluations is highly desirable. The expense of the optimisation must also be compared to the number of designs that will be manufactured and additional cost benefits outside of material purchase for reducing the mass of the structure.

An advantage of tabu search is a semi-deterministic nature as it acts both as a local and global search method. With the current implementation, from an initial starting point, tabu search always results in the same solution. It was seen though that the performance of tabu search improved by using an initial scatter search but, this eliminated the deterministic nature. A problem with non-deterministic methods is, while solution quality is improved overall, this can only be assessed based on a number of optimisation runs. A common rule of thumb used with simulated annealing is to take the best solution found from three runs.

The steepest descent and tabu search methods employed a hard constraint for stress and buckling so the constraints were never violated at any point in the search. In contrast, simulated annealing used dynamic penalty functions such that constraint violations were allowed throughout the search but were penalised in the cost function. The inclusion of soft constraints allows a method to track through regions of infeasible space.

As both methods are heuristic a number of search parameters are required. To have a search technique that is robust over a wide range of design problems it is advantageous to reduce the number of parameter adjustments required from problem to problem. The simulated annealing implementation used has 6 temperature schedule parameters, 15 parameters associated with the selection of moves, and 8 parameters associated with the dynamic constraint weights. While the majority of these parameters are robust over a wide range of problems, some fine tuning is generally necessary. In comparison, the tabu search method has a total of 5 control parameters for which an empirically derived heuristic has been developed that relates the values required to the number of design parameters.

Many structural optimisation methods allow members to reduce to a minimum area, which is generally very small, with the implication that these members can be removed from the structure. As was shown in Section 4, this is generally not a simple task. Skill and understanding are necessary to transform the solutions found by the optimisation algorithms into sensible designs. Deciding which members can be removed without the structure collapsing or inducing stress and buckling violations can be difficult, particularly for large structures. This emphasises that optimisation can only be an aid for designers and engineering judgement is still imperative.

To reduce the mass of a structure further than shown in this study, topology changes are required. Simulated annealing has been applied to this problem in the past and a best mass of 853 kg was reported for the same problem [12]. A goal of this study was to investigate the advantages of tabu search over simulated annealing for topology optimisation of discrete structures. General advantages are a decreased number of evaluations, potential for parallel evaluation, fewer search parameter adjustments, possibilities for incorporating design knowledge and potential for learning. These could lead to a more appropriate search technique for topology design problems.

6. Conclusions

This paper compares the relative merits of two mature implementations of different search algorithms to a structural optimisation problem. While the maturity of each implementation may have removed any bias in results due to differing implementation effort, it has also led to difficulties in comparison of the two methods in more than a very general sense. Both methods perform considerably better than a local search method, with simulated annealing finding higher quality solutions than tabu search at the cost of increased computational expense.

The performance of each method is as much dependent on the low level functionality of each implementation as to the actual method. The dynamic soft constraints of the simulated annealing approach have more merit than the hard constraints implemented in the tabu search. While the tabu search has the advantage of requiring less effort to set the control variables of the search, the inclusion of dynamic soft constraints will lead to the introduction of more control variables. One of the main conclusions of this preliminary study is that there is an inevitable trade off between enhancing the representation of the problem to aid the search and the complexity of controlling the search.

The aim of future work is to combine the strengths of each method to produce a hybrid approach. This may involve the use of memory cycles to escape local optima, the incorporation of domain knowledge and the utilisation of search history. The use of probabilistic moves will be investigated as an attempt to remove the inherent disadvantage of tabu search in evaluating all possible moves. The intent of the resulting method is an efficient and effective technique for topology synthesis problems in engineering design.

References

1. Glover G and Laguna M, 1997. *Tabu Search*, Kluwer Academic Publishers

2. Kirkpatrick S, Gelatt Jr., C D, Vecchi M P, 1983. Optimization by simulated annealing, *Science*, 220:4598:671-679.

3. Connor A M, Tilley D G, 1998. A tabu search algorithm for the optimisation of fluid power circuits, *Journal of Systems and Control* 212(5):373-381

4. Shea K, Cagan J, Fenves S J, 1997. A shape annealing approach to optimal truss design with dynamic grouping of members, *ASME Journal of Mechanical Design*, 119:388-394.

5. Arostegui, M, Kadipasaoglu, S N, Khumawala, B M, 1998. Empirical evaluation of tabu search, simulated annealing, and genetic algorithms on facilities location problem, Annual Meeting of the Decision Sciences Institute, Vol. 3, pp. 1091

6. Pirlot M, 1996. General local search methods, *European Journal of Operational Research*, 92(3):493-511

7. Sinclair M, 1993. Comparison of the performance of modern heuristics for combinatorial optimization of real data, *Computers in Operations Research*, 20(7):687-695

8. Metropolis N, Rosenbluth A, Rosenbluth M, Teller A, Teller M, 1953. Equation of state calculations by fast computing machines, *Journal of Chemical Physics*, 21:1087-1092.

9. Swartz W, Sechen C, 1990. New algorithms for the placement and routing of macro cells, Proceedings of the IEEE Conference on Computer-Aided Design, Santa Clara, CA, November 11-15, IEEE proceedings: Cat No. 90CH2924-9, pp. 336-339.

10. Hustin S. (1988), "Tim, a new standard cell placement program based on the simulated annealing algorithm," Master of Science, University of California, Berkeley, Department of Electrical Engineering and Computer Science.

11. Ochotta E S, 1994. Synthesis of high-performance analog cells in ASTRX/OBLX. Ph.D. Thesis, Carnegie Mellon University.

12. Shea K, Cagan J, 1998. Topology Design of Truss Structures by Shape annealing, Proceedings of DETC98: 1998 ASME Design Engineering Technical Conferences, September 1998, Atlanta, GA, DETC98/DAC-5624.

A Multi-population Approach to Dynamic Optimization Problems

Jürgen Branke, Thomas Kaussler, Christian Smidt and Hartmut Schmeck

Institute AIFB
University of Karlsruhe
D-76128 Karlsruhe, Germany
Email: {branke | Schmeck}@aifb.uni-karlsruhe

Abstract. Time-dependent optimization problems pose a new challenge to evolutionary algorithms, since they not only require a search for the optimum, but also a continuous tracking of the optimum over time. In this paper, we will will use concepts from the "forking GA" (a multi-population evolutionary algorithm proposed to find multiple peaks in a multi-modal landscape) to enhance search in a dynamic landscape. The algorithm uses a number of smaller populations to track the most promising peaks over time, while a larger parent population is continuously searching for new peaks. We will show that this approach is indeed suitable for dynamic optimization problems by testing it on the recently proposed Moving Peaks Benchmark.

Keywords: evolutionary algorithm, forking, dynamic optimization problem, time-dependent optimization

1   Introduction

Most research in evolutionary computation focuses on optimization of static, non-changing problems. However, many real-world optimization problems are actually dynamic, and optimization methods capable of continuously adapting the solution to a changing environment are needed. Applications include for example scheduling, where new jobs have to be added all the time, or manufacturing, where the quality of raw material is changing over time.

Although evolutionary algorithms seem to be a natural candidate to solve dynamic optimization problems, this area has only recently attracted significant research interest. A comprehensive survey can be found in [1].

As has been argued by Branke [2], continuous adaptation only makes sense when the landscapes before and after the change are sufficiently correlated, otherwise it would be at least as efficient to restart the search from scratch. Therefore it is valid to assume only small to moderate changes. But although after a small change local hill-climbing might often be sufficient, even a slight change might move the optimum to a totally different location, for example when the heights of the peaks change such that a different peak becomes the maximum peak. In these cases the EA basically has to "jump", or cross a valley, to reach the new maximum peak.

The main problem with standard evolutionary algorithms used for dynamic optimization problems appears to be that EAs eventually converge to an optimum and thereby lose their diversity necessary for efficiently exploring the search space and consequently also their ability to adapt to a change in the environment when such a change occurs.

In this paper, we adapt the concept of Forking Genetic algorithms (FGAs) as introduced by Tsutsui, Fujimoto and Ghosh [3] to time varying multimodal optimization problems. The FGA is based on the idea of dividing the search space into several parts, each exclusively explored by one of several subpopulations. A parent population continuously searches for new peaks, while a number of child populations try to exploit previously detected promising areas. Here, we use this general idea to maintain individuals on several peaks simultaneously, which should be helpful in the context of dynamic optimization problems. A number of adaptations were necessary to adapt the FGA to allow it to react on changes of the fitness function.

The only other multi-population approach to dynamic optimization problems the authors are aware of is the recently published Shifting Balance GA proposed by Oppacher and Wineberg [4]. Again, that algorithm tries to maintain the EAs exploratory power by dividing the population into one core population and a number of smaller colony populations. But there, the task of the core population is to exploit the best solution found, while the colony populations are forced to search in different areas of the fitness landscape, i.e. they are responsible for exploration. In the approach presented here, a number of smaller populations try to exploit several peaks simultaneously, while a large search population is searching for new peaks. Also, the mechanisms to maintain separated populations are completely different. At the current state we are unable to tell which of these two independently developed paradigms is more promising.

The paper's outline is as follows: First, Section 2 will present our approach in more detail and point out the major differences compared to the forking GA. Section 3 will report on some preliminary experiments. The paper concludes with a summary and some remarks on possible future work.

2 Concept of Forking and Adaption to Dynamic Problems

The original Forking Genetic Algorithm (FGA) as introduced by Tsutsui, Fujimoto and Ghosh is rather complex, therefore only a brief summary can be given here. The interested reader is referred to [3] for more details.

As pointed out before, the FGA has originally been designed to find multiple peaks of a multi-modal landscape, and is based on the idea of dividing up the search space. It uses a parent population, continuously searching for new peaks, and a number of child populations, which are restricted to search in some promising areas identified previously. Whenever the parent population has converged "sufficiently" to one region, the FGA separates this region from the parent population's search space and assigns a child population to it for further exploitation, i.e. the parent population and the child populations operate on disjoint parts of the search space (cf. Figure 1). If the maximally allowed number of forking populations is reached, the oldest forking population is discarded.

To make the FGA suitable for dynamic problems, at least two issues have to be addressed:

- the subpopulations have to be enabled to follow "their" peak through time
- we need to efficiently distribute our individuals (and thereby search efforts) between the different child populations and the parent population. This is done by adjusting the number of individuals in each population according to its assumed optimization potential. The least promising child populations may be erased completely when that seems appropriate.

Overall, the child populations should exploit the knowledge gained previously and follow the most promising peaks in the search space, while the parent population should constantly explore and search for new peaks.

The new approach introduced in this paper will be called "Self-Organizing Scouts" (SOS), since the individuals here act as scouts, that may divide the search space into different regions and distribute their search efforts onto the most promising of these regions.

The algorithm starts just as a simple EA with a single population searching through the entire search space. In regular intervals, this population (called the parent population) is analyzed, and it is checked whether the conditions for forking are fulfilled. If that is the case, a child population is split off from the parent population and henceforth independently explores the corresponding subspace. The parent population continues to search in the remaining search space. Note that the total number of individuals is never affected, the individuals are just assigned to different (sub)populations. Each child population is able to adapt its search space to the changing fitness landscape by moving it through the phenotypical feature space. The number of individuals in each child population is adjusted regularly to reflect its estimated optimization potential and quality. The overall algorithm is depicted below.

Algorithm: Self Organizing Scouts
    REPEAT
        Compute the next generation of child populations and adjust search spaces
        Compute the next generation of parent population
        IF (forking generation)
            Create new child population when possible
            Adjust size of base and child populations
        UNTIL termination criterion

302

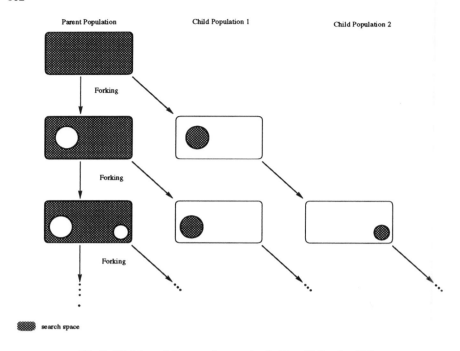

Fig. 1: Division of the search space by forking (following [3])

In the following, the different steps are explained in more detail.

Computing the next generation: Generally, computing the next generation of the parent or child population is equivalent to a single generation of an ordinary EA.

The mutation step size is adjusted to the diameter of the corresponding search space, i.e. child populations generally have much smaller mutation step sizes than the parent population.

Since the fitness function is dynamic, all individuals have to be reevaluated in every generation. It is always ensured that no individual from the parent population lies within any of child populations' search spaces.

Creating a new child population: A child population is an independent subpopulation working on a part of the phenotypic feature space. It is defined by a center (the most fit individual in the subpopulation) and a distance (range), and consists of all individuals whose distance (phenotypical manhattan distance) from the center is smaller than or equal to the given range. The population fitness is defined as the fitness of its best individual.

Child populations underly a number of restrictions:

- minimum and maximum number of individuals relative to overall amount

- minimum and maximum diameter of the subspaces relative to the size of the total search space
- minimum fitness of new forking populations relative to current overall best individual
- minimum fitness of existing forking populations relative to current overall best individual.

The original FGA only considers forking when certain convergence criteria are fulfilled and the best individual did not change over some number of generations. Since we are dealing with a dynamic fitness here, convergence may never occur, and a regular attempt for forking seemed more appropriate.

Therefore, at specific generations, called forking generations, the parent population is analyzed for the existence of a group of individuals that would fulfill the above constraints for child populations. If more than one group is found, the one with the maximum ratio of number of individuals to diameter is selected. All individuals in that group are split off from the parent population and assigned the smallest subspace encompassing all individuals as search space (note that this allows a variable size of the subspace, while the original FGA always uses a fixed size subspace).

Moving the child population's search space: As the peaks of the fitness landscape may move, the search space of each child population is allowed to move as well. This is achieved by defining the best individual in a child population as the center of its search space. The diameter of each search space, however, is kept constant in the current implementation.

When a child population's search space moves, some of the parent population's individuals may become invalid, as they lie within a child's population search space. These individuals are replaced by valid randomly generated individuals. Individuals that drop out of a moving child population's search space are kept for the next generation.

In addition, it may happen that the search spaces of child populations overlap. Usually this is tolerated, only when a center individual falls into the search space of another child population, its whole child population is removed.

Adjusting the population sizes: Since the overall computing power is limited, it should be distributed efficiently over the different populations. Generally, more effort should be devoted to areas with high quality and high dynamics. On the other hand, when a child population has converged to a peak and that peak has not changed for several generations, it may be sufficient to maintain a very small "outpost" on that peak in order to be able to detect when that peak becomes interesting again, i.e. when it changes height and/or position.

For that purpose, in each forking generation, first of all, any child population with a fitness smaller than the minimum required fitness is discarded. Then, for all remaining child populations as well as the parent population a quality measure Q_i is calculated.

The quality Q_i of population i is simply a linear combination of its fitness F_i and dynamism measure D_i which depends on the difference between a population's current and previous fitness (see equation below).

$$F_i(t) = \text{fitness of best individual in population } i \text{ at time } t$$

$$D_i(t) = \max\{0, \frac{F_i(t) - F_i(t-1)}{F_i(t-1)}\}$$

$$Q_i(t) = \begin{cases} \alpha \frac{D_i(t)}{\sum_j D_j(t)} + (1-\alpha)\frac{F_i(t)}{\sum_j F_j(t)} & : \quad \sum_j D_j(t) > 0 \\ \frac{F_i(t)}{\sum_j F_j(t)} & : \quad otherwise \end{cases}$$

The desired population size S_i is then chosen proportionally to each population's relative quality:

$$S_i = \frac{Q_i}{\sum_j Q_j} \cdot (\text{total number of individuals})$$

Of course, the restrictions on minimum and maximum population size of each child population and the parent population always have to be respected.

When the size of a population is increased, new random individuals are generated within the corresponding search space. If individuals have to be removed as the new population size is smaller than the old one, the worst individuals are removed.

We consider that the above described attempt to measure quality is a rather straightforward and preliminary approach, in future studies other indicators like e.g. convergence may be included.

3 Empirical Evaluation

For a preliminary empirical evaluation, we compare the SOS algorithm as presented above to a simple evolutionary algorithm (EA) using the Moving Peaks Problem [2]. This benchmark consists of a number of peaks, randomly changing their height, location and width from time to time. The step size by which a peak is moved can be set explicitly and thus allows to define the severity of a change.

As standard settings for SOS as well as for the simple EA we use a total number of 100 individuals, real-valued encoding, rank based selection, generational replacement with elitism of one individual, mutation probability of 0.2, and crossover probability of 0.6. The α-Parameter to set the relative importance of dynamism and quality in SOS ha For the Moving Peaks benchmark, each of 5 dimensions was restricted to values between 0 and 100, the step size for a peak has been set to 1, and the landscape changes every 5000 evaluations (50 generations). All reported values are averages over 20 runs with different random seeds but identical fitness function.

Since for dynamic fitness functions it is not useful to report the best solution found, we will use the offline performance as quality measure. The offline performance is defined as the average of the best solutions at each time step, i.e. $x^*(T) = \frac{1}{T}\sum_{t=1}^{T} e_t^*$ with e_t^* being the best solution at time t (cf. [5]). Note that the number of values that are used for the average grows with time, thus the curves tend to get smoother and smoother.

Figure 2 depicts three different settings of the SOS with differing restrictions to cluster size, and compares them to the standard EA. As can be seen, SOS always significantly outperforms the standard EA. With increasing cluster size, the performance of SOS improves slightly. Nevertheless, the other results reported in this paper have been computed with the smallest cluster size restrictions, which leaves some potential for improvement.

The effect of changing the total number of individuals can be seen in Figure 3, again for the standard EA (lower 3 curves) and SOS (upper 3 curves). Overall, the effect of varying the population size seems to be small.

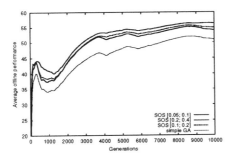
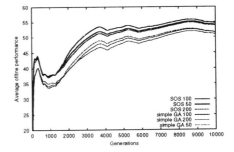

Fig. 2: Offline performance of standard EA compared to SOS with different restrictions for cluster size.

Fig. 3: Offline performance of standard EA compared to SOS with different total population size

Figure 4 shows the results obtained with standard settings but a higher change frequency, namely a change of the fitness landscape occurring every 20 generations. The change frequency is one aspect of dynamism, and it seems that with growing dynamism SOS is gaining additional relative advantage compared to the standard EA.

Finally, we examined the effect of changing the step size, or severity of the changes. Figure 5 reports the offline performance after 5000 generations for different values of severity. As can be seen, SOS is much less affected by an increased change severity than the standard EA, which means that again the difference between standard EA and SOS is growing when dynamism is increased. This is particularly noteworthy in comparison to the results reported in [2], as the memory based approaches presented there have proven to be very sensitive to changes of the peak locations, i.e. the severity.

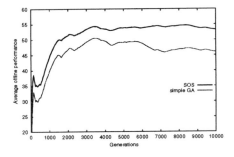

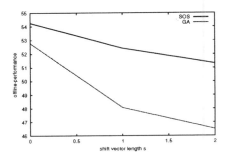

Fig. 4: Offline performance of standard EA compared to SOS with higher change frequency (every 20 generations).

Fig. 5: Offline performance of standard EA and SOS after 5000 generations, varying shift length s.

4 Conclusion and Future Work

We have presented a new way to tackle dynamic optimization problems by means of evolutionary algorithms. The newly proposed Self Organizing Scouts approach is based on the forking GA presented earlier in [3].

The basic idea behind this work was to design an evolutionary algorithm that is capable of tracking several different promising parts of the search space simultaneously.

Although some of the design decisions have been made rather ad hoc in order to quickly prove feasibility, in the preliminary tests performed so far, our approach significantly outperformed the simple GA on the Moving Peaks problem.

Furthermore, the approach is largely independent of the total number of individuals or the severity of the changes (as opposed to memory based approaches, which are very sensitive to changes of the peak locations).

For future work, it would certainly be valuable to examine more closely the effect of the different modifications made to the original FGA. In particular, the resizing of the populations should be refined, not only by refining the quality rating, but also by allowing the size of the search spaces to adapt.

Testing the approach on problems with more peaks than the maximally allowed number of child populations will reveal whether the resizing procedure is still capable of identifying the most promising search regions.

A comparison with other approaches, especially the shifting balance GA, should also be done.

References

1. J. Branke. Evolutionary algorithms for dynamic optimization problems - a survey. Technical Report 387, Insitute AIFB, University of Karlsruhe, February 1999.
2. J. Branke. Memory enhanced evolutionary algorithms for changing optimization problems. In *Congress on Evolutionary Computation CEC99*, volume 3, pages 1875–1882. IEEE, 1999.
3. S. Tsutsui, Y. Fujimoto, and A. Ghosh. Forking genetic algorithms: GAs with search space division schemes. *Evolutionary Computation*, 5(1):61–80, 1997.
4. F. Oppacher and M. Wineberg. The shifting balance genetic algorithm: Improving the ga in a dynamic environment. In W. Banzhalf et al., editor, *Genetic and Evolutionary Computation Conference*, volume 1, pages 504–510. Morgan Kaufmann, 1999.
5. K. De Jong. *An analysis of the behavior of a class of genetic adaptive systems*. PhD thesis, University of Michigan, Ann Arbor MI, 1975.

Short Term Memory in Genetic Programming

K Bearpark and A J Keane
School of Engineering Sciences, University of Southampton

Abstract. The recognition of useful information, its retention in memory, and subsequent use plays an important part in the behaviour of many biological species. Information gained by experience in one generation can be propagated to subsequent generations by some form of teaching. Each generation can then supplement its taught learning by its own experience. In this paper we explore the role of memorized information in the performance of a Genetic Programming (GP) system that uses a tree structure as its representation. Memory is implemented in the form of a set of sub-trees derived from successful members of each generation. The memory is used by a genetic operator similar to the mutation operator but with the following difference. In a tree-structured system the mutation operator replaces randomly selected sub-trees by new randomly-generated sub-trees. The memory operator replaces randomly selected sub-trees by sub-trees randomly selected from the memory. To study the memory operator's impact a GP system is used to evolve a well-known expression from classical kinetics using fitness-based selection. The memory operator is used together with the common crossover and mutation operators. It is shown that the addition of a memory operator increases the probability of a successful evolution for this particular problem. At this stage we make no claim for its impact on other problems that have been successfully addressed by Genetic Programming.

1. Introduction

Genetic Programming (GP) is one of the most recently developed fields in the study of Evolutionary Computation (EC). Koza [1] has shown that GP can be successfully applied to a wide range of problems across a number of technical and social disciplines. Included in these problems is Symbolic Regression - finding a function, in symbolic form, that fits a given finite sample of data. This paper is concerned with the application of a short-term memory operator to symbolic regression.

Genetic Programming uses a Genetic Algorithm (GA) to evolve a progressively improving solution to a problem. In a conventional GA the solution is represented in vector form with the simplest GAs operating on a binary string.

The main difference between the pure GA and the GA-based GP approaches lies in the complexity of the representation. A GP representation generally has a hierarchical rather than a linear structure. This hierarchical structure requires some modification to the genetic operators defined for GA systems.

A typical GA operates in parallel on a set or population of solutions. Each member of the population is a potential solution and must be represented in such a way that its suitability, or fitness, as a solution can be measured. An initial population is created by random selection from a pool of components to give the first generation. Successive generations are produced by the application of genetic operators that mirror those responsible for the evolution of biological species. Individuals are selected to participate in the creation of the next generation according to their fitness. Some selected individuals pass into the next generation unchanged while others are subject to a crossover operation in which two parents produce children by exchanging genes. A further operation may occur in which a small amount of genetic material in a child is randomly mutated.

The three operations of selection, crossover and mutation occur in nature and in most GAs. Fitness-based selection ensures that highly fit individuals are well represented in the mating pool for the next generation while individuals with lower fitness tend to disappear. Crossover attempts to produce better individuals by incorporating genes from each parent. Random mutation of some genes in a child introduces new genetic material and extends the search for a solution to different regions of the search space. Many GA and GP applications have shown that all three operators, individually and in combination, are beneficial in improving the average fitness of each generation and evolving better solutions to the problem in hand or, in some cases, the best solution.

Evolution in nature or in EC relies on stochastic processes. The average ability of members of a generation to perform a particular task in a given environment tends to improve as the fitter members survive and reproduce at the expense of the less fit members. This ability may also improve through accumulated experience handed down from one generation to the next, i.e. if this experience is recorded it may be used by future generations to augment their development. In the GP system described here experience is recorded in the form of a set of sub-trees derived from an analysis of the more successful members of each generation. We call this set the 'memory'. The memory is used in a mutation-like process in which replacement material is selected randomly from the memory as a complete sub-tree. This should be contrasted with the conventional mutation operator that produces replacement sub-trees by randomly selecting operators and operands.

The use of a memory operator has some similarity to the encapsulation operator defined by Koza [1], automatically defined functions (ADF), also defined by Koza [2], and module acquisition, described by Angeline and Pollack [3].

Banzhaf *et. al.*[4] point out that these modularization techniques essentially preserve sub-sets of genetic material against the potential disruption of crossover. On the other hand, the memory operator defined here is used after crossover has occurred and as the final process in creating a new generation. It is akin to children learning from their grandparents and earlier generations.

2. The Objective Function

An elementary result from classical kinetics states that the distance travelled by an object in time t subject to constant acceleration a and with an initial velocity u is given by the expression

$$ut + \tfrac{1}{2}at^2.$$

The GP system described here uses symbolic regression to evolve this expression given sets of terminals and arithmetic operators and conventional GP techniques. The system is then used as a vehicle to explore the use of memory to improve its performance.

The fitness of an individual test expression is measured by evaluating the expression for values of t from 1 to 10, two values of u (20 and 200) and a single value of a (980). All units are arbitrary. These values are compared with the corresponding values of the true expression $ut + \tfrac{1}{2}at^2$ to give two error values

$$E20 = \sum_{t=1}^{10} ABS(\text{test }(t) - \text{true }(t)) \text{ for } u = 20$$

and

$$E200 = \sum_{t=1}^{10} ABS(\text{test }(t) - \text{true }(t)) \text{ for } u = 200.$$

If both E20 and E200 are zero the evolved expression is a 'hit' and is given an error value of zero. Otherwise the expression is given the error value

$$(E20+E200)/2.$$

3. The GP System

The GP system makes use of largely standard techniques with the exception of the memory operator. Its various components are described in the following sections.

3.1 The function and terminal sets

The functions available to the system are here restricted to the simple arithmetic binary operators resulting in the set

$$\{+,-,*,/\}.$$

The implementation ensures that the division operator is protected against a zero divisor.

The terminals are restricted to the variables u, a and t and integer constants from 1 to 9. The structure of the operand set has some bearing on the success of the evolutionary process. All the results reported here were achieved using the set

$$\{u,1,a,2,t,3,u,4,a,5,t,6,u,7,a,8,t,9\}.$$

These two sets are used in producing the first generation and in selecting replacement sub-strings during mutation.

We have also explored the use of a terminal set from which integer constants are removed thus relying on the emergence of rational numbers (e.g. $x/(x+x)$) during evolution. Our conclusions still hold although the performance of the system is reduced but can be restored by increasing the number of tests.

3.2 Representation

Expressions are represented as Reverse Polish (RP) strings except in the first randomly produced generation where it is simpler to generate conventional (unbracketed) arithmetic expressions and convert them to RP form. The string length is controlled by imposing a limit of 5 functions (and hence a maximum total string length of 11 characters) in the first generation and a maximum length of 19 characters (or 9 functions) in subsequent generations. Modification of the RP strings by genetic operators here makes use of a tree representation.

3.3 Production of the first generation

The first generation is produced by alternately selecting terminals and functions from the respective sets subject to a maximum of 5 functions. Each expression is converted to RP form and evaluated to give an error value as described earlier.

3.4 Fitness-based selection

The fitness of an expression when considering it for entry into the mating pool for the next generation is given by

$$fitness_i = maxerr/(error_i+1)$$

where $fitness_i$ and $error_i$ are respectively the fitness value and the error value of the ith expression and $maxerr$ is the maximum error value in the generation.

The mating pool for the next generation is then populated by a conventional roulette-wheel method. Generation $n+1$ is evolved from the generation n mating pool by applying the conventional genetic operators of crossover and mutation and also the memory operator described below.

3.5 Crossover

Crossover is applied according to a probability between 0% (no sexual reproduction) and 100% (full reproduction). Two members are selected at random from the mating pool and crossover points randomly selected in each member. The sub-trees anchored to these points are extracted and exchanged between the two members to provide two children. Iteration of this operation produces an interim generation of the same size as the previous generation which may then be subject to further genetic modification. Crossover may result in an RP string that exceeds the maximum length of 19 characters. If this occurs for a given pair of parents, 3 further attempts are made with alternative parents. If an acceptable string is still not obtained the crossover attempt is abandoned and the final pair of parents copied directly from the mating pool to the interim generation before proceeding to the next pair. The choice of 3 retry attempts is a compromise between allowing crossover to occur and ensuring that the system does not require excessive computing cycles.

3.6 Mutation

Mutation is also governed by a probability parameter. In common with most GP systems, members of the interim generation are selected randomly with a given probability and one randomly selected sub-tree in each selected member is replaced by a sub-tree generated by random selection of functions and terminals from the sets defined in Section 3.1. The length of the replacement and hence the length of the mutated member is controlled so as not to exceed the maximum of 19 characters.

It should be noted that if a mutation probability of $x\%$ is quoted it means that on average $x\%$ of the members of a generation are subject to mutation. The limit on RP string length of 19 characters implies a limit of 9 functions and hence 9 sub-trees. On average an individual member has between 4 and 5 sub-trees and a quoted mutation rate of $x\%$ translates to between $x/4\%$ and $x/5\%$ of the genetic material in the generation.

3.7 The memory operator

In a conventional genetic algorithm the quality of a generation is reflected in the next generation through the selection mechanism which ensures that highly fit members are well represented in the mating pool. The best member of a generation may also be copied one or more times into the next generation, following genetic modification, by an elitist strategy. Both operations ensure that good genetic material is available to succeeding generations. The memory mechanism introduced here also ensures that good material is preserved by keeping it in memory and explicitly introducing it into each generation as evolution proceeds.

Fitness-based selection, elitism and the use of memory are similar in that they all contribute to the propagation of successful genetic material through the generations. Selection ensures that highly-fit members of a population survive to reproduction at the expense of less-fit members. Elitism counters the potential disruptive impact of crossover, mutation and, indeed, the memory operator. The use of memory provides a pool of component sub-trees by breaking down good solutions in each generation. The memory is used as a source of replacement sub-trees to improve fitness by the introduction of material previously shown to be successful.

When the best member of generation n is an improvement on the best member of generation $n-1$ (or when $n=1$) the best member of generation n, together with all its sub-trees, is recorded in the memory. The memory is then used as a source of replacement sub-trees during the memory operation by randomly selecting the replacement from the memory. For example, if the expression *ut + at,* represented by the Reverse Polish string *ut*at*+,* is the best-so-far member of a generation, the strings *ut*at*+,* *ut** and *at** are added to the memory. The note relating to the meaning of a percentage probability in section 3.6 also applies to the memory operator. *x%* use of the memory operator means that *x%* of members have one sub-tree replaced by one from the memory.

3.8 Elitism

An elitist strategy is employed whereby a single copy of the best member of generation *n* replaces a randomly selected member of generation *n+1* after all genetic operations have been performed.

4. Results

In the presentation of results from the GP system, the following parameters are used:

MAXGEN
the maximum number of generations in a run
MAXPOP
the maximum size of the population
MAXRUN
the maximum number of runs in a set
RXOVER
the crossover operator probability (%)
RMUTATE
the mutation operator probability (%)
RMEM
the memory operator probability (%)

4.1 Run parameters

A run consists of *MAXGEN* x *MAXPOP* individual tests. A set consists of *MAXRUN* runs. Results for several combinations of these parameter, each consisting of 1000 runs with 40000 tests per run, are presented. The other important parameters are the probabilities with which the 3 genetic operators are applied and denoted by *RXOVER* for crossover, *RMUTATE* for mutation and *RMEM* for memory.

A set of runs is characterised by these 6 parameters and represented by the notation

$$\{1000,2000,20,95,20,20\}$$

where

1000 = runs in the set
2000 = population size
20 = generations per run
95 = crossover probability (%)
20 = mutation probability (%)
20 = memory probability (%)

and the operator probabilities represent the percentage of the population that, on average, have one sub-tree replaced. Note that the 3 genetic operators are listed in the order in which they are applied.

4.2 Measuring the success of the system

The success of the system is measured by the number of runs in a set in which a hit is achieved. When a hit is achieved the run is terminated and the system re-initialised for the next run. In particular, the memory is emptied, i.e. the runs are all statistically independent.

The number of runs in a set is a trade-off between statistical significance and running time. Table 1 shows 3 sets of 100, 250 and 1000 runs, respectively, with each set repeated 5 times with different random number generator seeds. A single run has 40,000 tests. The spread is defined as *standard deviation/mean* and expressed as a percentage.

Runs		100		250		1000
	Seed	Hits	Seed	Hits	Seed	Hits
	12345	91	12345	214	12345	832
	23456	73	23456	192	23456	782
	34567	84	34567	194	34567	811
	45678	78	45678	206	45678	824
	56789	90	56789	204	56789	816
Mean		81.2		202		813
Spread		7.45%		4.01%		2.10%

Table 1 The effect of different random seeds and numbers of runs

All results in this paper were produced on a 300MHz personal computer capable of 40,000 tests/minute. A set of 1000 runs takes between 60 and 100 minutes depending on the number of hits. Taking these figures into account all results presented here were accumulated with 1000 runs per set.

4.3 Preliminary scan of the parameter space

A series of 1000-run sets was made with a population of 2000, a maximum of 20 generations per run and a fixed mutation rate of 20%. The purpose of this series was to explore the effect of different combinations of crossover and memory rates. The results are presented in Figure 1 which shows a contour map of this region of the search space. The figure plots the number of runs with hits in 1000 runs at different levels of crossover and memory use and a fixed level of mutation. The data shows that the system performs best at high crossover (90%-95%) and medium memory use (30%-40%). The gathering of the data for Figure 1 represents 148 hours of continuous running of the system.

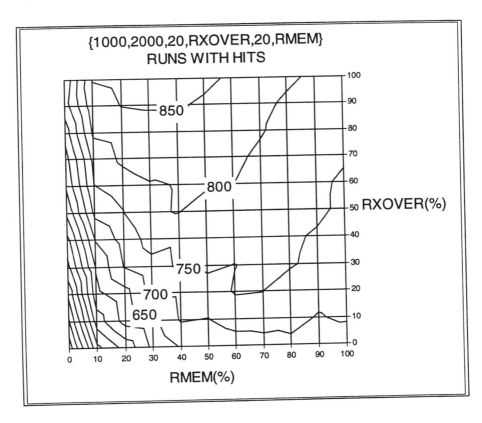

Figure 1 Runs with hits (out of 1000) at different crossover and memory probabilities and fixed mutation probability

It is worth noting here that the parameter set

$$\{1000,80000,1,*,*,*\}$$

i.e. 1000 runs with a population size of 80000 for 1 generation, representing an entirely random search over 80M tests yields no hits. The genetic operator probabilities (shown as '*') are then not relevant since no run proceeds beyond the first generation.

4.4 Variation of the genetic operators

In this section we present the results of applying the crossover, mutation and memory operators, singly or together, for a number of *population x generation* combinations each totalling 40000 tests per run. Table 2 and Figure 2 show the following situations:

100% crossover alone
100% crossover followed by 20% mutation
100% crossover followed by 35% memory
100% crossover followed by 20% mutation followed by 35% memory

for *population x generation* combinations ranging from 500 x 80 to 4000 x 10. Table 2 shows the total number of hits in a set of 1000 runs.

	4000 x 10	2000 x 20	1000 x 40	500 x 80
{100,0,0}	489	721	681	415
{100,20,0}	484	709	807	737
{100,0,35}	824	793	606	471
{100,20,35}	855	894	845	800

Table 2 Runs with hits (out of 1000) for different combinations of operators and different population/generation values

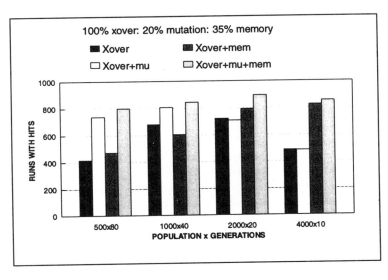

Figure 2

The impact of the different operator combinations may be summarised as follows:

Crossover alone

Reasonably good for population sizes 1000 (681 hits) and 2000 (721 hits).
Falls off at population size 4000 (489 hits) because 10 generations are not sufficient to achieve convergence.
Falls off at population size 500 (415 hits) because of insufficient genetic variety in the initial population with no means of adding to this.

Crossover + mutation

Mutation improves the hit rate at the lower populations, when the initial genetic material lacks variety, but has a slightly destructive effect at the higher population levels where it is not needed.

Crossover + memory

The memory operator is better than crossover alone in 3 of the 4 cases. It is also better than crossover + mutation at the two higher populations which are large enough to ensure that the memory contains good material. At the lower populations, with limited genetic diversity, the memory contents compete unfavourably with the random sub-trees produced during mutation alone.

Crossover + mutation + memory

The 3 operators together achieve high hit rates in all cases with little dependence on the population/generation mix. The reduced performance of the memory operator at lower populations is balanced by the improved performance of the mutation operator. The best result of 894 hits in 1000 runs was achieved with this combination although we believe that detailed tuning of the parameters is capable of a yielding a hit rate in excess of 90%.

These results show that the memory operator provides better performance than conventional mutation if the population size is sufficient to provide good memory material in the early generations. The two operators acting together further improve the hit rate and provide good results in remarkably few generations. For example 855 hits in 1000 runs result from a population size of 4000 evolved for only 10 generations.

5. Summary

We have proposed and investigated the use of a genetic operator based on short-term memory. It is used either as an alternative to mutation or as a supplement to mutation in restoring diversity in a population. When used instead of mutation it improves the hit rate of the GP system provided that the population size is sufficient to provide good genetic material for the memory . When used in combination with mutation the system performance is further improved and is maintained over a range of population sizes.

The objective function $ut + \frac{1}{2}at^2$ used as a vehicle to demonstrate the memory operator is a function of three variables and a non-trivial task for a symbolic regression system. The paper shows that the use of a memory operator, particularly when combined with crossover and mutation, improves the performance of the GP system for this particular problem. We plan to examine the generalization of this approach to other functions.

References

1. Koza J R, 1992. Genetic Programming: On the programming of computers by means of natural selection. MIT Press, Cambridge, MA.
2. Koza J R, 1994. Genetic Programming II: Automatic discovery of reusable programs. MIT Press, Cambridge, MA.
3. Angeline P J and Pollack J B, 1993. Proceedings of the 5th International Conference on Genetic Algorithms pp 264-270
4. Banzhaf W, Nordin P, Keller R E and Francone F D,1998. Genetic Programming - an introduction Morgan Kaufmann, San Francisco, CA.

Chapter 7

Neural Nets and Hybrid System

Application of Bayesian Neural Networks to Non-linear Function Modelling
for Powder Metals
R.P. Cherian, A.G. Pipe, P.S. Midha

Development of an Iterative Neural Network and GeneticAlgorithm
Procedure for Shipyard Scheduling
J. Manuputty, P. Sen, D. Todd

Adaptive Radial Basis Function Emulators for Robust Design
R.A. Bates, H.P. Wynn

An Evolutionary Neural Network Controller for Intelligent Active Force
Control
S.B. Hussein, A.M.S. Zalzala, H. Jamaluddin, M. Mailah

Quality Inspection of Veneer Using Soft-computing Methods
A. Stolpmann, J. Angel

Application of Bayesian Neural Networks to non-linear function modelling for Powder Metals.

RP Cherian AG Pipe and PS Midha
Faculty of Engineering,
University of the West of England,
Bristol,BS16 1QY,UK.
E-mail: Roy.Cherian@uwe.ac.uk /Anthony.Pipe@uwe.ac.uk

Abstract. Mechanical properties of Powder Metal parts depend on a number of material and process related factors and this relationship is of a Multi-dimensional highly non-linear nature. A neural network based model for predicting Tensile Strength of a commercially available ferrous powder grade is described in this paper. Bayesian techniques are used to optimise network parameters and this allows use of large networks without overfitting the noise. Experimental results confirm better generalisation properties than conventional neural networks and the method allows predictions to be linked with an indication of the uncertainty in the network output.

Introduction

Powder Metallurgy (PM) is a production process offering many economic and ecological benefits for the manufacture of small and medium sized parts. Parts manufactured using PM often require very little post-processing, and material utilisation is high compared to other competing manufacturing processes. Energy efficiency is also high and variants of the PM process, such as Hot Iso-Static Processing (HIP), are used in making a number of high performance components in aerospace and defence applications.

In PM, parts are manufactured by compacting metal powder to the required shape in a die and then 'sintering' the compacted part to achieve strength. During sintering the specimen is heated in a controlled atmosphere to promote particle bonding through diffusion. Strength of the sintered part depends on the pressure applied during compaction, chemical composition of the metal powder and additives in addition to sintering parameters such as atmosphere, time and temperature. The highly non-linear nature of the sintering process makes it difficult to develop a comprehensive model taking all these process variables into account as variation in particle size and shape, presence of impurities, crystal defects, changes in processing conditions, sintering atmosphere effects etc., can all influence the sintering process and properties of the end product [1,2]. Previous work done in this area used linear regression techniques and a review of different such models is given in [8]. These methods try to fit regression models using experimental data either to relate some of the process parameters that are linked linearly or approximate these relationships in regions of the parameter space where they are known to be linear [9]. The stochastic nature of the PM process causes the experimental data to contain a high level of noise (scatter in test results), making it difficult to use regression techniques. Current models relating sintered density and chemical composition are of limited practical use as other process variables, e.g. compacting and sintering are not directly related to the mechanical properties.

One of the advantages in using the PM process is its ability to 'engineer' properties of the product during the manufacturing stage using appropriate process parameters. In order to select the right material and process conditions for a given PM application, manufacturers either need to rely on expert knowledge or use commercial databases storing experimental data marketed by powder manufacturers [4]. However such databases are not of much use in choosing the optimum process settings for a specific application, as they are incapable of relating the effect of each variable on the final product.

In today's competitive market, a growing number of manufacturing firms are implementing a 'design for manufacture' (DFM) philosophy to improve lead-time and reduce the cost of design and manufacture of products. In the case of parts designed for PM, implementation of DFM practices requires automated systems that can advise part designers on the material and process constraints at an early stage in design. For a flexible process like PM, this necessitates models that relate the effect of various process-related parameters on the final properties of the part, so that part designers can use them during design.

The present work reports the results of using an Artificial Neural Network (ANN) for modelling the effect of various process parameters on the tensile strength of plain iron powder. The network is trained using a Bayesian framework and network parameters are optimised using Monte-Carlo simulation techniques. The Bayesian approach simplifies network selection and training as complex networks can be trained without 'overfitting' the data. In data modelling problems where there is little prior knowledge about the type of non-linear relationships, this is a major advantage. In the experiments reported here, Networks with differing complexities are tried with the same data and no overfitting was noticed, confirming superiority of the Bayesian approach compared to conventional neural network training methods. Bayesian Neural Networks (BNN) can also give an indication of the 'uncertainty' in the prediction.

ANNs have been successfully used in a number of data modelling problems, control and in other industrial applications [5]. Though impressive results are reported in many of these cases, selection and training of the right type of network is often a difficult task. A number of ANN settings must be made by trial and error including selecting the right number of neurones, number of layers, type of transfer functions, learning rate etc. As modelling ability of the network increases with the number of neurones, a complex network may mistakenly model the noise present in the data as part of the non-linear relationship, leading to 'overfitting' of the data.

At the same time, a network with few neurones may not be able to learn the intended mapping from input to output as its ability to model non-linear relationships is limited. As the objective of the training process in an ANN is to model the function from data without over-fitting the noise, selection of suitable network parameters and architecture is difficult. A number of techniques are used either to control the network complexity and hence its modelling ability or modify the training process to improve the 'generalisation' performance of the network. These include growing and pruning algorithms, generalised cross validation techniques, regularisation and training with noise [5]. These techniques are computationally demanding and require large amounts of data to satisfactorily model the data. Bayesian Neural Networks are proposed as a means to overcome some of the above limitations in the conventional neural network design and training process.

Bayesian Neural Networks

Unlike conventional training methods where a single set of weight vectors that minimises an error function is the objective of training, the Bayesian approach attempts to model the probability distribution function over a Multi-dimensional weight space. This probability distribution encodes the 'degree of belief' in different values of weight vector to model the given data. The distribution function for the network weights are initially set to some 'prior' distribution and then converted to a posterior distribution using Bayes theorem once the network performance on data is evaluated. The posterior probability for the weights is given by:

$$P(W \mid D, H) = \frac{P(D \mid W, H) P(W \mid H)}{P(D \mid H))} \ldots \ldots (1)$$

Here 'W' refers to the weight vector of the network, 'D' is the data set (input-output pairs) used for training the network and 'H' is the modelling assumptions like network architecture etc. Two sets of hyper-parameters are used in Bayesian methods to control the distribution of 'noise' in the data and the network weights, denoted by 'β' and 'α' respectively.

A modified formulation of the Eqn. 1 then will be -

$$P(W \mid D, \beta, \alpha, H) = \frac{P(D \mid W, \beta, H) P(W \mid \alpha, H)}{P(D \mid \beta, \alpha, H))} \ldots \ldots (2)$$

Assuming Gaussian noise in the data, Gaussian 'prior' distribution for network weights, and removing the denominator term as it acts only as a normalisation factor, the above can be written as: -

$$P(W \mid D, \beta, \alpha, H) \cong \frac{1}{Z_D(\beta)} \frac{1}{Z_W(\alpha)} \exp\left(-\left(\frac{1}{2}\beta(y-t)^2 + \frac{1}{2}\alpha \cdot W^2\right)\right) \ldots \ldots (3)$$

Where $Z_D(\beta)$ and $Z_W(\alpha)$ are the normalisation factors for 'data error' and 'prior term' respectively in the above equation and are given by $\int exp\ (-\beta\ E_D)\ dD$ and $\int exp\ (-\alpha\ E_W)\ dW$. The inverse of '$\beta$' and '$\alpha$' control the noise in the data and weight prior distribution respectively.

Learning in Neural Networks using a Bayesian method is aimed at estimating the above posterior probability distribution. Maximising this probability is similar to minimising the Sum Squared Error with a regulariser function ('weight decay factor') in conventional neural network training. This can be verified by taking the 'natural log' of Eqn.3, after eliminating the constant factors $Z_D(\beta)$ and $Z_W(\alpha)$. Eliminating this constant multiplying factor, the equation becomes: -

$$\left(\frac{\beta}{2} \sum (y-t)^2 + \frac{\alpha}{2} \cdot \sum W^2 \right) \dots\dots (4a)$$

which can be modified to: -

$$\left(\frac{1}{2} \sum (y-t)^2 + \frac{\alpha}{2*\beta} \cdot \sum W^2 \right) \dots\dots (4b)$$

However in conventional training, estimation of the right 'penalty' or 'decay' factor, (given by 'α/β' in Eqn 4b.) is done through time consuming cross validation processes. For ANNs with a large number of weights, this will require a large amount of data in addition to the computational burden of training and testing the networks. Bayesian learning allows optimisation of these parameters as part of training and facilitates setting of these decay factors for individual weights or groups of weights with much less computational burden and training data.

The predictive distribution for a new input data $x_{(n+1)}$ is obtained from this posterior distribution of 'W' using: -

$$P\left(y_{(n+1)} \mid x_{(n+1)}, D\right) = \int P\left(y_{(n+1)} \mid x_{(n+1)}, W\right) P\left(W \mid D, H\right) dW \dots\dots (5)$$

Network prediction $y_{(n+1)}$ on an unknown input $x_{(n+1)}$ can then be made with:-

$$y_{(n+1)} = \int f\left(x_{(n+1)}, W\right) P\left(W \mid D, H\right) dW \dots\dots (6)$$

When network predictions are made in this manner, an expected value for the prediction is arrived at by integrating over the 'probable weight' space, i.e. no single weight vector value will dominate the network output. Contribution of each probable weight vector value is weighed by the second term in the above equation and this is how a Bayesian framework handles complex network models without unduly overfitting the data.

Training ANNs using Bayesian methods requires prior beliefs in the noise present in the data and the type of function to be approximated in the form of prior probability distributions, i.e. $P(\beta)$ and $P(\alpha)$. These hyper-parameters are given broad prior distributions and are then optimised as part of the training process.

Evaluation of the above integrals for a typical network with multiple modes can be difficult in practice and is typically done with some simplifying assumptions. One approach is to use a Gaussian approximation for the posterior distribution of weights and it is reported to give good results when the number of training examples are much higher than the number of parameters in the network [6,7]. The network parameters are optimised using gradient descent methods and the hyper-parameters 'α and β' are fixed to values that maximise the probability of the data [6]. In this approach, these values are found by integrating out network parameters using Gaussian approximation for the distribution.

Bayesian methods for neural networks can also be implemented using numerical integration techniques like Monte-Carlo methods [3]. In this method, a complex integral of the type given in Eqn.6 is approximated by generating enough samples of 'W_i' from the multi-dimensional W (weight) space, that are representative of $P(W|D,H)$ and approximate the expected value of the function by: -

$$y_{(n+1)} = \frac{1}{N} \sum f\left(x_{(n+1)}, W_i\right) \dots \dots (7)$$

where 'N' is the number of samples. In this way, the overall output is calculated by averaging the outputs of the group of networks with weight vectors that have higher probability of representing the data. An advantage of this approach is that no Gaussian assumption need be made. This makes it possible to fit complex models, where required, using limited data sets.

Straightforward application of Monte-Carlo methods for evaluating integrals of the type given by Eqn.5 and 6 is infeasible due to high dimensionality and complex shape of the density function, with multiple minima. Therefore many modifications have been suggested to make the algorithm suitable for neural network problems. Neal [3] discusses the implementation of 'Hybrid Monte-Carlo' methods as an alternative to those based on Gaussian approximation. Here representative samples from the posterior distribution of network parameters are obtained using a method analogous to gradient descent, as used in conventional training. This reduces the amount of random sampling required compared to conventional

Monte-Carlo techniques. In this approach, the distributions of hyper-parameters controlling the noise levels, and the prior distribution for weights are each sampled using Gibbs sampling techniques [12]. The network parameters, i.e. weights and biases are updated using the Hybrid Monte-Carlo method. These two techniques are alternated to obtain a Markov chain that explores the entire posterior distribution of weights 'W' and hyper-parameters. A number of samples of network parameters are obtained from the posterior distribution, once the Markov chain reaches an equilibrium state. These sampled values are then used to make network predictions on new data using Eqn 6. More details of these techniques can be found in (3,6).

Experimental Results

Experiments were carried out using data comprising of 189 sets of experimental results taken from CASIP [4], for water atomised plain iron powder (ABC100.30), with apparent density 3.0 g/cc and a mean particle size of 100 microns. Input data to the network consists of compacting pressure in MPa, sintering temperature in DegC, as well as carbon, phosphorous, copper and nickel additions in weight percentages. Network output is the predicted tensile strength of the sintered specimen. Sintering time is kept constant at 30 minutes for all cases studied here. Sintering Atmosphere is either endothermic or dissociated ammonia to suit the material composition. Sintering temperature varies from 1120 °C to 1220 °C and compaction pressure is in the range of 350 to 800 MPA. Carbon, Phosphorous, Copper and Nickel addition to the part is changed from 0 to 0.7 weight %age, 0 to 0.45 weight %age, 0 to 2.5 weight %age and 0 to 3.0 weight %age respectively.

Before training the network the training data was normalised in the range of -1 and +1 as follows:

$$x_n = 2 * \frac{(x - x_{min})}{(x_{max} - x_{min})} - 1 \ldots \ldots (8)$$

where x_n is the normalised value for the variable, and x_{min} and x_{max} are the minimum and maximum of each variable 'x'.

The Neural Networks were implemented using the Bayesian Neural Network software available from [10]. The implementation details of the software, written in 'C' for UNIX systems, are described in Neal [3] and the accompanying documentation. The package allows implementation of complex Bayesian network architectures with sophisticated priors for network parameters, hyper-parameters and data models. A 486DX PC (33 MHz) running LINUX was used in the experiments. Training involved minimising the Sum of Squared Error with the regularising function given by Eqn 4a. Networks with 4, 8, 12, 16, 24 and 32 hidden layer neurones (single hidden layer) were used in different training runs. The networks used 'tanh' hidden layer neurones and 'linear' output neurones. The hyper-parameter controlling the noise is given a prior distribution with mean 0.05 and variance 0.5. Hyper-parameters controlling the weights are in four groups, i.e. Input to Hidden layer weights, Hidden layer biases, Hidden layer to Output Unit weights and Output Bias. Each of these parameters, except output bias, is given a Gaussian prior with mean 0.05 and variance 0.5. Output Bias is given a Gaussian prior with mean zero and standard deviation 100.

The 189 sets of input/output pairs form the data set and are split randomly into two groups; one is used for training and the other for testing. Training of the network proceeds in two stages. In stage 1, training involves Gibbs sampling for noise level and then Hybrid Monte-Carlo (HMC) updates for network weights using 100 (n) 'leapfrog' steps with a stepsize adjustment factor(ε) 0.4. These 'leapfrog' steps explore the weight space, using gradient information with a stochastic element. This is repeated 10 times (m) to reach a rough approximation to equilibrium state corresponding to posterior distribution of network parameters. Network hyper-parameters were kept fixed at this stage. The network parameters sampled at the end of this run are used for stage 2 where multiple samples are drawn. The training process in HMC is summarised below and details on the theoretical basis of the algorithms can be found in [3,10]: -

a. Initialise the network weights (W) using uniform random numbers and define a temporary variable x, which is set equal to W. Initialise hyper-parameters from the respective prior probability distributions.

b. Repeat 'm' times,
c. Initialise a momentum variable 'p' of dimension 'W' using uniform random numbers.
d. Calculate a *hypothetical Energy* function H given by the sum of Error function (E) at 'W' using Eqn 4a and *'Kinetic Energy'* of momentum component - $p^2/2$.
e. Repeat for 'n' times ('leapfrog' steps)
   f. Revise value for p as $p = p - (\text{stepsize}(\varepsilon) * \text{gradient of } E)/2$,
   g. Update 'x' as $x = x + \varepsilon * p$,
   h. Update 'p' using $p = p - (\varepsilon * \text{gradient of } E \text{ at } 'x')/2$,
   i. End

j. Evaluate Error function (Eqn 4a) E_{new} using x,
k. Evaluate new *Energy* function $H_{new} = E_{new} + p^2/2$,
l. Calculate change in H, $dH = H_{new} - H$,
m. Accept the new 'x' value as value for 'W' and E_{new} as E if,

n. $dH < 0$ or $\exp^{-dH}$ greater than a uniformly generated random number (rand()).

o. Update hyperparameters α and β using Gibbs sampling (if required),
p. end

During stage 2, multiple samples ('N' in Eqn 7) of the network parameters are generated in order to make predictions using Eqn 7. To generate these samples, hybrid Monte-Carlo (HMC) updates were carried out for the network parameters preceded by Gibbs sampling for hyper-parameters controlling the noise and priors. The number of 'leapfrog steps' is increased to 1000 (n) and step size adjustment factor is 0.2 during this stage. Of the 100 neural networks sampled during this stage (m), the last 80 are used for making predictions on test data. The first 20 samples can be discarded as they may not come from the equilibrium distribution. The results of a typical run are given in Fig. 1. The average of the outputs from the 80 networks (with 16 hidden layer neurones) sampled after an HMC run is shown as the network prediction as per Eqn7. The average sum squared error calculated for predictions made on test data for different networks (Single hidden layer networks with hidden layer neurones varying) using Bayesian techniques is plotted with those results from conventional network training error in Fig. 2. The ANNs were trained using backpropagation with momentum for 50,000 iterations in each case. In fact the situation with respect to conventional ANN is worse than that indicated by Fig 2 in some runs, probably due to different initial random seeds for the weight values. Results of conventional ANNs were always inferior to BNN and there was a wide variance from one run to another, compared to stable performance of BNN over a number of runs of this experiment.

The model was also tested with 289 data sets for a different grade of proprietary iron powder (SC100.26, apparent density 2.6 g/cc and mean particle size 100 microns [4]). The results compared with actual Tensile strength are given in Fig. 3. More than 90% of the network predictions were within +/- 15% of the actual value. As this powder grade has different compressibility and particle characteristics to those used for training, this variation is to be expected.

Predictions made by 80 samples of networks generated from the above Monte-Carlo runs for varying Nickel content from 0 to 7.5 Wt %age, keeping other process variables constant, is illustrated in Fig 4. Networks with 16 hidden layer neurones are used in making these predictions. The trend in predicted values of tensile strength with variation in process parameters was also found to be in agreement with the results reported in Salak [2]. In Fig 4., the variance of the network predictions increases, as the input parameter moves away from the four matching data in the training set. Although Nickel content above a Wt % of 3.0 is not of any practical importance in PM applications, this serves to demonstrate the wide variations in predictions where data density is low. This variance in predictions on new data can be used as a measure of uncertainty in network predictions in practical applications.

Conclusion

A Bayesian neural network based model for prediction of tensile strength of plain iron powder material is presented in this paper. The model is based on experimental data from the CASIP [4] database and tries to correlate tensile strength to compaction pressure, sintering temperature and alloy additions. Training of the network was carried out using Bayesian software implementing a Hybrid Monte-Carlo method.

The Bayesian based approach described here offers many advantages over conventional statistical methods. The difficult task of network model selection for function with large input spaces, in conventional neural networks is eliminated because users are free to use complex network architectures without affecting performance. As the model is therefore able to incorporate a large number of input parameters, it has potential as a new and powerful modelling tool for PM technologists to evaluate the effect that each of these variables has without having to physically try all possibilities. It can also be re-trained using new experimental data and thus reflect changes in production settings. The Bayesian approach also permits model fitting using less data than conventional neural network training methods. As prediction can be made using a sample of networks selected, the variance of these predictions can give an indication of the amount of uncertainty in the network model.

Training networks using Hybrid Monte-Carlo techniques took much more computer time than conventional training methods using gradient descent techniques. In the present case, it took 48 hours to train the 6 different network architectures compared to 4 hours using back-propagation training. However in many practical applications, such "one time" computer intensive training is worth the advantages that the Bayesian technique offers. Rasmussen [10] reports that Bayesian Neural Networks trained using Monte-Carlo techniques perform better compared to conventional networks even when allowed shorter simulation time than we have used for HMC.

As PM technology is being adopted more and more in mass-produced items, manufacturers are under pressure to improve quality, cut down lead-time and costs. For PM industry to adopt 'DFM' practices, material and process related issues need to be taken into account at an early stage in part design. A neural network based system of the type described here, could help to incorporate process related knowledge and facilitate material selection at the design stage.

REFERENCES

1. German R M, 1994, *Powder Metallurgy Science Second Edition*, Metal Powder Industries Federation, Princeton, NJ, USA, ISBN 1-878954-42-3.
2. Salak A, 1995, *Ferrous Powder Metallurgy*, Cambridge International Science Publishing, ISBN 1-898326-03-7.
3. Neal R M, 1996, *Bayesian Learning for Neural Networks*, Springer Verlag, New York. ISBN 0-387-94724-8.
4. CASIP, 1996, *Computer Aided Selection of Iron Powder*, Database and information package from Höganäs AB Sweden, (software copyright Polydata Ltd, Dublin, Ireland).
5. Bishop C M, 1995, *Neural Networks for Pattern Recognition*, Oxford University Press. ISBN 0-19-853849-9.
6. MacKay D J C, 1992, *Bayesian Methods for Adaptive Models*, PhD Thesis, California Institute of Technology, USA.
7. Thodberg H H, 1996, A review of Bayesian Neural Networks with an application to near infrared spectroscopy, *IEEE Transactions on Neural Networks*, 7(1), pp 56-72.
8. Salak A, V Miskovic, E Dudrova and E Rudnayova, 1974, The dependence of mechanical properties of sintered iron compacts upon porosity. *Powder Metallurgy international*, Vol 6. No.3 pp. 128-132.
9. Martini L L and Prucher T A, 1993, Predicting powder metal mechanical properties- a statistical approach, *Advances in Powder Metallurgy and Particulate Materials*, **3**, pp. 139-153.
10. Bayesian Software of Neal RM. URL: http://www.cs.toronto.ca/~radford/
11. Rasmussen C E, 1996, *Evaluation of Gaussian Processes and other methods for Non-Linear Regression*, PhD thesis, Deptt of Computer Science, University of Toronto.
12. Gelman Andrew, Carlin John B, Stern Hal S and Rubin Donald B, 1995, *Bayesian Data Analysis*, Chapman & Hall, UK, ISBN 0-412-03991 5.

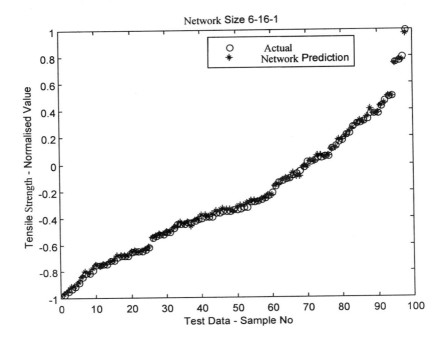

Fig 1. Predicted Tensile Strength (Normalised between -1 and +1) for the test data set.

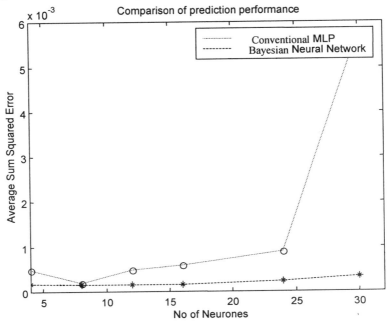

Fig 2. Network performance after training using conventional gradient descent methods and Bayesian approach.

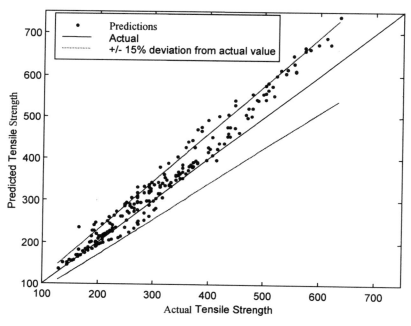

Fig 3. Average of Network predictions on data for a different Iron Powder grade (SC100.26).

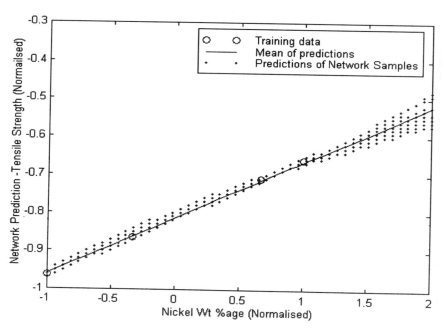

Fig 4. Variance in Network Predictions where data density is low demonstrated by the diverging predictions as Nickel content (one of the input) is varied from 0 to 7.5%.

Development of an Iterative Neural Network and Genetic Algorithm Procedure for Shipyard Scheduling

J Manuputty, P Sen and D Todd
Department of Marine Technology, University of Newcastle
NE1 7RU, United Kingdom
email: j.d.manuputty@ncl.ac.uk, pratyush.sen@ncl.ac.uk,
d.s.todd@ncl.ac.uk

Abstract. Scheduling has always been a key issue in the field of manufacturing and as such any tools that can improve the quality and efficiency of the processes involved have been welcomed. This paper demonstrates the development of an iterative technique involving the use of a neural network, single objective GA and multiple objective GA with a view to squeezing maximum benefits out of a given manufacturing environment. Additionally it also aims to identify key areas where improvements can be made to the facilities within the system. The system is demonstrated on a real world shipyard problem and is shown to give significant improvement on the shipyards' current schedules.

1. Introduction

Scheduling is understandably a very active area of research within the adaptive computing community. Scheduling problems of any size are combinatorially demanding and there are often several dispersed pockets of efficiency in the fitness landscape. In such a situation the use of adaptive and evolutionary algorithms are of great benefit [1], as the appropriate combination of exploration and convergence can be conveniently pursued. Not surprisingly, there is a large body of literature dealing with this subject. In practical terms the benefit of scheduling can be enormous as the consequent effects of such techniques, even on a local scale, can be felt throughout the whole business process. Hence any improvement should be strived for as it will improve overall organisational efficiency.

In the shipbuilding context, which deals mainly with a large number of sub-assembly and assembly tasks, scheduling is a key element of the overall product development strategy [2]. Many shipyards use formal planning tools to ensure that the product is capable of being constructed within the limitations of the production capacity. Planning tools are also used to manage resources and assess the influence of outsourcing. However, the pre-planning scheduling of tasks to get the most benefit out of a facility is often the realm of intuitive common sense approaches. Work is organised by planners on the basis of past experience and little if any use is made of formal scheduling tools [3]. The aim of this paper is to report on an ongoing project on shipyard management that combines facility modelling and scheduling to get the most out of a given shipyard. The stress is on the medium term, allowing for improvement in facilities if this is considered to be appropriate for the product mix in question.

The application is directed at a simplified shipyard model based on an actual yard in South East Asia.

2. System Architecture

The system is configured as shown in Figure 1. In the normal existing setup the user must provide a Product Work Breakdown Structure (PWBS) [4] and manufacturing environment configuration on the basis of which a set of sequences and routeings would be generated using standard techniques within the company's workplace. For a stable product base the planners are expected to do reasonably well even when unaided by scheduling algorithms.

The system loop can be entered in one of two ways. The first option is to go directly into simulation (Route 1) and the second is through the Multiple Criteria GA Scheduler (Route 2). Both of these routes are demonstrated later in the shipyard example. If route two is chosen the data is first put into the MCGA Scheduler. This is a Multiple Objective Genetic Algorithm (MOGA) based scheduling tool which optimises both sequences and routeings for a fixed manufacturing environment. The MCGA generates an optimal sequencing for the work which is then passed into the simulation model. A further and more complete explanation of the MCGA Scheduler can be found in [5] and [6], this includes information on the string construction and a full GA description. Alternatively the sequences and routeings provided by the user can be put straight into the simulation model (Route 1) without the initial optimisation by the MCGA.

The system model must be constructed by the user using a suitable simulation tool; in this case Simple++ was used (AESOP, Simple++ v.5.0). The model is then primed with the sequencing and routeing information from the user (or MCGA) giving a complete model of the manufacturing environment. This model can then be interrogated to find out finishing times and utilisations for the given sequencing and routeing of jobs for different configurations of manufacturing environments. Parameters such as capacity, quantity and throughput of machines can be modified and the input setup and corresponding output noted. This process is repeated several times in order to form a training set for the next stage of the optimisation process.

The set of input/output pairs from the simulator are used to train an Artificial Neural Network (ANN) (Cheshire Engineering Corporation, Neuralyst V. 1.41) [7]. The ANN based response surface can then be searched to find system configurations with the required properties in terms of finishing time and utilisation. This search is performed by a single objective GA (Paliside Corporation, Evolver V. 3.12). The GA experiments with different configurations of the manufacturing layout and tries to find configurations that achieve objectives of time and utilisation most closely. The resulting optimal configuration can then be taken and implemented along with the sequencing and routeing or can be passed back into the MCGA Scheduler to re-optimise the schedule with the new configuration.

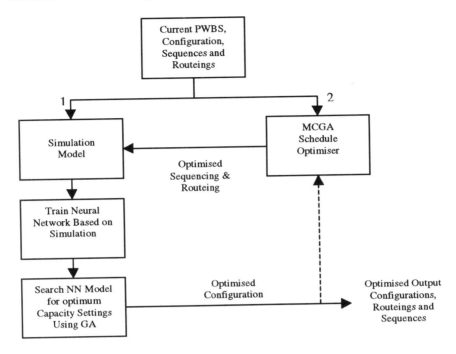

Figure 1 – System Configuration

3. Shipyard Problem

To test the system a simplified shipyard scheduling problem was used. In this paper the two routes through the system were tested to show their effectiveness and a full cycle of the system is reported below to show the potential improvements. Further tests are ongoing to show the benefits of iterative procedure and its underlying economic viability.

The shipyard problem is a real world example with 100 jobs, each going through a seven stage manufacturing process, are scheduled (i.e. 700 stages). In real terms this equates to about 195 days (4680 Working Hours) of total process time for the yard with the current system configuration and job sequence and routeing. The process is shown in Figure 2.

The data for the simulation include the ship's PWBS and its properties the building sequences of the parts included in the PWBS, the routeing of the parts, the process data, the facility (workstations) throughput capacity and layout and the process time required for each part. This data was provided by the shipyard. The

338

simulator (Simple++) produces the timings for a production schedule and utilisation values for each of the machines. The utilisation values are combined into a single value.

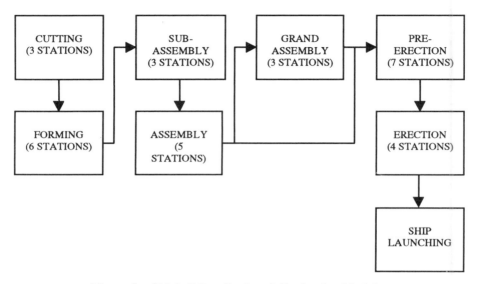

Figure 2 – Shipbuilding Steelwork Production Model

In these tests the capacity of two of the workstations (Sub-Assembly and Assembly) was investigated along with the modification of the current schedule to give maximum combined benefits. Two runs through the system were carried out, the first starting with the simulation (Route 1) and the second starting with the MCGA Scheduler (Route 2). This was done to compare the benefits of configuration optimisation with the combined effects of sequence/routeing optimisation and configuration.

In both cases the simulator was interrogated to provide 25 data pairs for the ANN, 20 for training and 5 for testing. The ANN was configured with two inputs (2x Workstation Capacities), 10 hidden layer nodes and two outputs (Production Finish Time and Resource Utilisation) using a sigmoid transfer function [8]. The single objective GA was then run with the workstation capacities as parameters and the difference between the ANN output and the desired finish time/utilisation as the fitness function. The GA was asked to configure the workstations such that a specific finish time and utilisation was achieved. Once the converged results were obtained from the GA the simulation was re-run with the best workstation configurations to give a more precise value for finishing time and utilisation. The output schedule can then be implemented or the new configuration can be returned to the MCGA Scheduler for further job rescheduling to take advantage of the new configuration improvements provided by the ANN and GA. The process can be repeated several times until no further improvement is achieved.

Work on this aspect of the system was still being carried out at the time of submission.

4. Results

Tables 1-5 shows the results from the various runs carried out on the system. The aim of the tests was to show the effectiveness of each of the components of the system and then to show the combined effects and potential of the entire system when applied to the shipyard problem.

Table 1 - Test 1 - Existing Configuration, Sequence and Routeing

SUB	ASS
16	20

Finish Time	Utilisation
194.22	39.99

Table 2 - Test 2 - GA Search for Best Configuration - Original Sequence and Routeing

GA SEARCH TARGET		GA SEARCH RESULT		GA SUGGESTIONS		RE-RUN SIM RESULT	
Finish Time	Utilisation	Finish Time	Utilisation	SUB	ASS	Finish Time	Utilisation
170	40	170.43	40.71	20	19	174.84	41.87

Table 3 - Test 3 - Existing Configuration with MCGA Optimised Sequence and Routeing

SUB	ASS
16	20

Finish Time	Utilisation
173.6311	40.00

Table 4 - Test 4 - GA Search for Best Configuration - New MCGA Optimised Sequence and Routeing

GA SEARCH TARGET		GA SEARCH RESULT		GA SUGGESTIONS		RE-RUN SIM RESULT	
Finish Time	Utilisation	Finish Time	Utilisation	SUB	ASS	Finish Time	Utilisation
170	40	171.03	40.33	15	15	177.24	40.93

Table 5 - Test 5 - Potential Performance with Configuration from Test 2 and
Sequence and Routeing from Test 3

SUB	ASS		Finish Time	Utilisation
20	19		150.38	43.54

Test 1 – Existing Configuration, Sequence and Routeing

In the first test the current configuration and original sequence and routeing were simulated with the standard shipyard setup in order to provide benchmark results for Finish Time and Utilisation. The SUB and ASS values are the capacity settings in tons/day for the Sub-Assembly and Assembly workstations. These are the search variables throughout these tests.

Test 2 - GA Search for Best Configuration - Original Sequence and Routeing

The second case follows route 1 (from Figure 1) where simulation is the first stage of the process followed by ANN training and GA search of the ANN model to find the optimum configuration based on the original sequence and routeing. The first two columns in the table shows the target Finish Time and Utilisation. The GA was set to run on the problem and produced as configuration of SUB=20 and ASS=19. These results are quite close to the goal values given. This configuration was re-run within the simulator to give more reliable estimates for the system performance. They are slightly higher than the ANN prediction but are still reasonable.

Test 3 - Existing Configuration with MCGA Optimised Sequence and Routeing

In this case the sequence and routeing was first optimised using the MCGA Scheduler based on the existing configuration. The optimised sequences were then put into the simulation model to assess the effects of the scheduler. As can be seen the Finish Time of the system has almost reached the goal of 170 days and the Utilisation is correct at a value of 40%. This means the current system can potentially achieve these goals without any modification to the configuration. This would be considerably more economic than changing the configuration as in Test 2 above and points to the potential advantage of this approach.

Test 4 - GA Search for Best Configuration - New MCGA Optimised Sequence and Routeing

The fourth case starts with a MCGA Scheduler optimisation of the original sequencing and routeing. Then the process goes on to the simulator and so on as in the second case. This run represents one complete iteration of the system. The goal is again to achieve a Finish Time=170 and Utilisation =40%. The current system can almost achieve this goal in its current SUB=16, ASS=20 configuration. However the GA still tries to find a more efficient configuration. In this case the workstations have been down sized due to the existing excess capacity (SUB=15, ASS=15). Again these results are resimulated to verify the results. In this case the inaccuracy of using the current ANN model is demonstrated as the simulated finish time is 6 days slower than the predicted value. This error can be virtually eliminated with a bigger training set. After further investigation the best simulated configuration of the system was found to be SUB=16, ASS=15 with values of Finish Time=171.29 and Utilisation = 40.72.

Test 5 - Potential Performance with Configuration from Test 2 and Sequence and Routeing from Test 3

The final experiment was carried out to examine the potential performance improvement of the system if the Test 2 configuration (SUB=20, ASS=19) and the Test 3 sequence and routeing were combined. These setting were simulated giving a Finish Time=150.38 and Utilisation = 43.54. This is an improvement of just under 44 days on the original configuration – a considerable gain.

5. Discussion of Results

As can be seen from the test cases each of the components of the system has an important role to play in terms of process and schedule improvement. Significant benefits in terms of project time and process management have been demonstrated.

During the use of the software several areas were identified for improvement. Firstly the collection of data from the simulator is quite time consuming even for modest problems. For large configuration search spaces many points are required to train the ANN and this slows down the process. Similarly the ANN becomes more difficult to train as increased function complexity means more neurons in all layers. As complexity increases so does the inaccuracy of prediction due to such a small training set. This was seen in test case 4 where the values of SUB and ASS found by the GA to be the closest to the goal were actually some distance away due to a value on the ANN surface that deviated modestly from the optimum.

Another feature is that the system requires significant data management between the various stages. Ideally the component parts would be linked together and the data file conversions would be carried out automatically so that the system models and schedules could be passed seamlessly from unit to unit to form a useful decision support tool for management. This work is in hand.

6. Conclusions

Although the system is in its early development stages it has shown itself to be capable of making significant improvements compared to the current experience based scheduling implementations in the shipyard. The system has shown the effects of configuration optimisation and schedule optimisation and has also demonstrated the synergy between the two. The benefits of using a full iterative procedure is expected to be significant on the basis of results obtained to date. For large scale facilities planning the benefits of such a procedure arise naturally from a combination of good facilities planning and efficient utilisation of the facilities themselves.

7. References

1. Goldberg, D E. 1989. *Genetic Algorithms in Search, Optimisation, and Machine Learning*. Addison-Wesley Publishing.

2. Okumoto Y, 1999. Optimising of Working Route Using Genetic Algorithm. *Proceedings of 1999 Ship Production Symposium and Expo*, July 29 – 30, 1999, Arlington, VA.

3. Aga S, Hatling J F, 1997. Simulation – A Powerful Tool for the Shipbuilding Industry. *Proceedings of ICCAS'97, 9th International Conference on Computer Applications in Shipbuilding*, pp. 8.103 – 8.117

4. Storch R L, Hammon C P, Bunch H M, 1995. *Ship Production*, 2nd edition , Cornell Maritime Press.

5. Todd D S, Sen P., 1997. Multiple Criteria Scheduling Using Genetic Algorithms in a Shipyard Environment. *Proceedings of ICCAS'97, 9th International Conference on Computer Applications in Shipbuilding*; 13-17 Oct 1997, Yokohama, Japan.

6. Todd D S, Sen P., 1998. Tacking Complex Job Shop Problems Using Operation Based Scheduling. *Proceedings of ACDM'98, Adaptive Computing in Design and Manufacture*. Springer, London, pp45-58.

7. Hurrion R D, 1997. An example of Simulation Optimisation Using a Neural Network Metamodel: Finding the Optimum Number of Kanbans in a Manufacturing System. *Journal of the Operational Research Society*, Vol. 48, No. 11, pp. 1105-1112

8. Rumelhart D E, McClelland J L, 1986. *Parallel Distributed Processing*, Volume 1, Cambridge, MA, MIT Press, 1986.

Adaptive radial basis function emulators for robust design

R A Bates and H P Wynn
Department of Statistics, University of Warwick
Coventry CV4 7AL, UK.

Abstract

In the field of engineering design, tradeoffs between competing design objectives can only be made if there is a good understanding of the product or process under development. To facilitate this, adaptive classes of models can be used to represent complex engineering systems and provide important information for design development. This paper describes a fast implementation of a radial basis function model, intended for this purpose. By exploiting the mathemtical form of the Gaussian basis function, a computationally efficient method of estimating smoothness is developed and used in the model fitting process. The method is applied to a set of existing experimental data and compared with two alternative modelling strategies involving polynomial and stochastic process models.

1 Background

Complex product optimisation requires fast, accurate models so that design tradeoffs can be explored, robust designs found and product families developed. Computational methods are used to find novel design solutions in several ways. There is a distinction between, say, structural optimisation, where the architecture of a product is changed radically in the search for a design solution, and Robust Engineering Design (RED), where an existing design architecture is optimised for robustness in manufacture and use.

In the case of RED, experiments are performed to determine the performance of a given design within set modes of operation, thus there is a continuous relationships between design factors and design performance criteria. Often, evaluations of design performance are costly and involve physical experimentation. Increasingly this experimentation is being replaced by computer simulation, the experiments being called *computer experiments*. But the computational burden remains high. By using methods developed in the field of Experimental Design, the number of evaluations, or *trials*, of a product prototype can be minimised and a framework provided for modelling performance with fast approximations, or *emulators*. These emulators describe the relationship between design factors and design performance and can be evaluated orders of magnitude faster than the original simulation or physical experiment. They can therefore be used in global multiobjective numerical optimisation of the product design. Experimental design, emulation and numerical optimisation provides a framework for overall design optimisation that can be implemented effectively in a computational environment.

In this paper an adaptive method for emulation of engineering designs based on radial basis function (RBF) networks [1] is described, based on a measure of smoothness. The method is compared with two other emulation methods.

2   Radial basis function networks

As emulators, RBF networks can be considered as one-layer neural networks with a radial activation function. The network provides a mapping between the various settings of the design factors, determined by a carefully chosen experimental design, and the corresponding measured response. In this paper it is assumed that there is a separate RBF emulator for each response.

In computer experiments there is no measurement error and the RBF emulator is an exact interpolator of the response, meaning that there is one basis function at each observation site.

There exist many basis functions, here the Gaussian function:

$$\phi(x) = \exp(-\frac{(x-s)^2}{2c^2})   \qquad (1)$$

is used, where s is centre of the basis function and c is a measure of the width of the basis function. Figure 2.1 shows three basis functions, all centred at $s = 0$, with three different width parameters $c = 0.5, c = 1$ and $c = 2$.

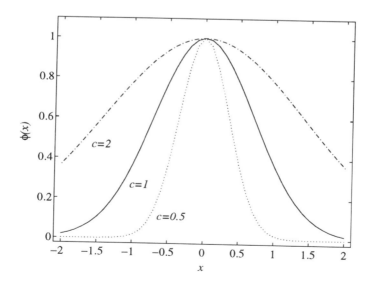

Figure 2.1: Gaussian basis function with three different width parameters

In this paper we shall adopt a particular class of radial basis functions in which each dimension $i = 1, \ldots, d$ of the input space is allowed its own width c_i:

$$\phi(x_1, \ldots, x_d) = \exp\left(-\sum_{i=1}^{d} \frac{(x_i - s_i)^2}{2c_i^2}\right).$$ (2)

This simple approach has proved effective in dealing with the different nature of the individual inputs x_i, for example in terms of units of measurement, while preserving computational simplicity.

3 Measures of curvature

We first specify the "ideal" form of the smoothness criteria we have adopted and then give a number of useful simplifications adopted for fast implementation.

The Gaussian curvature of a d-dimensional surface

$$y(x_1, \ldots, x_d)$$ (3)

is proportional to the determinant of the Hessian

$$H = \left\{ \frac{\partial^2 y}{\partial x_i \partial x_j} \right\}, \quad (i, j = 1, \ldots, d)$$ (4)

when the gradient $(\partial y/\partial x_1, \ldots, \partial y/\partial x_d)$ is zero. Several authors have suggested using the Hessian to measure the smoothness of the function y in the multivariate case. Use of the full Gaussian curvature is computationally difficult and the Hessian generalises $\partial^2 y/\partial x^2$ used in one-dimensional smoothing. We adopt here a generalised criteria

$$\begin{aligned} h_\alpha(x) &= \operatorname{trace}(H^\alpha), \quad \alpha > 0, \\ h_0(x) &= \det(H). \end{aligned}$$ (5)

In practice we have used mostly

$$\begin{aligned} h_2(x) &= \operatorname{trace}(H^2), \\ &= \sum_i \sum_j \left(\frac{\partial^2 y}{\partial x_i \partial x_j}\right)^2, \end{aligned}$$ (6)

being the natural generalisation of the much used $(\partial^2 y/\partial x^2)^2$ in one dimension. If $\phi_\alpha(x)$ is our criteria at a particular point then the overall measure of smoothness is taken to be

$$\int_X h_\alpha(x) dx$$ (7)

over the "design space" X, very often a hyperrectangle.

We introduce two further simplifcations.

(i) Replace the integral by the sum

$$\sum_S h_\alpha(x) \tag{8}$$

where $S \subseteq X$ is the experimental design.

(ii) Select $\alpha = 2$ and replace ϕ_2 by

$$h_2^* = \sum_i \left(\frac{\partial^2 y}{\partial x_i^2}\right)^2 \tag{9}$$

that is ignore the off-diagonal elements of the Hessian.

We now give some calculations which demonstrate the advantage of using h_2^*. Let us first compute the derivitives of a generic radial basis function

$$\phi = \exp(-\frac{(x-s)^2}{2c^2}),$$

$$\frac{\partial \phi}{\partial x} = -\frac{(x-s)}{c^2}\phi,$$

$$\frac{\partial^2 \phi}{\partial x^2} = \frac{(x-s)^2 - c^2}{c^4}\phi. \tag{10}$$

This shows that a term such as $(\partial^2 y/\partial x^2)$ is $O(1/c^4)$. Note that at $x = s$ we have $-\phi/c^2$

Now if our emulator is

$$\hat{y} = \sum_{j=1}^n w_j \phi_j(x), \tag{11}$$

its Hessian is

$$\left\{\frac{\partial^2 \hat{y}}{\partial x_r \partial x_s}\right\} = \sum_{j=1}^n w_j \frac{\partial^2 \phi_j(x)}{\partial x_r \partial x_s}, \tag{12}$$

and making the approximating assumptions (i) and (ii) above with $y = \hat{y}$, and taking the design $\{s^{(j)}\}_{j=1}^n$ we have

$$h_2^* = \sum_{k=1}^n \left(\sum_{j=1}^n w_j \sum_{i=1}^d \frac{(s_i^{(k)} - s_i^{(j)})^2 - c_i^2}{c_i^4}\phi_j(s^{(k)})\right)^2. \tag{13}$$

We seek a further approximation as $c = \max_i c_i \to 0$. The experimental terms are of a smaller order than $1/c_i^4, 1/c_i^2$ so we simply put, in the formula,

$$\phi_j(s^{(k)}) = \delta_{kj}, \quad \text{(Kronecker)},$$

$$w_j = y_j, \quad (j, k = 1, \ldots, n). \tag{14}$$

where y_j is the observation at $s^{(j)}$, $(j = 1, \ldots, n)$. Then the $O(1/c_i^4)$ terms are eliminated and

$$h_2^* = \left(\sum_{k=1}^n y_k^2\right)\left(\sum_{i=1}^d \frac{1}{c_i^2}\right)^2. \tag{15}$$

To this order of approximation, in fact, the contribution to the term h_2 from the off-diagonal Hessian terms are zero. In section 4 we shall fit with mean-adjusted y_i so that the first term in the above expression is proportional to the sample variance of the y_i.

4 Fitting RBF emulators

As previously stated, emulators are used to fit data derived from experiments on an engineering product or process, or a computer simulation thereof. In general, the product or process is considered as a model,

$$y = f(x),\tag{16}$$

where y is the simulator *response*, $x = \{x_1, \ldots, x_d\}$ is the vector of d design parameters, or *factors*. The response y is obtained by running the simulator at the factor values given by x. The emulator model, $\hat{f}$, approximates the response,

$$\hat{y} = \hat{f}(x) + \varepsilon,\tag{17}$$

where ε is the measurement error (set to zero in the case of computer experiments). The model is built by conducting an experiment which consists of running the simulator at a set of carefully selected *design points*, $s^{(j)} = s_1^{(j)}, \ldots, s_d^{(j)}$. This set of n design points, $S = s^{(1)}, \ldots, s^{(n)}$ is called an *experimental design*. The experimental design, S, and the corresponding set of responses, $Y = \{y_1, \ldots, y_n\}$, contain the information used to build the RBF emulator.

The RBF emulator therefore has n basis functions, centred at the design points $s^{(1)}, \ldots, s^{(n)}$. The model fitting process consists of finding suitable values for d basis function widths repeated at each site $s^{(j)}, j = 1, \ldots, n$. The accuracy of the RBF emulator is measured in terms of leave-one-out cross validation error (CVE):

$$\text{CVE} = \sqrt{\frac{1}{n} \sum_{j=1}^{n} (y_j - \hat{y}_j)^2}\tag{18}$$

and the model-fitting criteria is a weighted combination of this error and the estimate of curvature described in Section 3,

$$\min \text{CVE} + \alpha h_2^*,\tag{19}$$

where $\alpha > 0$ is set to reflect the relative importance of curvature in the model fitting process. In practice, the mean $\bar{y}$ of all observations, $y_j, j = 1, \ldots, n$, is subtracted from the observations prior to model fitting, and re-added afterwards.

The fitting procedure is summarised as follows:

1. subtract sample mean $\bar{y}$ from all $y_j, j = 1, \ldots, n$,

2. choose a value for α,

3. choose initial value for widths $c_i, i = 1, \ldots, d,$

4. calculate CVE and estimated curvature $h_2^*,$

5. choose new $c_i,$ to minimize $CVE + \alpha h_2^*,$

6. repeat Steps 4 and 5 until an acceptable minimum is found.

In practice, Steps 4, 5 and 6 are performed using a numerical optimisation algorithm.

5 Comparison of emulators

As an example of emulator fitting, an RBF emulator is fitted to a set of data from a diesel engine experiment and compared with two other methods of emulation: (i) DACE : Design and Analysis of Computer Experiments [2], and (ii) POLY : a polynomial emulator model fitting package [3]. The experiment consists of a 24 point Latin Hypercube experiment [4] conducted on a diesel engine with four design factors:

1. *fdel*, fuel delivery,

2. *ai*, injection timing,

3. *prex*, common rail pressure, and

4. *boost* boost pressure,

and a single response (power output). The diesel engine is placed on a testbed and the values for the four design factors are set using a computerised engine controller. The power output is measured at each setting once the engine has reached a steady state. For the purposes of this paper, the measurement error associated with this experiment is assumed to be zero. In practice, for real experiments with measurement error, one could fit a saturated RBF emulator (with a basis function at each design point) and then invoke a stepwise procedure to remove any undesirable basis functions.

The results of model fitting are given in Table 5.1 which shows the final cross validation error associated with each model and the time taken to fit.

Table 5.1: Comparison of results for 3 emulator types

	RBF	DACE	POLY
CVE	0.675	0.463	0.918
time(s)	133.3	12.7	782

Figures 5.1 and 5.2 show the cross validation and main effects plots respectively for the three different model types.

The cross validation plot shows the accuracy of the emulator in predicting the response at each design point $s^{(j)}$ with that point removed from the emulator. On the plots, the 'o' represents the prediction, 'x' represents the observed value and the error bar represents the 95% confidence interval associated with the prediction, based on all the cross validation residuals.

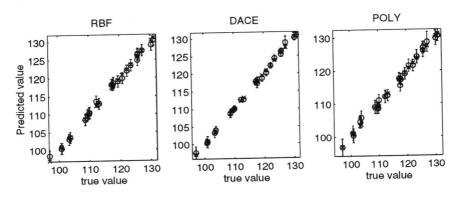

Figure 5.1: Cross validation plots for power output

The main effects plot describes the individual effect each factor has on the response, with the effects of all other factors integrated out. It is useful in displaying the relative importance of each factor and its complexity in the emulator. The values of each factor are normalised to lie in the range $[-0.5, +0.5]$ so that they can be plotted together. In this example, *fdel* is the most important factor, having a near-linear relationship with the response.

6   Discussion

The results of the example study show that the RBF emulator performs well in modelling the diesel engine data, better than the POLY emulator and almost as good as the DACE emulator. The main effects plots (Figure 5.2) show good agreement between the three emulator types, with *fdel* being the dominant factor in all cases.

The results also show the time taken to fit the emulators. The RBF emulator took 133.3 seconds, this is ten times longer than the DACE emulator. For RBF, the main computational burden is due to the cost of computing the CVE value wheras the DACE emulator is fitted using maximum likelihood. This highlights the need for fast, efficient model selection criteria, and it is important to note that the simplifications to the curvature calculations, introduced in Section 3, have greatly reduced computation time and give good model accuracy.

By improving the CVE calculation, or replacing it with a less computationally expensive estimate of model accuracy, the RBF emulator could be improved to make model fitting time more competitive with DACE. Another important point is that the RBF emulator is of a simpler form than the DACE emulator and so, once fitted, could be more useful in certain situations.

7   Conclusion

A method of fitting RBF emulators to experimental data has been described and implemented. By making simplifications to the way an estimate of curvature is cal-

350

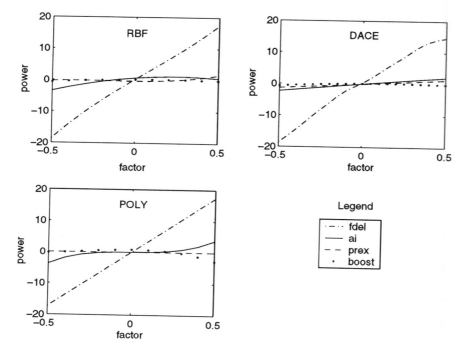

Figure 5.2: Main effects plots for power output

culated, an efficient method of fitting RBF emulators to experimental data has been developed and tested with data from a diesel engine study.

A distinction is made between fitting saturated RBF emulators to data from computer experiments and non-saturated emulators to data from real experiments. The latter case is only mentioned here and is the subject of ongoing research.

References

[1] D S Broomhead and D Lowe. Multivariate functional interpolation and adaptive networks. *Complex Systems*, 2:321–355, 1988.

[2] J Sacks, W J Welch, T J Mitchell, and H P Wynn. Design and analysis of computer experiments. *Statistical Science*, 4:409–435, November 1989.

[3] R A Bates, B Giglio, E M Riccomagno, and H P Wynn. Gröbner basis methods in polynomial modelling. In *Compstat 98*, 1998.

[4] M D McKay, W J Conover, and R J Beckman. A comparison of three methods for selecting values of input variables in the analysis of output from a computer code. *Technometrics*, 21:239–245, 1979.

An Evolutionary Neural Network Controller for Intelligent Active Force Control

S. B. Hussein[1], A.M.S. Zalzala[2], H. Jamaluddin[1], M. Mailah[1]

[1]Faculty of Mechanical Engineering, Universiti Teknologi Malaysia
Skudai, Johor Bahru 81310, Malaysia
E-mail: shamsul, hisahmj, musa@fkm.utm.my

[2]Department of Computing and Electrical Engineering, Heriot-Watt University
Edinburgh EH14 4AS, United Kingdom
E-mail: a.zalzala@hw.ac.uk

Abstract

In this paper, we examine the capability of an Evolutionary Neural Network Controller (ENNC) in estimating the inertia matrix of the two-arm rigid robot. The accurate estimation of the inertia matrix is very important in the active force control loop to calculate the disturbance torques which need to be compensated in order to control a robot subjected to unknown external forces. The proposed algorithm is a modification of the EPNET algorithm proposed by Xin Yao, where we emphasise the use of crossover to explore different offsprings which do not posses strong behavioural links to their parents but still perform better than them. At the same time, the mutation operations described in EPNET, i.e. hybrid training, node deletion and node addition, are still used to maintain the behavioural link between the strong parents and their offsprings. Therefore, the introduction of the crossover will create a kind of 'survival competition' scenario between the different offsprings. In addition, this algorithm also includes the evolution of transfer (activation) functions, which play an important role in the design of the neural network. The best offspring, which represents the optimum number of nodes and types of transfer functions, is selected as the optimum neural network design for the specified problem. Then, the selected network is once again trained using back propagation with adaptive learning rate and momentum to ensure global error convergence. Finally, The fast evolutionary programming (FEP) method, which is based on the Gaussian distribution and directional mutation scheme, is incorporated to fine tune the network parameters at the end of the training session. The trained network is implemented for the active force control problem of the two-arm robot with unknown external forces. Simulation is programmed in MATLAB/SIMULINK using the Neural Network and Geatbx toolboxes. Results show significant improvement in the performance of the evolving neural network as compared to the non-evolving network.

Keywords: Evolution, genetic algorithm, evolutionary programming, neural networks, robotics, active force control

1. Introduction

Researchers in robotic control have been trying to improve the performance of existing robot systems in order to cope with the increasing industrial demand and technological development. Some of the main industrial tasks involving robots are material handling, parts assembly, welding, spray painting, and polishing. Robots are also used in other areas such as in the autonomous robot, remote manipulation, and medical and tele-operation [1].

In general, robotic control can be divided into two main tasks, i.e. position control and force control. Some of the popular classical methods in robot position control are on-off control, independent joint control and computed torque control. In force control, the two popular methods are the impedance force control and the hybrid position-force control [2]. Another force control method is the active force control (AFC) first introduced by Hewit and Burdess [3]. Recently, robotic control involves the implementation of computational intelligence (i.e. neural network, evolutionary computation, and fuzzy logic) to function as a controller itself, or as part of the controller system. Some of the works on the implementation of ANN in robot control can be found in [4] and [5], while some researchers incorporate GA based algorithms [6].

In this paper, an evolutionary algorithm comprising a GA and an EP is used to optimise the architecture and weight of a neural network controller used for the estimation of the robot's inertia matrix. The estimated inertia matrix in the robot's active force controller loop is used to predict the external disturbance torque's acting on the robot arm. Description of the active force control strategy is described in the later section of this paper.

This paper is presented as follows. Section 2 describes a brief literature review on evolutionary artificial neural network (EANN). In section 3, some related works, i.e. the EPNet algorithm introduced by Yao and the fast EP algorithm by Kim, are discussed in brief. The active force control strategy is described in section 4. The new evolutionary neural network controller (ENNC) algorithm is explained in section 5. Simulation and result are shown in section 6 while discussions and conclusion are given in section 7.

2. Evolving Neural Networks – A Brief Literature Review

Evolving artificial neural network is an area of popular interest among the researchers in the field. Much work has been produced to improve the performance of the EANN in terms of its generalisation capability and its efficiency, i.e. time and space complexity. Whitley [7] and Fogel [8] have reported some of the early works in this area and a good review of the early development is reported by Yao [9]. In general, there are three approaches to EANNs: the evolution of connection weights, of architectures, and of learning rules. Some publications reports a fixed architecture, and evolve only the connection weights of the architecture [10], some evolve both the architecture and the connection weights [11], and others evolve only the architecture of the network and incorporate back-propagation or other methods to train the weight [12]. However, very few works deal with the evolving learning rules.

In the evolution of connection weights, two popular representations are binary strings (BS) and real numbers (RN). The problem with BS representation is that it requires large strings to capture precise resolution, and the training may require too long to compute. To overcome these shortcomings, RN representation is introduced. However, the classical techniques of crossover operation cannot be used for real number representation and new crossover techniques are proposed for use with real number representation [13].

In the evolution of EANN architectures, the chromosome representation can be divided into two different schemes, direct encoding (DE) and indirect encoding (IE). In DI, the number of nodes, the connection and the number of hidden layer are directly encoded as a BS representation. The drawback of this type of encoding is that it is only suitable for handling relatively small number of nodes since large ANNs need large represenatation matrices. On the other side, the IE scheme is a compact representation of the EANN architectures, where small chromosomes can grow massive neural network by some development rules during chromosome decoding. Some of the early work was reported by Kitano [14] and a comparison of both schemes has been reported recently [15].

3. Related Algorithms

a. The EPNet Algorithm

EPNet is an evolutionary system for evolving artificial neural networks [16]. The evolutionary algorithm used in EPNet is based on Fogel's EP. EP emphasises the behavioral link between parents and their offspring. EPNet actually emulates a kind of Lamarckian rather than Darwinian evolution. It is used to evolve feedforward ANN's with generalised multilayer perceptrons with unrestricted layer connection and hidden nodes. However, the transfer function in EPNet is not evolved but rather determined before the evolutionary program starts.

In EPNet, only the mutation operator is used in order to maintain the behavioural link between the offspring and their parents, and introducing a crossover operator may destroy parent ANNs. The major steps of EPNet are shown in figure 1. At an initial stage, the ANNs are trained partially to check weather or not the ANNs need to be mutated. If so, the selected ANNs will undergo the five-step mutation process, i.e. hybrid training, hidden node deletion, connection deletion, connection addition and hidden node addition.

354

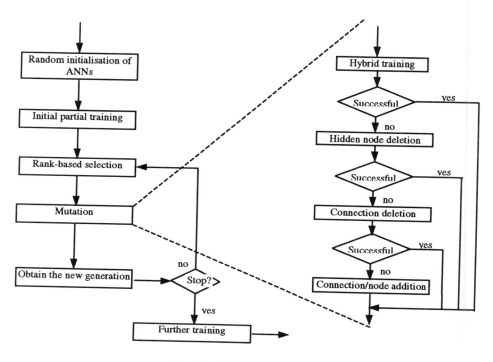

Fig. 1   Major steps in EPNet

b. The FEP Algorithm

Kim [17] proposed an evolutionary algorithm that converge faster than other evolutionary programming techniques without decreasing the diversity. The algorithm has two mutation operators, namely the direction operator and the Gaussian perturbation whose mean is zero. In order to enhance the diversity of the search and to prevent the individuals from remaining in the local minima, an additional variable "*age*" is used.

The mutation occurs according to the following two rules:

Rule 1:   If $f(\vec{x}^{\,i}[k]) < f(\vec{x}^{\,i}[k-1])$,

then: $dir((x_j^i)[k] = \mathrm{sgn}(x_j^i[k] - x_j^i[k-1])$,

$age^i[k] = 1;$

else: $age^i[k] = age^i[k-1] + 1,$

$\forall_i \in \{1,2,\dots,N_p\}, \ \forall_j \in \{1,2,\dots,n\}$

where $x_j^i[k]$ denotes the jth parameter in the ith vector among Np vectors at the kth generation, $dir(x_j^i)[k]$ denotes the evolving direction of $x_j^i[k]$ and "*sgn*" is a sign function. Based on rule 1, the mutation occurs as follows:

Rule 2: If $age^i[k] = 1$,

then: $\sigma^i = \beta_1 . f(\overrightarrow{x}^i[k])$,

$$x_j^i[k] = x_j^i[k-1] + dir(x_j^i).\left|N(0,\sigma^i)\right|;$$

else: $\sigma^i = \beta_1 . f(\overrightarrow{x}^i[k]). age^i$,

$$\forall_i \in \{1,2,...,N_p\}, \ \forall_j \in \{1,2,...,n\}$$

where $|.|$ denotes an absolute value and $dir(x_j^i).\left|N(0,\sigma^i)\right|$ is a realisation of Gaussian-distribution random variable which is polarized in the direction of $dir(x_j^i)$.

Rule 1 indicates that if the performance of a newly generated offspring is better than its parent, the previous search direction is retained toward the prospective regions in the search space by memorising the sign of each parameter's evolution direction. The age of the offspring is 1. On the other hand, if the performance of the parent is better or equal to its generated offspring, then the parent's age is increased by 1. As the age increases, its standard deviation becomes larger to give the individual a chance to get out of the local minima.

4. Active Force Control Strategy

Active force control (AFC) is a control method derived from the Newton's second law of motion for a rotating mass [18], i.e.

$$\sum T = I\alpha$$

where T is the sum of all torques acting on the body, I is the moment of inertia and α is the angular acceleration.

For a robot system which has a serial configuration, the equation of motion becomes

$$T + Q = I(\theta)\alpha$$

where T is the applied torque Q is the disturbance torques, $I(\theta)$ is the mass moment of inertia of the robot arm, θ is the robot joint angle, and α is the angular acceleration of the robot arm.

The idea of the AFC approach is that, if the value of disturbance torques Q' can be computed or measured within an acceptable accuracy, this value could be used to decouple the actual disturbance torques Q from the applied torque. This will make the system stays stable even under variable external forces. The estimated value of the disturbance torque can be calculated as follows

$$Q' = I'\alpha' - T'$$

where the superscript denotes a measured or computed quantity. The applied torque T' can be measured by using a current sensor and α' can be measured by using an accelerometer. On the other hand, I' can be obtained by several means such as by simple estimation or by assuming a perfect model. The more recent literature considered ANN to estimate the value of I' by implementing an offline training method [19]. Figure 2 shows the schematic diagram of the AFC method applied to a robot arm together with the resolved motion acceleration controller.

356

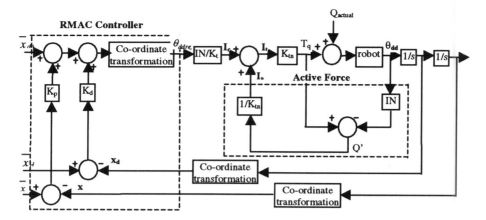

Fig. 2: Diagram of the active force controller applied to the robot force control

In figure 2, the estimated disturbance torques Q' can be estimated by the following equations:

$$Q' = IN\ \theta_{dd} - T_q$$

$$T_q = K_{tn}\ I_t$$

where θ_{dd} is the acceleration signal, IN is the estimated inertia matrix and T_q is the applied control torque, K_{tn} is the motor constant and I_t is the controlled current. The estimation of IN of the robot arm has been done previously by using traditional techniques such as crude estimation technique. Musa [19] incorporated an intelligent technique based on ANN to estimate the IN.

5. The ENNC Algorithm

The ENNC algorithm introduces some modifications to the EPNet algorithm described earlier. A GA procedure is incorporated into the algorithm for two reasons: (a) to explore better offspring, which acquires different identities (or behaviours) from that of their parents, and (b) to evolve the transfer function of the hidden nodes. By this approach, there is a possibility that an ANN converges faster than the previous method. The EP operator generates individuals that inherit the behaviour of their parents, but performs better.

5.1 ENNC Steps

The major steps in ENNC are shown in figure 3 and explained in the following:

1. Randomly generate M networks within certain ranges specified by the user, which represents the number of nodes, number of hidden layer, and types of transfer function at hidden nodes

2. Train each network using adaptive backpropagation for certain epochs, specified by the user. Evaluate the error of each network and calculate their cost function accordingly.

3. By using a rank-based selection method, rank the parent populations according to their cost function.

4. Use the EPNet operator, i.e. the 2 mutation steps, to produce a new offsprings. Delete one node from the hidden layer. Then, train and evaluate the network as in 2. If offspring is better than the parent, the offspring replaces the parent. Otherwise, add one node and then train and evaluate again as in 2. If the offspring is better than the parent, the offspring replaces the parent. Otherwise, the parent stays alive.

5. At the same time, the GA operators, i.e. crossover and mutation, are used to produce a new GA offsprings. Each of the GA offsprings is compared to their parent. Any offspring, which is better than its parent will represent the GA in the competition, section. If not, the parent itself will be the representative.

6. In the competition stage, all the selected nominees are combined and ranked from the best to the worst. The best one and a half of them are selected to live for the next generation cycle.

7. When the evolution stop, the best network selected undergoes the fine-tuning session using the FEP algorithms discussed in section 3. The procedure of the fine-tuning session is shown in figure 4.

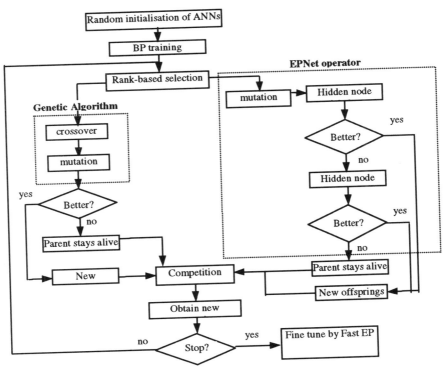

Fig. 3: The major steps in ENNC

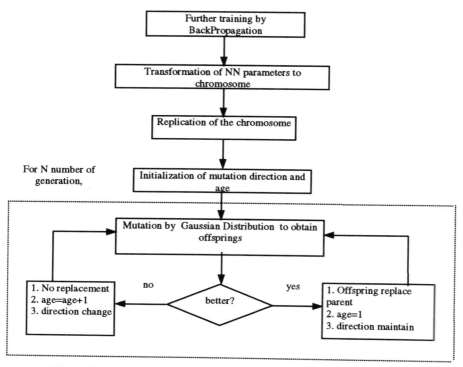

Fig.4: The steps in fine-tuning session by using the FEP algorithm

5.2 Chromosome Representation

In ENNC, both direct encoding and indirect encoding methods are used. The direct encoding method uses integer number representation to represents the network population. Each chromosome has the number of hidden layers, number of nodes and the types of transfer function. The advantage of using an integer representation over the real or binary number representation is that there is no need for further decoding. As for the transfer function (TF) types, the simple indirect encoding is used. The integer number represents the type of transfer function at hidden layers (HL) and output layers (OL) as shown in figure 5. Three types of TF are selected for the MLP network, i.e. tangent-sigmoid, log-sigmoid and pure-linear.

An example of a complete network encoding is shown in Figure 6. When the number of hidden layers is more than one, the number of nodes will be distributed equally for each layer. For the example shown in figure 6, the number of nodes for hidden layer 1 and 2 are 8 nodes each. If the network encounter extra nodes after the distribution, the first hidden layer will get the extra nodes. This is because the first hidden layer holds most of the generalised data, and therefore should have the same number of nodes as the other hidden layers or more.

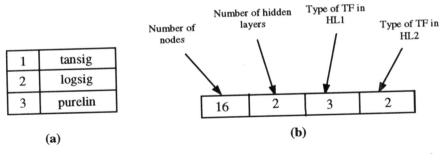

(a)   **(b)**

Fig. 5: (a) integer representing type of transfer function (TF), and (b) The complete representation of the MLP network in the form of chromosome

5.3 The Objective Variable and Cost Function

The objective variable in the ENNC is to minimise the error of each network output compared to the desired output. Therefore, the cost function to be minimised (the sum squared error) is calculated as follows

$$\text{Network Cost Function} = \sum_{i=1}^{n} (\hat{y}_i - \overline{y}_i)^2 + \beta.N$$

where $\hat{y}_i$ is the actual network output, $\overline{y}_i$ is the desired output, β is the node cost function constant, N is the total number of nodes and n is the number of output nodes. The value of $\beta.N$ means that if the network has more nodes, the efficiency of the network reduces and therefore the network cost function value is increased. The selection method used in ENNC is the linear-ranking method.

6. Simulation Results

The ENNC was programmed in MATLAB and applied to a two-link arm robot control using SIMULINK with ANN and GA toolboxes. The simulation is coded in two phases. The first phase is the offline training of the ENNC to produce the optimum ANN networks for the active force application. The second phase is the implementation of the ENNC for the two-link arm robot. A disturbance force is introduced to the robot end-effector to study the effectiveness of the active force controller. Suitable robot parameters are defined and the robot is planned to move in a circular trajectory. The robot position tracking error is plotted against time to get the performance result. For simplicity reason, some limitations have been imposed on the robot parameters.

6.1 Result of the ENNC Evolution

Figure 7 shows the evolution of ENNC for 30 generations. The result shows that the total sum-squared error of the network is reduced tremendously during early generations and settles down after 23 generations. It is observed that the evolution of the ENNC is very time consuming due to the fact that evolving networks has to go through two different procedures, i.e. GA and EPNet.

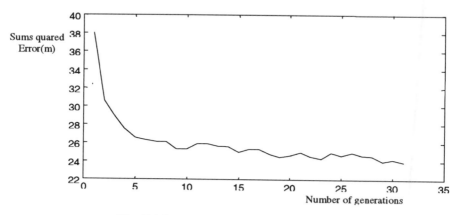

Fig. 7: The evolution of ENNC networks

The result of the ENNC controller applied to a 2-link arm robot is shown in figure 8(a), while the result of the non-evolving ANN controller is shown if figure 8(b). The ENNC controller provided better reduction in the tracking error.

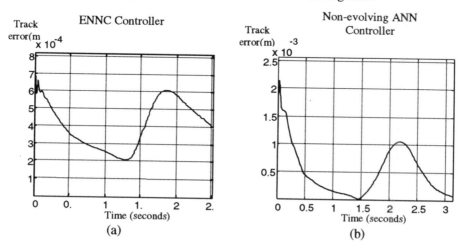

Fig. 8: The tracking error of (a) ENNC controller and (b) non-evolving ANN controller, when applied to the AFC controller of the 2-link arm robot.

7. Conclusions

Simulation results of a particular case study show that after 23 generations of evolution, the optimum design of MLP network has been obtained. After implementing the optimum ANN design obtained from the ENNC to the AFC problem of a 2-link arm robot, the results show that the sum-squared tracking error of the robot arm is reduced from 2.16×10^{-3} to 6.8×10^{-4} at time zero. The reduction of error is approximately the same for the rest of the simulation as shown in figure 8. In training the network, the number of epochs specified has to be more than a certain value. Otherwise, the offspring does not exactly represent the performance of the network.

It is concluded that the ENNC algorithm – which comprises a GA, EPNet, FEP and hybrid training – is a suitable searching algorithm for designing an optimum ANN for almost any kind of dynamic problems. The idea of introducing the GA in the EPNet algorithm is to widen the search space, without destroying the behavioural link of strong ANNs. Since the ENNC algorithm is relatively slow, it is only suitable for problems where training time is not a critical issue. Further work is progressing to improve the ENNC algorithm so that it can evolve faster and can be used for larger number of nodes and hidden layers.

References

1. D.R. Malcolm, *Robotics: An introduction*, PWS-Kent Publishing, 2nd ed., Boston, 1988
2. H. Asada and J.J.E. Slotine, *Robot Analysis and Control*, John Wiley and Son, 1986
3. J.R. Hewit and J.S. Burdess, Fast Dynamic Decoupled Control for Robotics Using Active Force Control, *Mechanism and Machine Theory*, Vol. 16, No.5, 1981, 535-542
4. D. T. Pham and S. J. Oh, Adaptive Control of a Robot Using Neural Networks, *Robotica*, vol. 12, 1994, pp. 553-561
5. A. Rana and A. M. S. Zalzala, A Neural Network Based Collision Detection Engine for Multi-Arm Robotic Systems, *Int'l Journal of Intelligent Control Systems*, vol. 2, No. 4, 1998, pp. 531-558
6. N. Chaiyaratana and A. M. S. Zalzala, Hybridisation of Neural Networks and Genetic Algorithms for Time-Optimal Control, *Congress on Evolutionary Computation*, vol. 1, July 1999, pp. 389-396
7. D. Whitley, T. Starkweather and C. Bogart, Genetic Algorithms and Neural Networks: Optimizing Connection and Connectivity, *Parralel Computing*, vol.14 No.3, 1990, pp. 347-361
8. D. B. Fogel, L.J. Fogel, and V. W. Porto, Evolving Neural Networks, *Biological Cybernetics*, 63, 1990, pp. 487-493
9. Xin Yao, A Review of Evolutionary Artificial Neural Networks, *International Journal of Intelligent Systems*, vol.8, No.4, April 1993, pp. 539-567

362

10. D. Fogel, Using evolutionary programming o create networks that are capable of playing tic-tac-toe, *in Proceedings of IEEE International Conference on Neural Networks*, San Francisco:IEEE, 1993, pp. 875-880

11. D. Dasgupta and D. McGregor, Designing application specific neural networks using the structured genetic algorithm, *in Proceedings of COGA NN-92 – IEEE International Workshop on Combinations of Genetic Algorithms and Neural Networks*, Baltimore: IEEE,1992, pp. 87-96

12. S. Oliker, M. Furst, and O. Maimon, A distributed genetic algorithm for neural network design and training, *Complex System*, vol. 6, no. 5, 1992, pp. 459-477

13. F. Herrera, M. Lozano, and J.L. Verdegay, Tackling Real-Coded Genetic algorithms: Operators and Tools for Behavioural Analysis, *Artificial Intelligence Review*, 12, 1998, pp. 265-319

14. H. Kitano, Designing Neural Networks using Genetic Algorithm with Graph Generation System, *Complex Systems*, vol.4, 1990, pp. 461-476

15. A. A. Siddiqi and S.M. Lucas, A Comparison of Matrix Rewriting versus Direct Encoding for Evolving Neural Networks, *Proceedings of IJCNN, FUZZ-IEEE,ICEC*, 1998, pp. 392-397

16. X. Yao, A New Evolutionary System for Evolving Artificial Neural Networks, *IEEE Transactions on Neural Networks*, vol. 8, No. 3, May 1997, pp. 694-713

17. J. H. Kim, J.Y. Jeon, H. K. Chae and K. Koh, A Novel Evolutionary Algorithm with Fast Convergence, *IEEE Conference on Evolutionary Computation*, vol. 1, Perth Australia, Dec 1995, pp. 228-233

18. J. R. Hewit and J.S. Burdess, An Active Method for the Control of Mechanical Systems in The Presence of Unmeasurable Forcing, *Transactions on Mechanism and Machine Theory*, vol. 21, No. 3, 1986, pp. 393-400

19. Musa Mailah, A simulation Study on the Intelligent Active Force Control of a Robot Arm using Neural Network, *Jurnal Teknologi UTM*, No.30(D), June 1999, pp. 55-78

Quality Inspection of Veneer Using Soft-Computing Methods

Alexander Stolpmann, Jürgen Angele
FH Braunschweig/Wolfenbüttel, Fachbereich Informatik
D-38302 Wolfenbüttel, Germany
{Stolpmann , Angele}@fh-wolfenbuettel.de

Laurence S. Dooley
Monash University
School of Computing and Information Technology
Gippsland Campus, Churchill, Victoria 3842, Australia
Laurence.Dooley@infotech.monash.edu.au

Abstract

This paper describes the use of a complex modular image processing system for veneer classification. An introduction into problems that arise when producing and processing veneer is given, namely flaws in the wood and faults in the final product, in this case spring boards for slatted frames.

A system has been developed that detects and classifies the faults. A line-scan camera is used for capturing the veneer images of which the features are extracted with statistical methods. The features are classified with either fuzzy clustering methods or neural networks. Additionally genetic algorithms are used for optimization purposes.

Keywords

System Optimization with Genetic Algorithms, Neural Networks and Fuzzy Clustering for Classification, Pattern Recognition and Image Processing, Application in Timber Industry.

1 Introduction

A major product in the timber industry is veneer 1(a). As wood is a natural product it contains a number of flaws that are in most cases unwanted and have to be detected during the production process. Among these shortcomings are checks 1(b), knottiness 1(c), knot holes 1(d), burn marks 1(e) and discolouring

1(f). In this project not the sheets of veneer themselves are inspected but the final product, spring boards for slatted frames. Therefore two more faults have to be detected, namely not properly applied coating 1(g) and missing veneer due to the end of a sheet 1(h).

Until now the spring boards are manually inspected. This is of course cost intensive and very tiring for the worker. As both sides have to be inspected at an approximate rate of one board per second some minor faults can easily slip through, effecting the overall quality of the production. Therefore a system is being developed that will automatically inspect and classify the spring board surfaces.

2 The Inspection System

The inspection system consists of specialized hardware and software. The system described in this paper is a laboratory system and therefore all the equipment which is necessary to use the system in a production line is not being mentioned.

2.1 The Hardware

The sensor element is a line-scan camera attached to a PCI framegrabber. The PC used is a dual processor Pentium Xeon machine with additional neural network hardware. Light comes from high frequency fluorescent tubes.

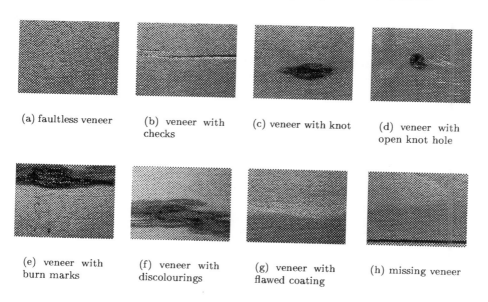

(a) faultless veneer

(b) veneer with checks

(c) veneer with knot

(d) veneer with open knot hole

(e) veneer with burn marks

(f) veneer with discolourings

(g) veneer with flawed coating

(h) missing veneer

Figure 1: Details of spring boards for slatted frames with faults.

2.2 The Software

The operating system used is Windows NT. The WiT image processing tool has been used for the implementation and testing of the application software.

The Classification System is composed of modules and layers. Figure 2 shows a simplified layout of the complete system. The centerpiece is the Core-System which consists of three modules: the preprocessing methods, the statistical methods and the soft-computing methods. The Core-System classifies the input images and is itself the center module of the Enhanced-System with the image preparation module before and the postprocessing module after it. The Enhanced-System is capable of identifying faulty regions within a larger image. Thus endless lamina coming directly from a veneer lathe can be classified and cut accordingly. The Automised Optimization System is made up of the genetic algorithm module and the fitness evaluation module. This subsystem can be used for optimization tasks throughout the complete system. The paper only describes the use for the feature vector size optimization.

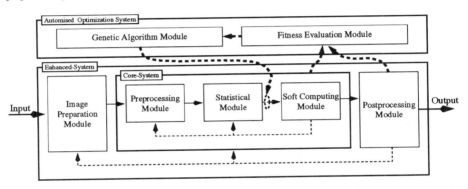

Figure 2: Texture Classification System

2.3 Statistical Feature Extraction

One way of classifying the quality of the veneer is by using statistical methods for feature extraction [16]. Such methods are used in this system.

Before the statistical methods are applied some preprocessing methods are used. These aim at the reduction of redundant information contained in the images so that the relevant information can be extracted more easily.

Currently used methods are a preliminary median filtering followed by a number of image analysis methods. Among these are different kinds of edge detection, gradient extraction, SOBEL filtering, surface and spectral analysis [12]. Next to those standard methods the images are processed through wavelets filters [5] and LAWS-measures [8]. The results of these computational steps are not used in the classical sense—it is for example not of interest where edges can be found—but are statistically evaluated.

Additionally to first order statistical calculations—like mean, variance, skewness, kurtosis—high order statistics are used as the neighbourhood relations are of importance. Very good results have been obtained concerning the orientation of similar grey-level pixels within an image. The spatial grey-level dependence (SGLD) matrices used yield potential features, among which are the entropy, correlation, inertia and homogenity [3].

All this results in a vast amount of data, but only a fraction of the extracted features provide information that is unique to a specific fault in the veneer. Therefore the number of features has to be reduced to avoid wasting computational resources.

2.4 Feature Selection using Genetic Algorithms

As the number of features generated in the previous modules of the system is very large it is necessary to select relevant features for the classification which takes place in the following modules. Doing this manually is not an option as the dimension of the feature plane is by far too large to be visualisable and the possible connections between features too complex. Therefore an automised feature selector has to be included into the system.

Genetic algorithms have the capability of finding very good local or even global optimal solutions in complex data-planes [11], [14], [17]. Therefore every feature is associated to one *gene*—a boolean element—and all genes compose the equivalent of a *DNA*. If the gene is set to zero the associated feature is not used in the following modules of the system and it is used if the gene is set to one. One half of the starting population is created randomly, the other half consists of the negated first half. In the next step all or some members of the population are used to create a new generation by exchanging parts of the DNA-string. This is called *crossover*. Depending on the way of selection and production of DNAs the population can grow rapidly. Additionally some DNAs can be mutated to avoid getting stuck in a local optimum.

Figure 3 shows graphically how the crossover works. The spot where the DNA-string is cut is chosen by random.

Next to the algorithm itself the fitness evaluation is most important. Fitness evaluation is the performance test of the system using every DNA of the population and thus a number of sets of selected features. After those tests only the better DNAs stay in the population and the production of a new population starts again.

This process continues as long as the fitness differences between parent and child population are significantly different for a specified number of generations.

In this case the fitness describes the ability to distinguish between different faults in the images. Figure 4 gives an idea how the features that describe a certain fault build a cluster in the feature plane. The aim is to find such a selection of features for which the cluster do not intersect with one another. The better the clusters are kept apart, the easier the classification is.

Parent Generation   Crossover   Child Generation

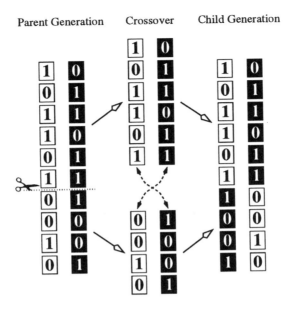

Figure 3: Crossover

2.5 Classification

As the aim is to find the relevant features to classify the possible faults a system had to be developed that handles the multi-dimensional feature vectors. Additionally it had to provide the possibility to judge the relevancy of the features.

2.5.1 Fuzzy Clustering

One possibility of classifying the feature vectors is using a clustering system [6], [9], [19]. For each (relevant) feature a new dimension is created. This results in an n-dimensional plane. All vectors that belong to a specific fault class make up a set, or cluster, which is diverse to all other fault clusters in this plane. With increasing numbers of fault classes the chance of clusters overlapping in one or more dimension increases. Therefore new methods had to be adopted to overcome this problem.

Fuzzy logic methods have a potential for handling uncertain knowledge [13], [20], [22]. Thus it is possible to classify veneer of which the feature vectors do not point to the center of a certain fault cluster but into an overlapping area of two or more clusters [1], [2], [4], [15], [18]. Such a veneer image has a membership value of a certain height for every cluster of the plane. Usually this value equals zero for almost all clusters and obtains a high value for the fault cluster in question. Veneer images that have membership values of about equal height have to be treated in a postprocessing system.

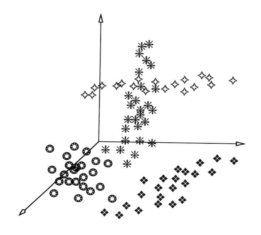

Figure 4: Cluster in a 3D-feature-plane

2.5.2 Neural Networks

An alternative to fuzzy clustering is to use neural networks. Widely known backpropagation (BP) networks have been successfully tested for this task [7], [10]. One problem with BP networks is that they are not retrainable, i.e. new classes can not be added without destroying the trained classes. This makes it more difficult and time consuming to expand the number of classes.

A neural network type which overcomes these problems is the adaptive resonance theory (ART) family [21]. Networks of this type can be retrained at any time. These networks are currently included into the system.

A general problem with neural networks is the rather long training time and the great need for computational resources, especially when input (number of features) and output (number of classes) vectors are large in size. In such cases the use of special neural network hardware is almost unavoidable, especially if the network has to be trained again and again to find the optimal feature vector as in the case of this classification system.

Therefore the above mentioned and some other neural network types are being adapted to run on the recently acquired Siemens-Nixdorf Synapse 3•PC neural network hardware.

3 Results

The results of the first set of tests are very promising. Using genetic algorithms to reduce the size of the feature vector proved to be a very good and fast approach for optimization and thus speed up of the system. At the moment 101 features are being extracted with the help of the preprocessing and statistical modules. This means that the optimal subset of features has to be found among all possible 2.510^{30} solutions. The optimization module finds an optimal solution within 500 tests, sometimes even significantly faster.

Until now only the fuzzy clustering has been used for classification. The fuzzy-c-means and Gustafsson-Kessel algorithms used will not give good results for any kind of clustering of the feature plane. In combination with the optimization module those feature subsets are found that build clusters that are separable. The introduction of further specialized fuzzy clustering methods aims at the exclusion of even more features, especially those that are computationally extensive.

A further speed up of the system can be achieved by compiling a subsystem once the optimal setup has been found.

4   Conclusions

This paper has introduced the use of a complex modular image processing system for veneer classification. The necessity and advantages of this approach have been delineated and discussed. Problems associated with the actual spring boards themselves and other modules of the system have also been addressed.

5   References

[1] J C Bezdek. *Fuzzy mathematics in pattern classification.* PhD Thesis, Cornell University, 1973.

[2] J C Dunn. *A Fuzzy Relative of the ISODATA Process and Its Use in Detecting Compact Well-Separated Clusters.* Journal of Cybernetics, 3/3:32–57, 1973.

[3] I M Elfadel, R W Picard. *Gibbs Random Fields, Co-occurrences, and Texture Modeling.* Technical Report # 204, MIT Media Laboratory Perceptual Computing Group, Januar 1993.

[4] I Gath, A B Geva. *Unsupervised Optimal Fuzzy Clustering.* IEEE Transactions on Pattern Analysis and Machine Intelligence, 11/7:773–781, 1989.

[5] A Graps. *An Introduction to Wavelets.* IEEE Computational Science and Engineering, 2(2), 1995.

[6] L Kaufmann, P J Rousseeuw. *Finding Groups in Data: An Introduction to Cluster Analysis.* John Wiley and Sons, New York, 1990.

[7] C Klevenhusen. *Texture Analysis with Artificial Neural Networks.* Diplomarbeit, Fachhochschule Braunschweig/Wolfenbüttel / University of Glamorgan, 1996.

[8] K I Laws. *Rapid texture identification*. SPIE Image Processing for Missile Guidance, 238:376–380, 1980.

[9] J MacQueen. *Some methods for classification and analysis of multivariate observations*. Proceedings of the 5th Berkeley Symposium, 1:281–297, 1967.

[10] S Malon. *Entwicklung eines Systems zur Texturanalyse mittels Wavelet-Transformation und neuronalem Netz für das Bildverarbeitungssystem WiT*. Diplomarbeit, Fachhochschule Braunschweig/Wolfenbüttel, 1997.

[11] E Schöneburg, F Heinzmann, S Feddersen. *Genetische Algorithmen und Evolutionsstrategien*. Addison-Wesley, Bonn/Paris/Reading (Mass.), 1994.

[12] M Sonka, V Hlavac, R Boyle. *Image Processing, Analysis and Machine Vision*. Chapman & Hall, London, 1993.

[13] A Stolpmann. *Fuzzy Logik: Eine Einführung ins Unscharfe*. Diplomarbeit, Fachhochschule Braunschweig/Wolfenbüttel, 1993.

[14] A Stolpmann, L S Dooley. *Genetic Algorithms for Automised Feature Selection in a Texture Classification System*. Proceedings of the 4th International Conference on Signal Processing, ICSP'98, Beijing, 1998.

[15] A Stolpmann, L S Dooley. *About the Use of Fuzzy Clustering for Texture Classification*. Proceedings of the 18th International Conference of the North American Fuzzy and Information Processing Scociety, NAFIPS'99, New York, 1999.

[16] A Stolpmann, L S Dooley. *A Texture Classification System Using Statistical and Soft-Computing Methods*. Proceedings of the 1st Infotech Oulu Workshop on Texture Analysis in Machine Vision, Oulu, Finland, 1999.

[17] A Stolpmann, L S Dooley. *A Texture Classification System with Automatic Feature Vector Optimization using Genetic Algorithms*. Proceedings of the 7th European Congress on Intelligent Techniques and Soft Computing, EUFIT'99, Aachen, 1999.

[18] T Tilli. *Mustererkennung mit Fuzzy-Logik: Analysieren, klassifizieren, erkennen und diagnostizieren*. Franzis-Verlag, München, 1993.

[19] R C Tryon. *Cluster Analysis*. Edwards Bros., Ann Arbor, 1939.

[20] L A Zadeh. *Fuzzy Sets*. Information Control, 8:338–353, 1965.

[21] A Zell. *Simulation Neuronaler Netze*. Addison-Wesley, Bonn/Paris/Reading (Mass.), 1994.

[22] H-J Zimmermann. *Fuzzy Set Theory – And Its Applications*. Kluwer Academic Publishers, Boston/Dordrecht/London, 1991.

AUTHOR INDEX